Manifold Learning
Theory and Applications

Manifold Learning Theory and Applications

Yunqian Ma and Yun Fu

CRC Press
Taylor & Francis Group
Boca Raton London New York

CRC Press is an imprint of the
Taylor & Francis Group, an **informa** business

CRC Press
Taylor & Francis Group
6000 Broken Sound Parkway NW, Suite 300
Boca Raton, FL 33487-2742

Printed in the United States of America on acid-free paper
Version Date: 20111110

International Standard Book Number: 978-1-4398-7109-6 (Hardback)

Visit the Taylor & Francis Web site at
http://www.taylorandfrancis.com

and the CRC Press Web site at
http://www.crcpress.com

Contents

List of Figures

List of Tables

Preface

Scientists and engineers working with large volumes of high-dimensional data often face the problem of dimensionality reduction: finding meaningful low-dimensional structures hidden in their high-dimensional observations. Manifold learning, as an important mathematical tool for high dimensionality analysis, has derived an emerging interdisciplinary research area in machine learning, computer vision, neural networks, pattern recognition, image processing, graphics, and scientific visualization, with various real-world applications.

Much research has been published since the most well-known, "A global geometric framework for nonlinear dimensionality reduction," by Joshua B. Tenenbaum, Vin de Silva and John C. Langford and "Nonlinear dimensionality reduction locally linear embedding" by Sam T. Roweis and Lawrence K. Saul, both in *Science*, Vol. 290, 2000. However, what is lacking in this field is a book grounded with the fundamental principles of existing manifold learning methodologies, and that provides solid theoretical and practical treatments for algorithmic and implementations from case studies.

Our purpose for this book is to systematically and uniquely bring together the state-of-the-art manifold learning theories and applications, and deliver a rich set of specialized topics authored by active experts and researchers in this field. These topics are well balanced between the basic theoretical background, implementation, and practical applications. The targeted readers are from broad groups, such as professional researchers, graduate students, and university faculties, especially with backgrounds in computer science, engineering, statistics, and mathematics. For readers who are new to the manifold learning field, the proposed book provides an excellent entry point with a high-level introductory view of the topic as well as in-depth discussion of the key technical details. For researchers in the area, the book is a handy tool summarizing the up-to-date advances in manifold learning. Readers from other fields of science and engineering may also find this book interesting because it is interdisciplinary and the topics covered synergize cross-domain knowledge. Moreover, this book can be used as a reference or textbook for graduate level courses at academic institutions. There are some universities already offering related courses or those with particular focus on manifold learning, such as the CSE 704 Seminar in Manifold and Subspace Learning, SUNY at Buffalo, 2010.

This book's content is divided by two criteria. Chapters 1 through 8 describe the manifold learning theory, with Chapters 9 through 11 presenting the application of manifold learning.

Chapter 1, as an introduction to this book, provides an overview of various methods in manifold learning. It reviews the notion of a smooth manifold using basic concepts from topology and differential geometry, and describes both linear and nonlinear manifold methods. Chapter 2 discusses how to use global information in manifold learning, particularly regarding Laplacian eigenmaps with global information. Chapter 3 describes manifold learning from the density-preserving point of view, and defines a density-preserving map on a Riemannian submanifold of a Euclidean space. Chapter 4 describes the sample complexity of classification on a manifold. It examines the informational aspect of manifold learning by studying two basic questions: first, classifying data that is drawn from a distribution supported on a submanifold of Euclidean space and, second, fitting a submanifold to data from a high-dimensional ambient space. It derives bounds on the amount of data required to perform these tasks that are independent of the ambient dimension, thus delineating two settings in which manifold learning avoids the curse of dimensionality. Chapter 5 deals with manifold learning in multiple datasets using manifold alignment, which constructs lower dimensional mapping between the multiple datasets by aligning their underlying learned model. Chapter 6 presents a large scale study on manifold learning using 18M sample data. Chapter 7 describes the heat kernel on a Riemannian manifold and focuses on the relation between the metric and heat kernel. Chapter 8 discusses the Ricci flow for designing Riemannian metrics by prescribed curvatures on surfaces and 3-dimensional manifolds.

Manifold learning applications are presented in Chapter 9 through Chapter 11. Chapter 9 describes manifold learning in the application of morphing in 2- and 3-dimensional shapes. Chapter 10 presents the application of manifold learning in visual recognition. It presents a framework of learning manifold representation from local features in images. Using the manifold representation, the visual recognition applications including object categorization, category discovery, feature matching can be fulfilled. Chapter 11 describes the application of manifold learning in human motion analysis. Manifold representation for the shape and appearance of moving objects is used in synthesis, pose recovery, reconstruction and tracking.

Overall, this book is intended to provide a solid theoretical background and practical guide of manifold learning to students and practitioners.

We would like to sincerely thank all the contributors of this book for presenting their research in an easily accessible manner, and for putting such discussion into a historical context. We would like to thank Mark Listewnik, Richard A. O'Hanley, and Stephanie Morkert of Auerbach Publications/CRC Press of Taylor & Francis Group for their strong support of this book.

Editors

Yunqian Ma received his PhD in electrical engineering from the University of Minnesota at Twin Cities in 2003. He then joined Honeywell International Inc., where he is currently senior principal research scientist in the advanced technology lab at Honeywell Aerospace. He holds 12 U.S. patents and 38 patent applications. He has authored 50 publications, including 3 books. His research interests include inertial navigation, integrated navigation, surveillance, signal and image processing, pattern recognition and computer vision, machine learning and neural networks. His research has been supported by internal funds and external contracts, such as AFRL, DARPA, HSARPA, and FAA. Dr. Ma received the International Neural Network Society (INNS) Young Investigator Award for outstanding contributions in the application of neural networks in 2006. He is currently associate editor of *IEEE Transactions on Neural Networks*, on the editorial board of the *Pattern Recognition Letters Journal*, and has served on the program committee of several international conferences. He also served on the panel of the National Science Foundation in the division of information and intelligent system and is a senior member of IEEE. Dr. Ma is included in *Marquis Who's Who in Engineering and Science.*

Yun Fu received his B.Eng. in information engineering and M.Eng. in pattern recognition and intelligence systems, both from Xi'an Jiaotong University, China. His M.S. in statistics, and Ph.D. in electrical and computer engineering were both earned at the University of Illinois at Urbana-Champaign. He joined BBN Technologies, Cambridge, Massachusetts, as a scientist in 2008 and was a part-time lecturer with the Department of Computer Science, Tufts University, Medford, Massachusetts, in 2009. Since 2010, he has been an assistant professor with the Department of Computer Science and Engineering, SUNY at Buffalo, New York. His current research interests include applied machine learning, human-centered computing, pattern recognition, intelligent vision system, and social media analysis. Dr. Fu is the recipient of the 2002 Rockwell Automation Master of Science Award, Edison Cups of the 2002 GE Fund Edison Cup Technology Innovation Competition, the 2003 Hewlett-Packard Silver Medal and Science Scholarship, the 2007 Chinese Government Award for Outstanding Self-Financed Students Abroad, the 2007 DoCoMo USA Labs Innovative Paper Award (IEEE International Conference on Image Processing 2007 Best Paper Award), the 2007–2008 Beckman Graduate Fellowship, the 2008 M. E. Van Valkenburg Graduate Research Award, the ITESOFT Best Paper Award of 2010 IAPR International Conferences on the Frontiers of Handwriting Recognition (ICFHR), and the 2010 Google Faculty Research Award. He is a lifetime member of the Institute of Mathematical Statistics (IMS), a senior member of IEEE, and a member of ACM and SPIE.

Contributors

Pierre-Antoine Absil
Department of Mathematical Engineering
Université Catholique de Louvain
Louvain-la-neuve, Belgium

Ahmed Elgammal
Department of Computer Science
Rutgers University
Piscataway, New Jersey

Joydeep Ghosh
Department of Electrical and Computer Engineering
University of Texas at Austin
Austin, Texas

Alexander Gray
College of Computing
Georgia Institute of Technology
Atlanta, Georgia

David Gu
Computer Science Department
SUNY at Stony Brook
Stony Brook, New York

Ren Guo
Department of Mathematics
Oregon State University
Corvallis, Oregon

Alan Julian Izenman
Department of Statistics
Temple University
Philadelphia, Pennsylvania

Peter Krafft
Department of Computer Science
University of Massachusetts Amherst
Amherst, Massachusetts

Sanjiv Kumar
Google Research
New York, New York

Chan-Su Lee
Electronic Engineering Department
Yeungnam University
Gyeongsan, Gyeongbuk, Korea

Feng Luo
Department of Mathematics
Hill Center-Busch Campus
Rutgers University
Piscataway, New Jersey

Sridhar Mahadevan
Department of Computer Science
University of Massachusetts Amherst
Amherst, Massachusetts

Mehryar Mohri
Courant Institute (NYU)
and
Google Research
New York, New York

Hariharan Narayanan
Laboratory for Information and Decision Systems
Massachusetts Institute of Technology
Cambridge, Massachusetts

Arkadas Ozakin
Georgia Tech Research Institute
Atlanta, Georgia

Henry Rowley
Google Research
Mountain View, California

Shounak Roychowdhury
Oracle Corporation
Austin, Texas

Chafik Samir
ISIT
Faculty of Medicine
Clermont-Ferrand, France

Jian Sun
Mathematical Science Center
Tsinghua University
Beijing, China

Ameet Talwalkar
Computer Science Division
University of California at Berkeley
Berkeley, California

Marwan Torki
Department of Computer Science
Rutgers University
Piscataway, New Jersey

Paul Van Dooren
Department of Mathematical Engineering
Université Catholique de Louvain
Louvain-la-neuve, Belgium

Nikolaos Vasiloglou II
College of Computing
Georgia Institute of Technology
Atlanta, Georgia

Chang Wang
Department of Computer Science
University of Massachusetts Amherst
Amherst, Massachusetts

Shing-Tung Yau
Department of Mathematics
Harvard University
Cambridge, Massachusetts

Wei Zeng
Computer Science Department
SUNY at Stony Brook
Stony Brook, New York

Chapter 1

Spectral Embedding Methods for Manifold Learning

Alan Julian Izenman

1.1 Introduction

Manifold learning encompasses much of the disciplines of geometry, computation, and statistics, and has become an important research topic in data mining and statistical learning. The simplest description of manifold learning is that it is a class of algorithms for recovering a low-dimensional manifold embedded in a high-dimensional ambient space. Major breakthroughs on methods for recovering low-dimensional nonlinear embeddings of high-dimensional data (Tenenbaum, de Silva, and Langford, 2000; Roweis and Saul, 2000) led to the construction of a number of other algorithms for carrying out nonlinear manifold learning and its close relative, nonlinear dimensionality reduction. The primary tool of all embedding algorithms is the set of eigenvectors associated with the top few or bottom few eigenvalues of an appropriate random matrix. We refer to these algorithms as *spectral embedding methods*. Spectral embedding methods are designed to recover linear or nonlinear manifolds, usually in high-dimensional spaces.

Linear methods, which have long been considered part-and-parcel of the statistician's toolbox, include PRINCIPAL COMPONENT ANALYSIS (PCA) and MULTIDIMENSIONAL SCALING (MDS). PCA has been used successfully in many different disciplines and applications. In computer vision, for example, PCA is used to study abstract notions of shape, appearance, and motion to help solve problems in facial and object recognition, surveillance, person tracking, security, and image compression where data are of high dimensionality (Turk and Pentland, 1991; De la Torre and Black, 2001). In astronomy, where very large digital sky surveys have become the norm, PCA has been used to analyze and classify stellar spectra, carry out morphological and spectral classification of galaxies and quasars, and analyze images of supernova remnants (Steiner, Menezes, Ricci, and Oliveira, 2009). In bioinformatics, PCA has been used to study high-dimensional data generated by genome-wide, gene-expression experiments on a variety of tissue sources, where scatterplots of the top principal components in such studies often show specific classes of genes that are expressed by different clusters of distinctive biological characteristics (Yeung and Ruzzo, 2001; Zheng-Bradley, Rung, Parkinson, and Brazma, 2010). PCA has also been used to select an optimal subset of single nucleotide polymorphisms (SNPs) (Lin and Altman, 2004). PCA is also used to derive approximations to more complicated nonlinear subspaces, including problems involving data interpolation, compression, denoising, and visualization.

MDS, which has its origins in psychology, has recently been found most useful in bioinformatics, where it is known as "distance geometry." MDS, for example, has been used to display a global representation (i.e., a map) of the protein-structure universe (Holm and Sander, 1996; Hou, Sims, Zhang, and Kim, 2003; Hou, Jun, Zhang, and Kim, 2005; Lu, Keles, Wright, and Wahba, 2005; Kim, Ahn, Lee, Park, and Kim, 2010). The idea is that points that are closely positioned to other points provide important information on the shape and function of proteins within the same family and so can be used for prediction and classification purposes. See Izenman (2008, Table 13.1) for a list of many diverse application areas and research topics in MDS.

There are other linear methods available, such as *projection pursuit (PP)*, which seeks linear projections of high-dimensional data that expose specific non-Gaussian features, and *independent component analysis (ICA)*, which looks for independent linear projections of the data by maximizing non-Gaussianity. Although PP and ICA are closely related, and PP has had a significant influence on the development of ICA, they approach the same type of problem from two very different directions. See, for example, Izenman (2008, Chapter 15). We will not be discussing PP and ICA here.

When linear manifold learning does not result in a good low-dimensional representation of high-dimensional data, then we are led to consider the possibility that the data may lie on or near a nonlinear manifold. Nonlinear manifold-learning algorithms include ISOMAP, LOCAL LINEAR EMBEDDING, LAPLACIAN EIGENMAPS, DIFFUSION MAPS, HESSIAN EIGENMAPS, and different interpretations of what NONLINEAR PCA means.

Whereas linear manifold-learning methods seek to preserve global structure on the manifold (mapping close points on a manifold to close points in low-dimensional space, and distant points to distant points), most nonlinear manifold-learning methods are considered to be local methods, which seek to preserve local structure in small neighborhoods on the manifold (Lee and Verleysen, 2007). ISOMAP is considered to be a global method because the embedding it derives builds upon MDS (itself a global method), and because it is based upon computing the geodesic distance between all pairs of data points. On the other hand, LOCAL LINEAR EMBEDDING, LAPLACIAN EIGENMAPS, DIFFUSION MAPS, and HESSIAN EIGENMAPS are all considered to be local methods because the embedding process of each method deals only with the relationship between a data point and a small number of its neighbors. There is a LOCAL ISOMAP algorithm in which ISOMAP is applied to many small neighborhoods of the manifold $\mathcal{M}$ and then the local mappings are patched together to form a global reconstruction of $\mathcal{M}$ (Donoho and Grimes, 2003a).

It is difficult to find real data that lie exactly on a nonlinear manifold. As a result, nonlinear manifold-learning algorithms are usually compared using simulated data, typically data that have been sampled from manifolds having specific quirks (e.g., S-curved manifold, Swiss-roll manifold, open box, torus, sphere, fishbowl) that are designed to expose any weaknesses of the various algorithms. Perhaps not surprisingly, it appears that no one method outperforms the rest over all types of manifolds, but some have been shown to be better in certain situations than the others. In an important paper, Goldberg, Zakai, Kushnir, and Ritov (2008) obtained necessary conditions on manifolds for these algorithms to be successful (i.e., recover the underlying manifold); they also showed that there exist simple manifolds where these conditions are violated and where the algorithms fail to recover the underlying manifold.

Although manifold learning is usually considered to be a topic in unsupervised learning, it has also been studied in the contexts of supervised-learning and semi-supervised learning. For example, variable selection in multiple regression is made a lot more difficult when the predictors live on a low-dimensional nonlinear manifold of a higher-dimensional sample space; one proposed solution is to learn the manifold first and then carry out a regularization of the regression problem (Aswani, Bickel, and Tomlin, 2011). In nonparametric regression, it was found that a nonparametric estimator of a regression function with a large number of predictors can automatically adapt to situations in which the predictors lie on or close to a low-dimensional smooth manifold (Bickel and Li, 2007). In semi-supervised learning, additional information takes the form of either a mixture of labeled and unlabeled points or a continuous function value that is known only for some of the data points. If such data live on a low-dimensional nonlinear manifold, it has been shown that classical methods will adapt automatically, and improved learning rates may be achieved even if one knows little about the structure of the manifold (Belkin, Niyogi, and Sindhwani, 2006; Lafferty and Wasserman, 2007). This raises the following question: under what circumstances is knowledge of the underlying manifold beneficial when carrying out supervised or semi-supervised learning, where the data lie on or close to a nonlinear manifold? See Niyogi (2008) for a theoretical discussion of this issue.

This chapter is organized as follows. In Section 1.2, we outline the basic ideas behind topological spaces and manifolds. Section 1.3 deals with linear manifold learning, and Section 1.4 deals with nonlinear manifold learning.

1.2 Spaces and Manifolds

Manifold learning involves concepts from general topology and differential geometry. Good introductions to topological spaces include Kelley (1955), Willard (1970), Bourbaki (1989), Mendelson (1990), Steen (1995), James (1999), and several of these have since been reprinted. Books on differential geometry include Spivak (1965), Kreyszig (1991), Kühnel (2000), Lee (2002), and Pressley (2010).

Manifolds generalize the notions of curves and surfaces in two and three dimensions to higher dimensions. Before we give a formal description of a manifold, it will be helpful to visualize the notion of a manifold. Imagine an ant at a picnic, where there are all sorts of items from cups to doughnuts. The ant crawls all over the picnic items, but because of its tiny size, the ant sees everything on a very small scale as flat and featureless. Similarly, a human, looking around at the immediate vicinity, would not see the curvature of the earth. A *manifold* (also referred to as a *topological manifold*) can be thought of in similar terms, as a topological space that locally looks flat and featureless and behaves like Euclidean space. Unlike a metric space, a topological space has no concept of distance. In this Section, we review specific definitions and ideas from topology and differential geometry that enable us to provide a useful definition of a manifold.

1.2.1 Topological Spaces

Topological spaces were introduced by Maurice Fréchet (1906) (in the form of metric spaces), and the idea was developed and extended over the next few decades. Amongst those who contributed significantly to the subject was Felix Hausdorff, who in 1914 coined the phrase "topological space" using Johann Benedict Listing's German word *Topologie* introduced in 1847.

A *topological space* $\mathcal{X}$ is a nonempty collection of subsets of $\mathcal{X}$ which contains the empty set, the space itself, and arbitrary unions and finite intersections of those sets. A topological space is often denoted by $(\mathcal{X}, \mathcal{T})$, where $\mathcal{T}$ represents the topology associated with $\mathcal{X}$. The elements of $\mathcal{T}$ are called the *open sets* of $\mathcal{X}$, and a set is *closed* if its complement is open. Topological spaces can also be characterized through the concept of neighborhood. If $\mathbf{x}$ is a point in a topological space $\mathcal{X}$, its *neighborhood* is a set that contains an open set that contains $\mathbf{x}$.

Let $\mathcal{X}$ and $\mathcal{Y}$ be two topological spaces, and let $U \subset \mathcal{X}$ and $V \subset \mathcal{Y}$ be open subsets. Consider the family of all cartesian products of the form $U \times V$. The topology formed from these products of open subsets is called the *product topology* for $\mathcal{X} \times \mathcal{Y}$. If $W \subset \mathcal{X} \times \mathcal{Y}$, then W is open relative to the product topology iff for each point $(x, y) \in \mathcal{X} \times \mathcal{Y}$ there are open neighborhoods, U of x and V of y, such that $U \times V \subset W$. For example, the *usual topology* for d-dimensional Euclidean space $\Re^d$ consists of all open sets of points in $\Re^d$, and this topology is equivalent to the *product topology* for the product of d copies of $\Re$.

One of the core elements of manifold learning involves the idea of "embedding" one topological space inside another. Loosely speaking, the space $\mathcal{X}$ is said to be *embedded* in the space $\mathcal{Y}$ if the topological properties of $\mathcal{Y}$ when restricted to $\mathcal{X}$ are identical to the topological properties of $\mathcal{X}$. To be more specific, we state the following definitions. A function $g : \mathcal{X} \rightarrow \mathcal{Y}$ is said to be *continuous* if the inverse image of an open set in $\mathcal{Y}$ is an open set in $\mathcal{X}$. If g is a *bijective* (i.e., one-to-one and onto) function such that g and its inverse g^{-1} are continuous, then g is said to be a *homeomorphism*. Two topological spaces $\mathcal{X}$ and $\mathcal{Y}$ are said to be *homeomorphic* (or *topologically equivalent*) if there exists a homeomorphism from one space onto the other. A topological space $\mathcal{X}$ is said to be *embedded* in a topological space $\mathcal{Y}$ if $\mathcal{X}$ is homeomorphic to a subspace of $\mathcal{Y}$.

If $A \subset \mathcal{X}$, then A is said to be *compact* if every class of open sets whose union contains A has a finite subclass whose union also contains A (i.e., if every *open cover* of A contains a finite *subcover*). This definition of compactness extends naturally to the topological space $\mathcal{X}$, and is itself a generalization of the celebrated *Heine–Borel theorem* that says that closed and bounded subsets of $\Re$ are compact. We note that subsets of a compact space need not be compact; however, closed subsets will be compact. *Tychonoff's theorem* that the product of compact spaces is compact is said to be "probably the most important single theorem of general topology" (Kelley, 1955, p. 143). One of the properties of compact spaces is that if $g : \mathcal{X} \rightarrow \mathcal{Y}$ is continuous and $\mathcal{X}$ is compact, then $g(\mathcal{X})$ is a compact subspace of $\mathcal{Y}$.

Another important idea in topology is that of a *connected space*. A topological space $\mathcal{X}$ is said to be *connected* if it cannot be represented as the union of two disjoint, nonempty, open sets. For example, $\Re$ itself with the usual topology is a connected space, and an interval in $\Re$ containing at least two points is connected. Furthermore, if $g : \mathcal{X} \rightarrow \mathcal{Y}$ is continuous and $\mathcal{X}$ is connected, then its image, $g(\mathcal{X})$, is connected as a subspace of $\mathcal{Y}$. Also, the product of any number of nonempty connected spaces, such as $\Re^d$ for any $d \geq 1$, is connected. The space $\mathcal{X}$ is *disconnected* if it is not connected.

A topological space $\mathcal{X}$ is said to be *locally Euclidean* if there exists an integer $d \geq 0$ such that around every point in $\mathcal{X}$, there is a local neighborhood which is homeomorphic to an open subset in Euclidean space $\Re^d$. A topological space $\mathcal{X}$ is a *Hausdorff space* if every pair of distinct points has a corresponding pair of disjoint neighborhoods. Almost all spaces are Hausdorff, including the real line $\Re$ with the standard metric topology. Also, subspaces and products of Hausdorff spaces are Hausdorff. $\mathcal{X}$ is *second-countable* if its topology has a countable basis of open sets. Most reasonable topological spaces are second countable, including the real line $\Re$, where the usual topology of open intervals has rational numbers as interval endpoints; a finite product of $\Re$ with itself is second countable if its topology is the product topology where open intervals have rational endpoints. Subspaces of second-countable spaces are again second countable.

1.2.2 Topological Manifolds

It was not until 1936 that the first clear description of the nature of an abstract manifold was provided by Hassler Whitney. We regard a "manifold" as a generalization to higher dimensions of a curved surface in three dimensions. A topological space $\mathcal{M}$ is a *topological manifold of dimension* d (sometimes written as $\mathcal{M}^d$) if it is a second-countable Hausdorff space that is also locally Euclidean of dimension d. The last condition says that at every point on the manifold, there exists a small local region such that the manifold enjoys the properties of Euclidean space. The Hausdorff condition ensures that distinct points on the manifold can be separated (by neighborhoods), and the second-countability condition ensures that the manifold is not too large. The two conditions of Hausdorff and second countability, together with an embedding theorem of Whitney (1936), imply that any d-dimensional manifold, $\mathcal{M}^d$, can be embedded in $\Re^{2d+1}$. In other words, a space of at most $2d+1$ dimensions is required to embed a d-dimensional manifold. A *submanifold* is just a manifold lying inside another manifold of higher dimension. As a topological space, a manifold can have topological structure, such as being compact or connected.

1.2.3 Riemannian Manifolds

In the entire theory of topological manifolds, there is no mention of the use of calculus. However, in a prototypical application of a "manifold," calculus enters in the form of a "smooth" (or differentiable) manifold $\mathcal{M}$, also known as a Riemannian manifold; it is usually defined in differential geometry as a submanifold of some ambient (or surrounding) Euclidean space, where the concepts of length, curvature, and angle are preserved, and where smoothness relates to differentiability. The word manifold (in German, *Mannigfaltigkeit*) was coined in an "intuitive" way and without any precise definition by Georg Friedrich Bernhard Riemann (1826–1866) in his 1851 doctoral dissertation (Riemann, 1851; Dieudonné, 2009); in 1854, Riemann introduced in his famous *Habilitations* lecture the idea of a topological manifold on which one could carry out differential and integral calculus.

A topological manifold $\mathcal{M}$ is called a *smooth* (or *differentiable*) *manifold* if $\mathcal{M}$ is continuously differentiable to any order. All smooth manifolds are topological manifolds, but the reverse is not necessarily true. (Note: Authors often differ on the precise definition of a "smooth" manifold.)

We now define the analogue of a homeomorphism for a differentiable manifold. Consider two open sets, $U \in \Re^r$ and $V \in \Re^s$, and let $g : U \to V$ so that for $\mathbf{x} \in U$ and $\mathbf{y} \in V$, $g(\mathbf{x}) = \mathbf{y}$. If the function g has finite first-order partial derivatives, $\partial y_j / \partial x_i$, for all $i = 1, 2, \ldots, r$, and all $j = 1, 2, \ldots, s$, then g is said to be a *smooth* (or *differentiable*) mapping on U. We also say that g is a $\mathcal{C}^1$-function on U if all the first-order partial derivatives are continuous. More generally, if g has continuous higher-order partial derivatives, $\partial^{k_1+\cdots+k_r} y_j / \partial x_1^{k_1} \cdots \partial x_r^{k_r}$, for all $j = 1, 2, \ldots, s$ and all nonnegative integers $k_1, k_2, \ldots, k_r$ such that $k_1 + k_2 + \cdots + k_r \leq r$, then we say that g is a $\mathcal{C}^r$-function, $r = 1, 2, \ldots$. If g is a $\mathcal{C}^r$-function for all $r \geq 1$, then we say that g is a $\mathcal{C}^\infty$-function.

If g is a homeomorphism from an open set U to an open set V, then it is said to be a $\mathcal{C}^r$*-diffeomorphism* if g and its inverse g^{-1} are both $\mathcal{C}^r$-functions. A $\mathcal{C}^\infty$-diffeomorphism is simply referred to as a *diffeomorphism*. We say that U and V are *diffeomorphic* if there exists a diffeomorphism between them. These definitions extend in a straightforward way to manifolds. For example, if $\mathcal{X}$ and $\mathcal{Y}$ are both smooth manifolds, the function $g : \mathcal{X} \to \mathcal{Y}$ is a *diffeomorphism* if it is a homeomorphism from $\mathcal{X}$ to $\mathcal{Y}$ and both g and g^{-1} are smooth. Furthermore, $\mathcal{X}$ and $\mathcal{Y}$ are *diffeomorphic* if there exists a diffeomorphism between them, in which case, $\mathcal{X}$ and $\mathcal{Y}$ are essentially indistinguishable from each other.

Consider a point $\mathbf{p} \in \mathcal{M}$. The set, $T_{\mathbf{p}}(\mathcal{M})$, of all vectors that are tangent to the manifold at the point $\mathbf{p}$ forms a vector space called the *tangent space* at $\mathbf{p}$. The tangent space has the same dimension as $\mathcal{M}$. Each tangent space $T_{\mathbf{p}}(\mathcal{M})$ at a point $\mathbf{p}$ has an inner-product, $g_{\mathbf{p}} = \langle \cdot, \cdot \rangle : T_{\mathbf{p}}(\mathcal{M}) \times T_{\mathbf{p}}(\mathcal{M}) \to \Re$, which is defined to vary smoothly over the manifold with $\mathbf{p}$. For $\mathbf{x}, \mathbf{y}, \mathbf{z} \in T_{\mathbf{p}}(\mathcal{M})$, the inner-product $g_{\mathbf{p}}$ is

bilinear: $g_{\mathbf{p}}(a\mathbf{x} + b\mathbf{y}, \mathbf{z}) = a g_{\mathbf{p}}(\mathbf{x}, \mathbf{z}) + b g_{\mathbf{p}}(\mathbf{y}, \mathbf{z})$, for $a, b \in \Re$,

symmetric: $g_{\mathbf{p}}(\mathbf{x},\mathbf{y}) = g_{\mathbf{p}}(\mathbf{y},\mathbf{x})$,

positive-definite: $g_{\mathbf{p}}(\mathbf{x},\mathbf{y}) \geq 0$ and $g_{\mathbf{p}}(\mathbf{x},\mathbf{x}) = 0$ iff $\mathbf{x} = \mathbf{0}$.

The collection of inner-products $g = \{g_{\mathbf{p}} : \mathbf{p} \in \mathcal{M}\}$ is a *Riemannian metric* on $\mathcal{M}$, and the pair $(\mathcal{M}, g)$ defines a *Riemannian manifold.*

Suppose $(\mathcal{M}, g^{\mathcal{M}})$ and $(\mathcal{N}, g^{\mathcal{N}})$ are two Riemannian manifolds that have the same dimension, and let $\psi : \mathcal{M} \to \mathcal{N}$ be a diffeomorphism. Then, ψ is an *isometry* if for all $\mathbf{p} \in \mathcal{M}$ and any two points $\mathbf{u}, \mathbf{v} \in T_{\mathbf{p}}(\mathcal{M})$, $g^{\mathcal{M}}(\mathbf{u},\mathbf{v}) = g^{\mathcal{N}}(\psi(\mathbf{u}), \psi(\mathbf{v}))$; in other words, ψ is an *isometry* if ψ "pulls back" one Riemannian metric to the other.

1.2.4 Curves and Geodesics

If the Riemannian manifold $(\mathcal{M}, g)$ is connected, it is a metric space with an induced topology that coincides with the underlying manifold topology. We can, therefore, define a function $d^{\mathcal{M}}$ on $\mathcal{M}$ that calculates distances between points on $\mathcal{M}$ and determines its structure.

Let $\mathbf{p}, \mathbf{q} \in \mathcal{M}$ be any two points on the Riemannian manifold $\mathcal{M}$. We first define the length of a (one-dimensional) curve in $\mathcal{M}$ that joins $\mathbf{p}$ to $\mathbf{q}$, and then the length of the shortest such curve.

A *curve* in $\mathcal{M}$ is defined as a smooth mapping from an open interval Λ (which may have infinite length) in $\Re$ into $\mathcal{M}$. The point $\lambda \in \Lambda$ forms a *parametrization* of the curve. Let $c(\lambda) = (c_1(\lambda), \cdots, c_d(\lambda))^{\tau}$ be a curve in $\Re^d$ parametrized by $\lambda \in \Lambda \subseteq \Re$. If we take the *coordinate functions*, $\{c_h(\lambda)\}$, of $c(\lambda)$ to be as smooth as needed (usually, $\mathcal{C}^{\infty}$, functions that have any number of continuous derivatives), then we say that c is a *smooth* curve. If $c(\lambda + \alpha) = c(\lambda)$ for all $\lambda, \lambda + \alpha \in \Lambda$, the curve c is said to be *closed.* The *velocity* (or *tangent*) *vector* at the point λ is given by

$$c'(\lambda) = (c'_1(\lambda), \cdots, c'_d(\lambda))^{\tau}, \tag{1.1}$$

where $c'_j(\lambda) = dc_j(\lambda)/d\lambda$, and the "speed" of the curve is

$$\| c'(\lambda) \| = \left\{ \sum_{j=1}^{d} [c'_j(\lambda)]^2 \right\}^{1/2}. \tag{1.2}$$

Distance on a smooth curve c is given by arc-length, which is measured from a fixed point λ_0 on that curve. Usually, the fixed point is taken to be the origin, $\lambda_0 = 0$, defined to be one of the two endpoints of the data. More generally, the arc-length $L(c)$ along the curve $c(\lambda)$ from point λ_0 to point λ_1 is defined as

$$L(c) = \int_{\lambda_0}^{\lambda_1} \| c'(\lambda) \| \; d\lambda. \tag{1.3}$$

In the event that a curve has unit speed, its arc-length is $L(c) = \lambda_1 - \lambda_0$.

Example: The Unit Circle in $\Re^2$. The unit circle in $\Re^2$, which is defined as $\{(x_1, x_2) \in \Re^2 : x_1^2 + x_2^2 = 1\}$, is a one-dimensional curve that can be parametrized as

$$c(\lambda) = (c_1(\lambda), c_2(\lambda))^{\tau} = (\cos\lambda, \sin\lambda)^{\tau}, \quad \lambda \in [0, 2\pi). \tag{1.4}$$

The unit circle is a closed curve, its velocity is $c'(\lambda) = (-\sin\lambda, \cos\lambda)^{\tau}$, and its speed is $\| c'(\lambda) \| = 1$.

One of the reasons that we study the topic of geodesics is because we are interested in finding the minimal-length curve that connects any two points on $\mathcal{M}$. Let $\mathcal{C}(\mathbf{p},\mathbf{q})$ be the set of all differentiable curves in $\mathcal{M}$ that join up the points $\mathbf{p}$ and $\mathbf{q}$. We define the distance between $\mathbf{p}$ and $\mathbf{q}$ as

$$d^{\mathcal{M}}(\mathbf{p},\mathbf{q}) = \inf_{c\in\mathcal{C}(\mathbf{p},\mathbf{q})} L(c), \tag{1.5}$$

where $L(c)$ is the arc-length of the curve c as defined by (1.3). One can show that the distance (1.5) satisfies the usual axioms for a metric. Thus, $d^{\mathcal{M}}$ finds the shortest curve (or *geodesic*) between any two points $\mathbf{p}$ and $\mathbf{q}$ on $\mathcal{M}$, and $d^{\mathcal{M}}(\mathbf{p},\mathbf{q})$ is the *geodesic distance* between the points. One can show that the geodesics in $\Re^d$ are straight lines.

1.3 Data on Manifolds

All the manifold-learning algorithms that we will describe in this chapter assume that finitely many data points, $\{\mathbf{y}_i\}$, are randomly drawn from a smooth t-dimensional manifold $\mathcal{M}$ with a metric given by geodesic distance $d^{\mathcal{M}}$. These data points are (linearly or nonlinearly) embedded by a smooth map ψ into high-dimensional input space $\mathcal{X} = \Re^r$, where $t \ll r$, with Euclidean metric $\|\cdot\|_{\mathcal{X}}$. This embedding results in the input data points $\{\mathbf{x}_i\}$. In other words, the embedding map is $\psi : \mathcal{M} \to \mathcal{X}$, and a point on the manifold, $\mathbf{y}_i \in \mathcal{M}$, can be expressed as $\mathbf{y} = \phi(\mathbf{x})$, $\mathbf{x} \in \mathcal{X}$, where $\phi = \psi^{-1}$.

The goal is to recover $\mathcal{M}$ and find an explicit representation of the map ψ (and recover the $\{\mathbf{y}\}$), given either the input data points $\{\mathbf{x}_i\}$ in $\mathcal{X}$, or the proximity matrix of distances between all pairs of those points. When we apply these algorithms, we obtain estimates $\{\widehat{\mathbf{y}}_i\} \subset \Re^{t'}$ that provide reconstructions of the manifold data $\{\mathbf{y}_i\} \subset \Re^t$, for some t'. Clearly, if $t' = t$, we have been successful. In general, we expect $t' > 3$, and so the results will be impractical for visualization purposes. To overcome this difficulty, while still providing a low-dimensional representation, we take only the points of the first two or three of the coordinate vectors of the reconstruction and plot them in a two- or three-dimensional space.

1.4 Linear Manifold Learning

Most statistical theory and applications that deal with the problem of dimensionality reduction are focused on linear dimensionality reduction and, by extension, linear manifold learning. A *linear manifold* can be visualized as a line, a plane, or a hyperplane, depending upon the number of dimensions involved. Data are observed in some high-dimensional space and it is usually assumed that a lower-dimensional linear manifold would be the most appropriate summary of the relationship between the variables. Although data tend not to live *on* a linear manifold, we view the problem as having two kinds of motivations. The first such motivation is to assume that the data live close to a linear manifold, the distance off the manifold determined by a random error (or noise) component. A second way of thinking about linear manifold learning is that a linear manifold is really a simple linear approximation to a more complicated type of nonlinear manifold that would probably be a better fit to the data. In both scenarios, the *intrinsic dimensionality* of the linear manifold is taken to be much smaller than the dimensionality of the data.

Identifying a linear manifold embedded in a higher-dimensional space is closely related to the classical statistics problem of linear dimensionality reduction. The recommended way of accomplishing linear dimensionality reduction is to create a reduced set of linear transformations of the input variables. Linear transformations are projection methods, and so the problem is to derive a sequence of low-dimensional projections of the input data that possess some type of optimal properties.

There are many techniques that can be used for either linear dimensionality reduction or linear manifold learning. In this chapter, we describe only two linear methods, namely, principal component analysis and multidimensional scaling. The earliest projection method was principal component analysis (dating back to 1933), and this technique has become the most popular dimensionality-reducing technique in use today. A related method is that of multidimensional scaling (dating back to 1952), which has a very different motivation. An adaptation of multidimensional scaling provided the core element of the ISOMAP algorithm for nonlinear manifold learning.

1.4.1 Principal Component Analysis

PRINCIPAL COMPONENT ANALYSIS (PCA) (Hotelling, 1933) was introduced as a technique for deriving a reduced set of orthogonal linear projections of a single collection of correlated variables, $\mathbf{X} = (X_1, \cdots, X_r)^\tau$, where the projections are ordered by decreasing variances. The amount of information in a random variable can be measured by its variance, which is a second-order property. PCA has also been referred to as a method for "decorrelating" $\mathbf{X}$, and so several researchers in different fields have independently discovered this technique. For example, PCA is also called the *Karhunen–Loève transform* in communications theory and *empirical orthogonal functions* in atmospheric science. As a technique for dimensionality reduction, PCA has been used in lossy data compression, pattern recognition, and image analysis. In chemometrics, PCA is used as a preliminary step for constructing derived variables in biased regression situations, leading to *principal component regression.*

PCA is also used as a means of discovering unusual facets of a data set. This can be accomplished by plotting the top few pairs of principal component scores (those having largest variances) in a scatterplot. Such a scatterplot can identify whether $\mathbf{X}$ actually lives on a low-dimensional linear manifold of $\Re^r$, as well as provide help identifying multivariate outliers, distributional peculiarities, and clusters of points. If the bottom set of principal components each have near-zero variance, then this implies that those principal components are essentially constant and, hence, can be used to identify the presence of collinearity and possibly outliers that might distort the intrinsic dimensionality of the vector $\mathbf{X}$.

Population Principal Components

Suppose that the input variables are the components of a random r-vector,

$$\mathbf{X} = (X_1, \cdots, X_r)^\tau, \tag{1.6}$$

where $\mathbf{A}^\tau$ denotes the transpose of the matrix $\mathbf{A}$. In this chapter, all vectors will be column vectors. Further, assume that $\mathbf{X}$ has mean vector $\mathrm{E}\{\mathbf{X}\} = \boldsymbol{\mu}_X$ and $(r \times r)$ covariance matrix $\mathrm{E}\{(\mathbf{X} - \boldsymbol{\mu}_X)(\mathbf{X} - \boldsymbol{\mu}_X)^\tau\} = \boldsymbol{\Sigma}_{XX}$. PCA replaces the input variables $X_1, X_2, \ldots, X_r$ by a new set of derived variables, $\xi_1, \xi_2, \ldots, \xi_t$, $t \leq r$, where

$$\xi_j = \mathbf{b}_j^\tau \mathbf{X} = b_{j1}X_1 + \cdots + b_{jr}X_r, \quad j = 1, 2, \ldots, r. \tag{1.7}$$

The derived variables are constructed so as to be uncorrelated with each other and ordered by the decreasing values of their variances. To obtain the vectors $\mathbf{b}_j$, $j = 1, 2, \ldots, r$, which define the principal components, we minimize the loss of information due to replacement. In PCA, "information" is interpreted as the "total variation" of the original input variables,

$$\sum_{j=1}^{r} \mathrm{var}(X_j) = \mathrm{tr}(\boldsymbol{\Sigma}_{XX}). \tag{1.8}$$

From the spectral decomposition theorem, we can write

$$\Sigma_{XX} = \mathbf{U}\Lambda\mathbf{U}^{\tau}, \quad \mathbf{U}^{\tau}\mathbf{U} = \mathbf{I}_r, \tag{1.9}$$

where the diagonal matrix Λ has as diagonal elements the eigenvalues $\lambda_1, \ldots, \lambda_r$ of Σ_{XX}, and the columns of the matrix $\mathbf{U}$ are the eigenvectors of Σ_{XX}. Thus, the total variation is $\text{tr}(\Sigma_{XX}) = \text{tr}(\Lambda) = \sum_{j=1}^{r} \lambda_j$.

The jth coefficient vector, $\mathbf{b}_j = (b_{1j}, \cdots, b_{rj})^{\tau}$ in (1.7), is chosen to have the following properties: (1) The top t linear projections $\xi_j, j = 1, 2, \ldots, t$, of $\mathbf{X}$ are ranked in importance through their variances, which are listed in decreasing order of magnitude: $\text{var}(\xi_1) \geq \text{var}(\xi_2) \geq \cdots \geq \text{var}(\xi_t)$, (2) ξ_j is uncorrelated with all $\xi_k, k < j$. The first t linear projections of (1.7) are known as the *top t principal components* of $\mathbf{X}$.

There are several derivations of the set of principal components of $\mathbf{X}$. We give only one here, through the least-squares optimality criterion.

Let $\mathbf{B} = (\mathbf{b}_1, \cdots, \mathbf{b}_t)^{\tau}$ denote the $(t \times r)$-matrix of weights, $t \leq r$, and let $\boldsymbol{\xi} = \mathbf{BX}$ be the t-vector of linear projections of $\mathbf{X}$, where $\boldsymbol{\xi} = (\xi_1, \cdots, \xi_t)^{\tau}$. We wish to find an r-vector $\boldsymbol{\mu}$ and an $(r \times t)$-matrix $\mathbf{A}$ such that the projections $\boldsymbol{\xi}$ have the property that $\mathbf{X} \approx \boldsymbol{\mu} + \mathbf{A}\boldsymbol{\xi}$ in some least-squares sense. Our least-squares criterion for finding $\boldsymbol{\mu}$ and $\mathbf{A}$ is given by

$$\text{E}\{(\mathbf{X} - \boldsymbol{\mu} - \mathbf{A}\boldsymbol{\xi})^{\tau}(\mathbf{X} - \boldsymbol{\mu} - \mathbf{A}\boldsymbol{\xi})\}. \tag{1.10}$$

If we substitute $\mathbf{BX}$ for $\boldsymbol{\xi}$ in (1.10), then the least-squares criterion becomes

$$\text{E}\{(\mathbf{X} - \boldsymbol{\mu} - \mathbf{ABX})^{\tau}(\mathbf{X} - \boldsymbol{\mu} - \mathbf{ABX})\}, \tag{1.11}$$

and the problem becomes one of finding $\boldsymbol{\mu}$, $\mathbf{A}$, and $\mathbf{B}$ to minimize (1.11). The following motivation for this minimization problem was suggested by Brillinger (1969). Suppose we have to send a message based upon the r components of a vector $\mathbf{X}$. Suppose further that such a message can only be transmitted using t channels, where $t \leq r$. We can first encode $\mathbf{X}$ into a t-vector $\boldsymbol{\xi} = \mathbf{BX}$, where $\mathbf{B}$ is a $(t \times r)$-matrix, and then, on receipt of the coded message, decode it using an $(r \times t)$-matrix $\mathbf{A}$ and a constant r-vector $\boldsymbol{\mu}$. The decoded message will then be the r-vector $\boldsymbol{\mu} + \mathbf{A}\boldsymbol{\xi}$, which we hope would be as "close" as possible to the original message $\mathbf{X}$.

We can think about this minimization problem in another way. Because $\mathbf{A}$ is an $(r \times t)$-matrix and $\mathbf{B}$ is a $(t \times r)$-matrix, where $t \leq r$, then $\mathbf{C} = \mathbf{AB}$ is an $(r \times r)$-matrix of multivariate regression coefficients obtained by regressing $\mathbf{X}$ on itself while requiring $\mathbf{C}$ to have "reduced-rank" t; that is, the rank of $\mathbf{C}$ is $r(\mathbf{C}) = t < r$. The rank condition on $\mathbf{C}$ implies that there may be a number of linear constraints on the set of regression coefficients. However, the value of the rank t and also the number and nature of those constraints may not be known prior to statistical analysis. We distinguish between the full-rank case when $t = r$ and the reduced-rank case when $t < r$. Note that $\mathbf{A}$ and $\mathbf{B}$ are not uniquely determined because the substitutions $\mathbf{A} \to \mathbf{AT}$ and $\mathbf{B} \to \mathbf{T}^{-1}\mathbf{B}$, where $\mathbf{T}$ is a nonsingular $(t \times t)$-matrix, give a different decomposition of $\mathbf{C}$. The matrix $\mathbf{T}$ can be used to rotate the least-squares solutions for $\mathbf{A}$ and $\mathbf{B}$, perhaps to allow a better interpretation of the results. The rotation idea is a popular method used in exploratory factor analysis.

The least-squares criterion (1.11) is minimized by

$$\mathbf{A}^{(t)} = (\mathbf{v}_1, \cdots, \mathbf{v}_t) = \mathbf{B}^{(t)\tau}, \tag{1.12}$$

$$\boldsymbol{\mu}^{(t)} = (\mathbf{I}_r - \mathbf{A}^{(t)}\mathbf{B}^{(t)})\boldsymbol{\mu}_X, \tag{1.13}$$

where $\mathbf{v}_j = \mathbf{v}_j(\Sigma_{XX})$ is the eigenvector associated with the jth largest eigenvalue, λ_j, of Σ_{XX}. Thus, the best rank-t reconstruction of the original $\mathbf{X}$ is given by

$$\widehat{\mathbf{X}}^{(t)} = \boldsymbol{\mu}^{(t)} + \mathbf{C}^{(t)}\mathbf{X} = \boldsymbol{\mu}_X + \mathbf{C}^{(t)}(\mathbf{X} - \boldsymbol{\mu}_X), \tag{1.14}$$

where

$$\mathbf{C}^{(t)} = \mathbf{A}^{(t)}\mathbf{B}^{(t)} = \sum_{j=1}^{t} \mathbf{v}_j \mathbf{v}_j^\tau \tag{1.15}$$

is the *multivariate reduced-rank regression coefficient matrix* with rank t. The minimum value of (1.11) is $\sum_{j=t+1}^{r} \lambda_j$, the sum of the smallest $r - t$ eigenvalues of $\boldsymbol{\Sigma}_{XX}$. The *first t principal components* of $\mathbf{X}$ are given by the linear projections. $\xi_1, \ldots, \xi_t$, where

$$\xi_j = \mathbf{v}^\tau \mathbf{X}, \quad j = 1, 2, \ldots, t. \tag{1.16}$$

The covariance between ξ_i and ξ_j is

$$\text{cov}(\xi_i, \xi_j) = \text{cov}(\mathbf{v}_i^\tau \mathbf{X}, \mathbf{v}_j^\tau \mathbf{X}) = \mathbf{v}_i^\tau \boldsymbol{\Sigma}_{XX} \mathbf{v}_j = \lambda_j \mathbf{v}_i^\tau \mathbf{v}_j = \delta_{ij} \lambda_j, \tag{1.17}$$

where δ_{ij} is the Kronecker delta, which equals 1 if $i = j$ and zero otherwise. Thus, λ_1, the largest eigenvalue of $\boldsymbol{\Sigma}_{XX}$, is var(ξ_1); λ_2, the second-largest eigenvalue of $\boldsymbol{\Sigma}_{XX}$, is var(ξ_2); and so on. Further, all pairs of derived variables are uncorrelated as required; that is, $\text{cov}(\xi_i, \xi_j) = 0$, $i \neq j$. We note that in the full-rank case, $t = r$, $\mathbf{C}^{(r)} = \mathbf{I}_r$, and $\boldsymbol{\mu}^{(r)} = \mathbf{0}$.

There are a number of stronger optimality results that can be obtained regarding the above least-squares choices of $\boldsymbol{\mu}^{(t)}$, $\mathbf{A}^{(t)}$, and $\mathbf{B}^{(t)}$. We refer the interested reader to Izenman (2008, Section 7.2).

Sample Principal Components

As we just saw, the population principal components are constructed using the eigenvalues and eigenvectors of the population covariance matrix $\boldsymbol{\Sigma}_{XX}$. However, in practice, $\boldsymbol{\Sigma}_{XX}$ (and $\boldsymbol{\mu}_X$) will be unknown and, therefore, has to be estimated by sample data. So, suppose we have n independent and identically distributed observations, $\{\mathbf{X}_i, i = 1, 2, \ldots, n\}$, on $\mathbf{X}$. First, we estimate $\boldsymbol{\mu}_X$ by

$$\widehat{\boldsymbol{\mu}}_X = \bar{\mathbf{X}} = n^{-1} \sum_{i=1}^{n} \mathbf{X}_i. \tag{1.18}$$

Now, let $\mathbf{X}_{ci} = \mathbf{X}_i - \bar{\mathbf{X}}$, $i = 1, 2, \ldots, n$, and set $\mathcal{X}_c = (\mathbf{X}_{c1}, \cdots, \mathbf{X}_{cn})$ to be an $(r \times n)$-matrix. We estimate $\boldsymbol{\Sigma}_{XX}$ by the sample covariance matrix,

$$\widehat{\boldsymbol{\Sigma}}_{XX} = n^{-1} \mathcal{X}_c \mathcal{X}_c^\tau. \tag{1.19}$$

The ordered sample eigenvalues of $\widehat{\boldsymbol{\Sigma}}_{XX}$ are given by $\widehat{\lambda}_1 \geq \widehat{\lambda}_2 \geq \cdots \geq \widehat{\lambda}_r \geq 0$, and the eigenvector corresponding to the jth largest sample eigenvalue $\widehat{\lambda}_j$ is the jth sample eigenvector $\widehat{\mathbf{v}}_j$, $j = 1, 2, \ldots, r$.

If r is fixed and n increases, then the sample eigenvalues and eigenvectors are consistent estimators[1] of the corresponding population eigenvalues and eigenvectors (Anderson, 1963). Furthermore, the sample eigenvalues and eigenvectors are approximately unbiased for their population counterparts, and their joint distribution is known. When both r and n are large, and they increase at the same rate (i.e., $r/n \to \gamma \geq 0$, as $n \to \infty$), then consistency depends upon γ in the following way: under certain moment assumptions on $\mathbf{X}$, if $\gamma = 0$, the sample eigenvalues converge to the population eigenvalues and, hence, are consistent; but if $\gamma > 0$, the sample eigenvalues will not be consistent (Baik and Silverstein, 2006).

Recent research regarding the statistical behavior of sample eigenvalues has been motivated by applications in which r is very large regardless of the sample size n. Examples include data obtained from microarray experiments where r can be in the tens of thousands

[1]An estimator $\widehat{\theta}$ is said to be *consistent* for a parameter θ if $\widehat{\theta} \to \theta$ in probability as $n \to \infty$.

while n would typically be fewer than a couple of hundred. This leads to the study of the eigenvalues and eigenvectors of large sample covariance matrices. *Random matrix theory*, which originated in mathematical physics during the 1950s and has now become a major research area in probability and statistics, is the study of the stochastic behavior of the *bulk* and the *extremes* of the spectrum of large random matrices. The bulk deals with most of the eigenvalues of a given matrix and the extremes refer to the largest and smallest of those eigenvalues. We refer the interested reader to the articles by Johnstone (2001, 2006) and the books by Mehta (2004) and Bai and Silverstein (2009).

We estimate $\mathbf{A}^{(t)}$ and $\mathbf{B}^{(t)}$ in (1.12) by

$$\widehat{\mathbf{A}}^{(t)} = (\widehat{\mathbf{v}}_1, \cdots, \widehat{\mathbf{v}}_t) = \widehat{\mathbf{B}}^{(t)\tau}. \tag{1.20}$$

The best rank-t reconstruction of $\mathbf{X}$ is given by

$$\widehat{\mathbf{X}}^{(t)} = \bar{\mathbf{X}} + \widehat{\mathbf{C}}^{(t)}(\mathbf{X} - \bar{\mathbf{X}}), \tag{1.21}$$

where

$$\widehat{\mathbf{C}}^{(t)} = \widehat{\mathbf{A}}^{(t)}\widehat{\mathbf{B}}^{(t)} = \sum_{j=1}^{t} \widehat{\mathbf{v}}_j \widehat{\mathbf{v}}_j^\tau \tag{1.22}$$

is the multivariate reduced-rank regression coefficient matrix of rank t. The jth *sample PC score* of $\mathbf{X}$ is given by $\widehat{\xi}_j = \widehat{\mathbf{v}}_j^\tau \mathbf{X}_c$, where $\mathbf{X}_c = \mathbf{X} - \bar{\mathbf{X}}$. The variance, λ_j, of the jth principal component is estimated by the sample variance, $\widehat{\lambda}_j$, $j = 1, 2, \ldots, t$. For diagnostic and data analytic purposes, it is customary to plot the first sample PC scores against the second sample PC scores, $(\widehat{\xi}_{i1}, \widehat{\xi}_{i2})$, $i = 1, 2, \ldots, n$, where $\widehat{\xi}_{ij} = \widehat{\mathbf{v}}_j^\tau \mathbf{X}_i$, $i = 1, 2, \ldots, n$, $j = 1, 2$. More generally, we could draw the scatterplot matrix to view all pairs of PC scores.

Note that PCA is not invariant under rescalings of $\mathbf{X}$. If we standardize the $\mathbf{X}$ variables by computing $\mathbf{Z} \leftarrow (\text{diag}\{\mathbf{\Sigma}_{XX}\})^{-1/2}(\mathbf{X} - \widehat{\boldsymbol{\mu}}_X)$, then PCA is carried out using the correlation matrix (rather than the covariance matrix). The lack of invariance implies that PCA based upon the correlation matrix could be very different from a PCA based upon the covariance matrix, and no simple relationship exists between the two sets of results. Standardization of $\mathbf{X}$ when using PCA is customary in many fields where the variables differ substantially in their variances; the variables with relatively large variances will tend to overwhelm the leading PCs with the remaining variables contributing very little.

So far, we have assumed that t is known. If t is unknown, as it generally is in practice, we need to estimate t, which is now considered a *metaparameter*. It is the value of t that determines the dimensionality of the linear manifold of $\Re^r$ in which $\mathbf{X}$ really lives. The classical way of estimating t is through the values of the sample variances. One hopes that the first few sample PCs will have large sample variances, while the remaining PCs will have sample variances that are close enough to zero for the corresponding subset of PCs to be declared essentially constants and, therefore, omitted from further consideration. There are several alternative methods for estimating t, some of them graphical, including the scree plot and the PC rank trace plot. See Izenman (2008, Section 7.2.6) for details.

1.4.2 Multidimensional Scaling

Multidimensional scaling (MDS) is a family of algorithms each member of which seeks to identify an underlying manifold consistent with a given set of data. A useful motivation for MDS can be viewed in the following way. Imagine a map of a particular geographical region, which includes several cities and towns. Such a map is usually accompanied by a two-way table of distances between selected pairs of those towns and cities. The number in each cell of that table gives the degree of "closeness" (or *proximity*) of the row city to the column city that identifies that cell. The general problem of MDS reverses that relationship between the map and table of proximities. With MDS, one is given only the table of proximities, and the problem is to reconstruct the map as closely as possible. There is one more wrinkle: the number of dimensions of the map is unknown, and so we have to determine the dimensionality of the underlying (linear) manifold that is consistent with the given table of proximities.

Proximity Matrices

Proximities do not have to be distances, but can be a more complicated concept. We can talk about the proximity of any two entities to each other, where by "entity" we might mean an object, a brand-name product, a nation, a stimulus, and so on. The proximity of a pair of such entities could be a measure of association (e.g., the absolute value of a correlation coefficient), a confusion frequency (i.e., to what extent one entity is confused with another in an identification exercise), or some other measure of how alike (or how different) one perceives the entities to be. A proximity can be a continuous measure of how physically close one entity is to another or it could be a subjective judgment recorded on an ordinal scale, but where the scale is sufficiently well-calibrated as to be considered continuous. In other scenarios, especially in studies of perception, a proximity will not be quantitative, but will be a subjective rating of "similarity" (how close a pair of entities are to each other) or "dissimilarity" (how unalike are the pair of entities). The only thing that really matters in MDS is that there should be a monotonic relationship (either increasing or decreasing) between the "closeness" of two entities and the corresponding similarity or dissimilarity value.

Suppose we have a particular collection of n entities to be compared. We represent the *dissimilarity* of the ith entity to the jth entity by δ_{ij}. A *proximity matrix* $\mathbf{\Delta} = (\delta_{ij})$ is an $(m \times m)$ square matrix of dissimilarities, where $m = n(n-1)/2$. In practice, the proximity matrix is stored (and displayed) as a lower-triangular array of nonnegative entries (i.e., $\delta_{ij} \geq 0,\ i, j = 1, 2, \ldots, n$), with the understanding that the diagonal entries are all zeroes (i.e., $\delta_{ii} = 0,\ i = 1, 2, \ldots, n$) and that the upper-triangular array of the matrix is a mirror image of the given lower triangle (i.e., $\delta_{ji} = \delta_{ij},\ i, j = 1, 2, \ldots, n$). Further, to be considered as a *metric* distance, it is usual to require that the triangle inequality be satisfied (i.e., $\delta_{ij} \leq \delta_{ik} + \delta_{kj}$, for all k). In some applications, we should not expect $\mathbf{\Delta}$ to be symmetric.

Classical Scaling

Although there are several different versions of MDS, we describe here only the classical scaling method. Other methods are described in Izenman (2008, Chapter 13).

So, suppose we are given n points $\mathbf{X}_1, \ldots, \mathbf{X}_n \in \Re^r$ from which we compute an $(n \times n)$-matrix $\mathbf{\Delta} = (\delta_{ij})$ of dissimilarities, where

$$\delta_{ij} = \| \mathbf{X}_i - \mathbf{X}_j \| = \left\{ \sum_{k=1}^{r} (X_{ik} - X_{jk})^2 \right\}^{1/2} \tag{1.23}$$

is the dissimilarity between $\mathbf{X}_i = (X_{i1}, \cdots, X_{ir})^\tau$ and $\mathbf{X}_j = (X_{j1}, \cdots, X_{jr})^\tau$, $i, j = 1, 2, \ldots, n$; these dissimilarities are the Euclidean distances between all $m = n(n-1)/2$ pairs of points in that space. Squaring both sides of (1.23) and expanding the right-hand side yields

$$\delta_{ij}^2 = \| \mathbf{X}_i \|^2 + \| \mathbf{X}_j \|^2 - 2\mathbf{X}_i^\tau \mathbf{X}_j. \tag{1.24}$$

Note that $\delta_{i0}^2 = \| \mathbf{X}_i \|^2$ is the squared distance from the point $\mathbf{X}_i$ to the origin. Let

$$b_{ij} = \mathbf{X}_i^\tau \mathbf{X}_j = -\frac{1}{2}(\delta_{ij}^2 - \delta_{i0}^2 - \delta_{j0}^2). \tag{1.25}$$

Summing (1.24) over i and over j gives the following identities:

$$n^{-1}\sum_i \delta_{ij}^2 = n^{-1}\sum_i \delta_{i0}^2 + \delta_{j0}^2 \tag{1.26}$$

$$n^{-1}\sum_j \delta_{ij}^2 = \delta_{i0}^2 + n^{-1}\sum_j \delta_{j0}^2 \tag{1.27}$$

$$n^{-2}\sum_i\sum_j \delta_{ij}^2 = 2n^{-1}\sum_i \delta_{i0}^2. \tag{1.28}$$

Let $a_{ij} = -\frac{1}{2}\delta_{ij}^2$. Using the usual "dot" notation, we define $a_{i\cdot} = n^{-1}\sum_j a_{ij}$, $a_{\cdot j} = n^{-1}\sum_i a_{ij}$, and $a_{\cdot\cdot} = n^{-2}\sum_i\sum_j a_{ij}$. Substituting (1.26)–(1.28) into (1.25) and then simplifying, we get

$$b_{ij} = a_{ij} - a_{i\cdot} - a_{\cdot j} + a_{\cdot\cdot}. \tag{1.29}$$

We can express this in matrix notation as follows. Set $\mathbf{A} = (a_{ij})$ and $\mathbf{B} = (b_{ij})$. Then, $\mathbf{A}$ and $\mathbf{B}$ are related through

$$\mathbf{B} = \mathbf{HAH}, \tag{1.30}$$

where $\mathbf{H} = \mathbf{I}_n - n^{-1}\mathbf{J}_n$ is a centering matrix and $\mathbf{J}_n$ is an $(n \times n)$-matrix of all ones. $\mathbf{HA}$ removes the row mean from each row of $\mathbf{A}$, while $\mathbf{AH}$ removes the column mean from each column of $\mathbf{A}$. The matrix $\mathbf{B}$ when expanded has the form,

$$\mathbf{B} = \mathbf{A} - n^{-1}\mathbf{AJ} - n^{-1}\mathbf{JA} + n^{-2}\mathbf{JAJ}. \tag{1.31}$$

As we can see from (1.31), $\mathbf{B}$ removes from each element a_{ij} of $\mathbf{A}$ the row mean $a_{i\cdot}$, and the column mean $a_{\cdot j}$ and then adds back in the overall mean $a_{\cdot\cdot}$. As a result, $\mathbf{B}$ is said to be a "doubly centered" version of $\mathbf{A}$.

We would now like to find a set of t-dimensional points, $\mathbf{Y}_1, \ldots, \mathbf{Y}_n \in \Re^t$ (called *principal coordinates*), that could represent the r-dimensional points $\mathbf{X}_1, \ldots, \mathbf{X}_n \in \Re^r$, with $t < r$, such that the interpoint distances in t-space "match" those in r-space. If we define dissimilarities as Euclidean distances between pairs of points, then the resulting representation will be equivalent to PCA because the principal coordinates are identical to the first t principal component scores of the $\{\mathbf{X}_i\}$.

The "catch" in all this is that we are not given the values of the $\mathbf{X}_1, \ldots, \mathbf{X}_n$. Instead, we are given only the dissimilarities $\{\delta_{ij}\}$ through the $(n \times n)$ proximity matrix $\mathbf{\Delta}$. The problem of constructing the $\mathbf{Y}_1, \ldots, \mathbf{Y}_n \in \Re^t$ given only the matrix $\mathbf{\Delta}$ is called *classical scaling* (Torgerson, 1952, 1958).

We form the matrix $\mathbf{A}$ from $\mathbf{\Delta}$, and then, using (1.30), we form the matrix $\mathbf{B}$. Next, we find a matrix $\mathbf{B}^* = (b_{ij}^*)$, with rank at most t, that minimizes

$$\text{tr}\{(\mathbf{B} - \mathbf{B}^*)^2\} = \sum_i\sum_j (b_{ij} - b_{ij}^*)^2. \tag{1.32}$$

If $\{\lambda_k\}$ are the eigenvalues of $\mathbf{B}$ and if $\{\lambda_k^*\}$ are the eigenvalues of $\mathbf{B}^*$, then the minimum of $\text{tr}\{(\mathbf{B} - \mathbf{B}^*)^2\}$ is given by $\sum_{k=1}^n (\lambda_k - \lambda_k^*)^2$, where $\lambda_k^* = \max(\lambda_k, 0)$ for $k = 1, 2, \ldots, t$, and zero otherwise (Mardia, 1978). Let $\mathbf{\Lambda} = \text{diag}\{\lambda_1, \cdots, \lambda_n\}$ be the diagonal matrix of the eigenvalues of $\mathbf{B}$ and let $\mathbf{V} = (\mathbf{v}_1, \cdots, \mathbf{v}_n)$ be the matrix whose columns are the eigenvectors of $\mathbf{B}$. By the spectral theorem, $\mathbf{B} = \mathbf{V\Lambda V}^\tau$. If $\mathbf{B}$ is nonnegative-definite with rank $r(\mathbf{B}) = t < n$, the largest t eigenvalues will be positive and the remaining $n - t$ eigenvalues will be zero. Let $\mathbf{\Lambda}_1 = \text{diag}\{\lambda_1, \cdots, \lambda_t\}$ be the $(t \times t)$ diagonal matrix of the positive eigenvalues of $\mathbf{B}$, and let $\mathbf{V}_1 = (\mathbf{v}_1, \cdots, \mathbf{v}_t)$ be the corresponding matrix of eigenvectors of $\mathbf{B}$. Then,

$$\mathbf{B} = \mathbf{V}_1\mathbf{\Lambda}_1\mathbf{V}_1^\tau = (\mathbf{V}_1\mathbf{\Lambda}_1^{1/2})(\mathbf{\Lambda}_1^{1/2}\mathbf{V}_1) = \mathbf{YY}^\tau, \tag{1.33}$$

where

$$\mathbf{Y} = \mathbf{V}_1 \mathbf{\Lambda}_1^{1/2} = (\sqrt{\lambda_1}\mathbf{v}_1, \cdots, \sqrt{\lambda_t}\mathbf{v}_t) = (\mathbf{Y}_1, \cdots, \mathbf{Y}_n)^\tau. \tag{1.34}$$

The *principal coordinates* are the columns, $\mathbf{Y}_1, \ldots, \mathbf{Y}_n \in \Re^t$, of the $(t \times n)$-matrix $\mathbf{Y}^\tau$, whose interpoint distances

$$d_{ij}^2 = \| \mathbf{Y}_i - \mathbf{Y}_j \|^2 = (\mathbf{Y}_i - \mathbf{Y}_j)^\tau (\mathbf{Y}_i - \mathbf{Y}_j) \tag{1.35}$$

are equal to the distances δ_{ij} in the matrix $\mathbf{\Lambda}$.

If the eigenvalues of $\mathbf{B}$ are not all nonnegative, then we can either ignore the negative eigenvalues (and associated eigenvectors) or we can add a suitable constant to the dissimilarities and then start the algorithm again. If t is too large for practical purposes, then the largest $t' < t$ positive eigenvalues and associated eigenvectors of $\mathbf{B}$ can be used to construct a reduced set of principal coordinates. In this case, the interpoint distances d_{ij} approximate the δ_{ij} from the matrix $\mathbf{\Delta}$.

Note that the classical-scaling algorithm automatically sets the mean $\bar{\mathbf{Y}}$ of all n points in the configuration to be the origin in $\Re^t$. This follows because $\mathbf{H}\mathbf{1}_n = \mathbf{0}$, and so $\mathbf{B}\mathbf{1}_n = \mathbf{0}$. Then,

$$n^2 \bar{\mathbf{Y}}^\tau \bar{\mathbf{Y}} = (\mathbf{Y}^\tau \mathbf{1}_n)^\tau (\mathbf{Y}^\tau \mathbf{1}_n) = \mathbf{1}_n^\tau \mathbf{B} \mathbf{1}_n = 0, \tag{1.36}$$

whence, $\bar{\mathbf{Y}} = \mathbf{0}$.

The classical-scaling solution is not unique. To see this, let $\mathbf{P}$ be an orthogonal matrix. Make an orthogonal transformation of two points obtained through the classical-scaling algorithm: $\mathbf{Y}_i \to \mathbf{P}\mathbf{Y}_i$ and $\mathbf{Y}_j \to \mathbf{P}\mathbf{Y}_j$. Then, $\mathbf{P}\mathbf{Y}_i - \mathbf{P}\mathbf{Y}_j = \mathbf{P}(\mathbf{Y}_i - \mathbf{Y}_j)$, whence, $\| \mathbf{P}(\mathbf{Y}_i - \mathbf{Y}_j) \|^2 = \| \mathbf{Y}_i - \mathbf{Y}_j \|^2$. This tells us that if we make a common orthogonal transformation of the points in the configuration found by classical scaling, we obtain a different solution to the classical-scaling problem.

The most popular way of assessing dimensionality of the classical-scaling configuration is to plot the ordered eigenvalues (from largest to smallest) of $\mathbf{B}$ against order number (dimension), and then identify a dimension at which the eigenvalues "stabilize." Eigenvalues become stable when they cease to change perceptively. At the dimension they become roughly constant, there will be an "elbow" in the plot where stability occurs. If $\mathbf{X}_i \in \Re^t$, $i = 1, 2, \ldots, n$, then we should see stability at dimension $t + 1$. Typically, one hopes that t is small, of the order of 2 or 3.

In a recent study of MDS, theoretical arguments were made for the presence of "horseshoe" patterns in plots of the first few principal coordinates (Diaconis, Goel, and Holmes, 2008). The study was illustrated with an application of MDS to the 2005 U.S. House of Representatives roll-call votes. Rather than use Euclidean distance (1.23) to construct a matrix $\mathbf{\Delta}$ of interpoint distances, the authors used an exponential-kernel distance based upon where bills (on which legislators vote) fall in a "liberal-conservative" policy space. The study showed that, when plotting the first three principal coordinates (i.e., $t = 3$ in (1.34)) for the roll-call votes, MDS produced a three-dimensional "horseshoe" pattern of legislators for each political party (Democrat and Republican), and that these two horseshoe patterns were well-separated from each other in the plot. Various explanations have been proposed for these horseshoe patterns, which are characteristic of MDS and many other manifold-learning techniques.

1.5 Nonlinear Manifold Learning

We next discuss some algorithmic techniques that proved to be innovative in the study of nonlinear manifold learning: ISOMAP, LOCAL LINEAR EMBEDDING, LAPLACIAN EIGENMAPS, DIFFUSION MAPS, HESSIAN EIGENMAPS, and the many different versions of NONLINEAR PCA. The goal of each of these algorithms is to recover the full low-dimensional

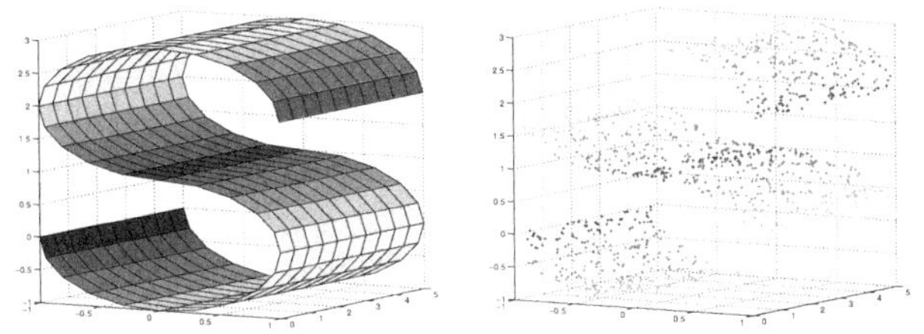

Figure 1.1: (See Color Insert.) Left panel: The S-curve, a two-dimensional S-shaped manifold embedded in three-dimensional space. Right panel: 2,000 data points randomly generated to lie on the surface of the S-shaped manifold. Reproduced from Izenman (2008, Figure 16.6) with kind permission from Springer Science+Business Media.

representation of an unknown nonlinear manifold, $\mathcal{M}$, embedded in some high-dimensional space, where it is important to retain the neighborhood structure of $\mathcal{M}$. When $\mathcal{M}$ is highly nonlinear, such as the S-shaped manifold in the left panel of Figure 1.1, these algorithms outperform the usual linear techniques. The nonlinear manifold-learning methods emphasize simplicity and avoid optimization problems that could produce local minima.

Assume that we have a finite random sample of data points, $\{\mathbf{y}_i\}$, from a smooth t-dimensional manifold $\mathcal{M}$ with metric given by the geodesic distance $d^{\mathcal{M}}$; see Section 1.2.4. These points are then nonlinearly embedded by a smooth map ψ into high-dimensional input space $\mathcal{X} = \Re^r$ $(t \ll r)$ with Euclidean metric $\|\cdot\|_{\mathcal{X}}$. This embedding provides us with the input data $\{\mathbf{x}_i\}$. For example, in the right panel of Figure 1.1, we randomly generated 20,000 three-dimensional points to lie uniformly on the surface of the two-dimensional S-shaped curve displayed in the left panel. Thus, $\psi : \mathcal{M} \rightarrow \mathcal{X}$ is the embedding map, and a point on the manifold, $\mathbf{y} \in \mathcal{M}$, can be expressed as $\mathbf{y} = \phi(\mathbf{x})$, $\mathbf{x} \in \mathcal{X}$, where $\phi = \psi^{-1}$. The goal is to recover $\mathcal{M}$ and find an implicit representation of the map ψ (and, hence, recover the $\{\mathbf{y}_i\}$), given only the input data points $\{\mathbf{x}_i\}$ in $\mathcal{X}$.

Each algorithm computes t'-dimensional estimates, $\{\widehat{\mathbf{y}}_i\}$, of the t-dimensional manifold data, $\{\mathbf{y}_i\}$, for some t'. Such a *reconstruction* is deemed to be successful if $t' = t$, the true (unknown) dimensionality of $\mathcal{M}$. In practice, t' will most likely be too large. Because we require a low-dimensional solution, we retain only the first two or three of the coordinate vectors and plot the corresponding elements of those vectors against each other to yield n points in two- or three-dimensional space. For all practical purposes, such a display is usually sufficient to identify the underlying manifold.

Most of the nonlinear manifold-learning algorithms that we discuss here are based upon different philosophies regarding how one should recover unknown nonlinear manifolds. However, they each consist of a three-step approach (except NONLINEAR PCA). The first and third steps are common to all algorithms: the first step incorporates neighborhood information at each data point to construct a weighted graph having the data points as vertices, and the third step is a spectral embedding step that involves an $(n \times n)$-eigenequation computation. The second step is specific to the algorithm, taking the weighted neighborhood graph and transforming it into suitable input for the spectral embedding step.

1.5.1 Isomap

The *isometric feature mapping* (or ISOMAP) algorithm (Tenenbaum, de Silva, and Langford, 2000) assumes that the smooth manifold $\mathcal{M}$ is a convex region of $\Re^t$ $(t \ll r)$ and that the embedding $\psi : \mathcal{M} \rightarrow \mathcal{X}$ is an isometry. This assumption has two key ingredients:

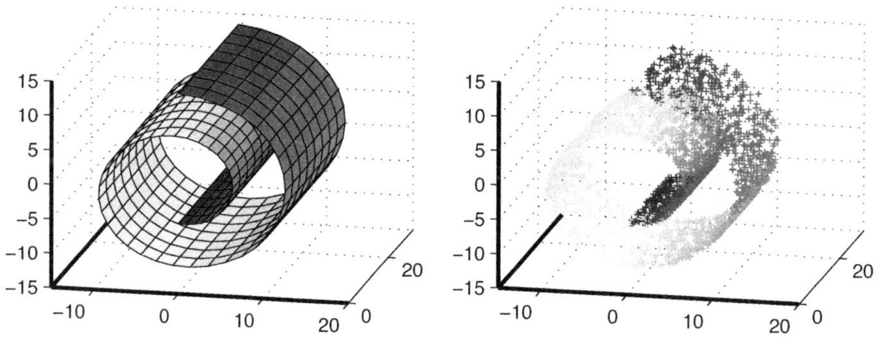

Figure 1.2: (See Color Insert.) Left panel: The Swiss roll: a two-dimensional manifold embedded in three-dimensional space. Right panel: 20,000 data points lying on the surface of the Swiss roll manifold. Reproduced from Izenman (2008, Figure 16.7) with kind permission from Springer Science+Business Media.

- *Isometry:* The geodesic distance is invariant under the map ψ. For any pair of points on the manifold, $\mathbf{y}, \mathbf{y}' \in \mathcal{M}$, the geodesic distance between those points equals the Euclidean distance between their corresponding coordinates, $\mathbf{x}, \mathbf{x}' \in \mathcal{X}$; i.e.,

$$d^{\mathcal{M}}(\mathbf{y}, \mathbf{y}') = \|\mathbf{x} - \mathbf{x}'\|_{\mathcal{X}}, \tag{1.37}$$

  where $\mathbf{y} = \phi(\mathbf{x})$ and $\mathbf{y}' = \phi(\mathbf{x}')$.

- *Convexity:* The manifold $\mathcal{M}$ is a convex subset of $\Re^t$.

ISOMAP considers $\mathcal{M}$ to be a convex region possibly distorted in any of a number of ways (e.g., by folding or twisting). The so-called *Swiss roll*,[2] which is a flat two-dimensional rectangular submanifold of $\Re^3$, is one such example; see Figure 1.2. Empirical studies show that ISOMAP works well for intrinsically flat submanifolds of $\mathcal{X} = \Re^r$ that look like rolled-up sheets of paper or "open" manifolds such as an open box or open cylinder. However, ISOMAP does not perform well if there are any holes in the roll, because this would violate the convexity assumption. The isometry assumption appears to be reasonable for certain types of situations, but, in many other instances, the convexity assumption may be too restrictive (Donoho and Grimes, 2003b).

ISOMAP uses the isometry and convexity assumptions to form a nonlinear generalization of multidimensional scaling (MDS). Recall that MDS looks for a low-dimensional subspace in which to embed input data while preserving the Euclidean interpoint distances (see Section 1.3.2). Unfortunately, working with Euclidean distances in MDS when dealing with curved regions tends to give poor results. ISOMAP follows the general MDS philosophy by attempting to preserve the global geometric properties of the underlying nonlinear manifold, and it does this by approximating *all* pairwise geodesic distances (i.e., lengths of the shortest paths between two points) on the manifold. In this sense, ISOMAP provides a *global* approach to manifold learning.

The ISOMAP algorithm consists of three steps:

1. *Nearest-neighbor search.* Select either an integer K or an $\epsilon > 0$. Calculate the distances,

$$d_{ij}^{\mathcal{X}} = d^{\mathcal{X}}(\mathbf{x}_i, \mathbf{x}_j) = \|\mathbf{x}_i - \mathbf{x}_j\|_{\mathcal{X}}, \tag{1.38}$$

[2]The Swiss roll is generated as follows: for $y_1 \in [3\pi/2, 9\pi/2]$ and $y_2 \in [0, 15]$, set $x_1 = y_1 \cos y_1$, $x_2 = y_1 \sin y_1$, $x_3 = y_2$.

between all pairs of data points $\mathbf{x}_i, \mathbf{x}_j \in \mathcal{X}$, $i, j = 1, 2, \ldots, n$. These are generally taken to be Euclidean distances but may be a different distance metric. Determine which data points are "neighbors" on the manifold $\mathcal{M}$ by connecting each point either to its K nearest neighbors or to all points lying within a ball of radius ϵ of that point. Choice of K or ϵ controls neighborhood size and also the success of ISOMAP.

2. Compute graph distances. This gives us a *weighted neighborhood graph* $\mathcal{G} = \mathcal{G}(\mathcal{V}, \mathcal{E})$, where the set of *vertices* $\mathcal{V} = \{\mathbf{x}_1 \ldots, \mathbf{x}_n\}$ are the input data points, and the set of *edges* $\mathcal{E} = \{e_{ij}\}$ indicate neighborhood relationships between the points. The edge e_{ij} that joins the neighboring points $\mathbf{x}_i$ and $\mathbf{x}_j$ has a weight w_{ij} associated with it, and that weight is given by the "distance" $d_{ij}^{\mathcal{X}}$ between those points. If there is no edge present between a pair of points, the corresponding weight is zero. Estimate the unknown true *geodesic distances*, $\{d_{ij}^{\mathcal{M}}\}$, between pairs of points in $\mathcal{M}$ by *graph distances*, $\{d_{ij}^{\mathcal{G}}\}$, with respect to the graph $\mathcal{G}$. The graph distances are the shortest path distances between all pairs of points in the graph $\mathcal{G}$ (see Equation (1.5)). Points that are not neighbors of each other are connected by a sequence of neighbor-to-neighbor links, and the length of this path (sum of the link weights) is taken to approximate the distance between its endpoints on the manifold.

If the data points are sampled from a probability distribution that is supported by the entire manifold, then, asymptotically (as $n \to \infty$), it turns out that the graph distance $d^{\mathcal{G}}$ converges to the true geodesic distance $d^{\mathcal{M}}$ if the manifold is flat (Bernstein, de Silva, Langford, and Tenenbaum, 2001).

An efficient algorithm for computing the shortest path between every pair of vertices in a graph is *Floyd's algorithm* (Floyd, 1962), which works best with dense graphs (graphs with many edges). However, *Dijkstra's algorithm* (Dijkstra, 1959) is preferred when the graph is sparse. Floyd's algorithm has a worst-case complexity of $O(n^3)$, while Dijkstra's algorithm with Fibonacci heaps has complexity $O(Kn^2 \log n)$, where K is the neighborhood size.

3. Spectral embedding via multidimensional scaling. Let $\mathbf{D}^{\mathcal{G}} = (d_{ij}^{\mathcal{G}})$ be the symmetric $(n \times n)$-matrix of graph distances. Apply the classical-scaling algorithm of MDS (see Section 1.3.2) to $\mathbf{D}^{\mathcal{G}}$ to give the reconstructed data points in a t-dimensional feature space $\mathcal{Y}$, so that the geodesic distances on $\mathcal{M}$ between data points are preserved as much as possible:

- Let $\mathbf{S}^{\mathcal{G}} = ([d_{ij}^{\mathcal{G}}]^2)$ denote the $(n \times n)$ symmetric matrix of squared graph distances. Form the "doubly centered" matrix,

$$\mathbf{A}_n^{\mathcal{G}} = -\frac{1}{2}\mathbf{H}\mathbf{S}^{\mathcal{G}}\mathbf{H}, \tag{1.39}$$

  where $\mathbf{H} = \mathbf{I}_n - n^{-1}\mathbf{J}_n$, and $\mathbf{J}_n$ is an $(n \times n)$-matrix of ones. The matrix $\mathbf{A}_n^{\mathcal{G}}$ will be nonnegative-definite of rank $t < n$.

- The embedding vectors $\{\widehat{\mathbf{y}}_i\}$ are chosen to minimize $\|\mathbf{A}_n^{\mathcal{G}} - \mathbf{A}_n^{\mathcal{Y}}\|$, where $\mathbf{A}_n^{\mathcal{Y}}$ is (1.39) with $\mathbf{S}^{\mathcal{Y}} = ([d_{ij}^{\mathcal{Y}}]^2)$ replacing $\mathbf{S}^{\mathcal{G}}$, and $d_{ij}^{\mathcal{Y}} = \|\mathbf{y}_i - \mathbf{y}_j\|$ is the Euclidean distance between $\mathbf{y}_i$ and $\mathbf{y}_j$. The optimal solution is given by the eigenvectors $\mathbf{v}_1, \ldots, \mathbf{v}_t$ corresponding to the t largest (positive) eigenvalues, $\lambda_1 \geq \cdots \geq \lambda_t$, of $\mathbf{A}_n^{\mathcal{G}}$.

- The graph $\mathcal{G}$ is embedded into $\mathcal{Y}$ by the $(t \times n)$-matrix

$$\widehat{\mathbf{Y}} = (\widehat{\mathbf{y}}_1, \cdots, \widehat{\mathbf{y}}_n) = (\sqrt{\lambda_1}\mathbf{v}_1, \cdots, \sqrt{\lambda_t}\mathbf{v}_t)^\tau. \tag{1.40}$$

  The ith column of $\widehat{\mathbf{Y}}$ yields the embedding coordinates in $\mathcal{Y}$ of the ith data point. The Euclidean distances between the n t-dimensional columns of $\widehat{\mathbf{Y}}$ are collected into the $(n \times n)$-matrix $\mathbf{D}_t^{\mathcal{Y}}$.

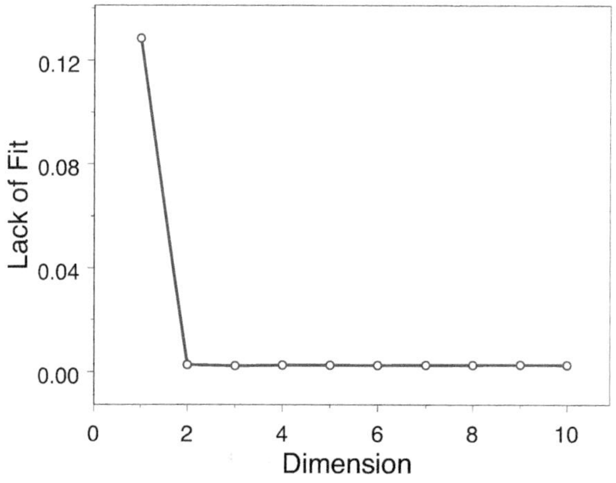

Figure 1.3: ISOMAP dimensionality plot for the first $n = 1{,}000$ Swiss roll data points. The number of neighborhood points is $K = 7$. The plotted points are $(t, 1 - R_t^2)$, $t = 1, 2, \ldots, 10$. Reproduced from Izenman (2008, Figure 16.8) with kind permission from Springer Science+Business Media.

The ISOMAP algorithm appears to work most efficiently with $n \leq 1,000$. To permit ISOMAP to work with much larger data sets, changes in the original algorithm were studied, leading to the LANDMARK ISOMAP algorithm (see below).

We can draw a graph that gives us a good idea of how closely the ISOMAP t-dimensional solution matrix $\mathbf{D}_t^{\mathcal{Y}}$ approximates the matrix $\mathbf{D}^{\mathcal{G}}$ of graph distances. We plot $1 - R_t^2$ against dimensionality t (i.e., $t = 1, 2, \ldots, t^*$, where t^* is some integer such as 10), and

$$R_t^2 = [\text{corr}(\mathbf{D}_t^{\mathcal{Y}}, \mathbf{D}^{\mathcal{G}})]^2 \tag{1.41}$$

is the squared correlation coefficient of all corresponding pairs of entries in the matrices $\mathbf{D}_t^{\mathcal{Y}}$ and $\mathbf{D}^{\mathcal{G}}$. The *intrinsic dimensionality* is taken to be that integer t at which an "elbow" appears in the plot.

Suppose, for example, 20,000 points are randomly and uniformly drawn from the surface of the two-dimensional Swiss roll manifold embedded in three-dimensional space. The 3D scatterplot of the data is given in the right panel of Figure 1.2. Using all 20,000 points as input to the ISOMAP algorithm proves to be overly computationally intensive, and so we use only the first 1,000 points for illustration. Taking $n = 1,000$ and $K = 7$ neighborhood points, Figure 1.3 shows a plot of the values of $1 - R_t^2$ against t for $t = 1, 2, \ldots, 10$, where an elbow correctly shows $t = 2$; the 2D ISOMAP neighborhood-graph solution is given in Figure 1.4.

As we remarked above, the ISOMAP algorithm has difficulty with manifolds that contain holes, have too much curvature, or are not convex. In the case of "noisy" data (i.e., data that do not necessarily lie *on* the manifold), it depends upon how the neighborhood size (either K or ϵ) is chosen; if K or ϵ are chosen neither too large (that it introduces false connections into $\mathcal{G}$) nor too small (that $\mathcal{G}$ becomes too sparse to approximate geodesic paths accurately), then ISOMAP should be able to tolerate moderate amounts of noise in the data.

LANDMARK ISOMAP

If a data set is very large (such as the 20,000 points on the Swiss roll manifold), then the performance of the ISOMAP algorithm is significantly compromised by having to store

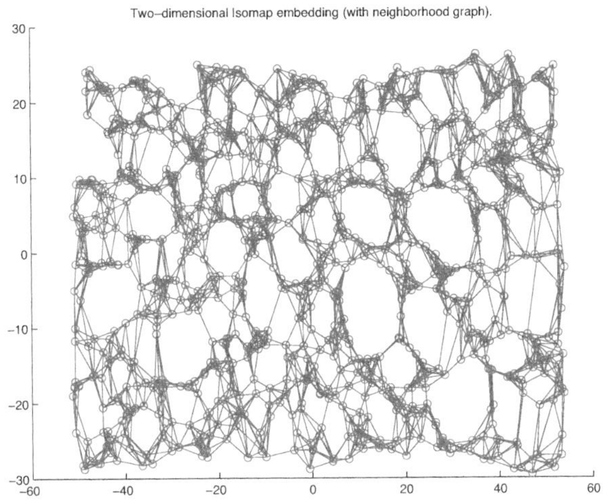

Figure 1.4: Two-dimensional ISOMAP embedding, with neighborhood graph, of the first $n = 1{,}000$ Swiss roll data points. The number of neighborhood points is $K = 7$. Reproduced from Izenman (2008, Figure 16.9) with kind permission from Springer Science+Business Media.

in memory the complete $(n \times n)$-matrix $\mathbf{D}_G$ (Step 2) and carry out an eigenanalysis of the $(n \times n)$-matrix $\mathbf{A}_n$ for the MDS reconstruction (Step 3). If the data are uniformly scattered all around a low-dimensional manifold, then the vast majority of pairwise distances will be redundant; to speed up the MDS embedding step, we eliminate as many of the redundant distance calculations as possible.

In LANDMARK ISOMAP (de Silva and Tenenbaum, 2003), we eliminate such redundancy by designating a subset of m of the n data points as "landmark" points. For example, if $\mathbf{x}_i$ is designated as one of the m landmark points, we calculate only those distances between each of the n points and $\mathbf{x}_i$. Input to the LANDMARK ISOMAP algorithm is, therefore, an $(m \times n)$-matrix of distances. The landmark points may be selected by random sampling or by a judicious choice of "representative" points. The number of such landmark points is left to the researcher, but $m = 50$ works well. In the MDS embedding step, the object is to preserve only those distances between all points and the subset of landmark points. Step 2 in LANDMARK ISOMAP uses *Dijkstra's algorithm* (Dijkstra, 1959), which is faster than Floyd's algorithm for computing graph distances and is generally preferred when the graph is sparse.

Applying LANDMARK ISOMAP to the first $n = 1{,}000$ Swiss roll data points with $K = 7$ and the first $m = 50$ points taken to be landmark points results in an elbow at $t = 2$ in the dimensionality plot; the 2D LANDMARK ISOMAP neighborhood-graph solution is given in Figure 1.5. This is a much faster solution than the one we obtained using the original ISOMAP algorithm. The main differences between Figure 1.4 and Figure 1.5 are roundoff error and a rotation due to sign changes.

Because of the significant increase in computational speed, we can apply LANDMARK ISOMAP to all 20,000 points (using $K = 7$ and $m = 50$); an elbow again correctly appears at $t = 2$ in the dimensionality plot, and the resulting 2D LANDMARK ISOMAP neighborhood-

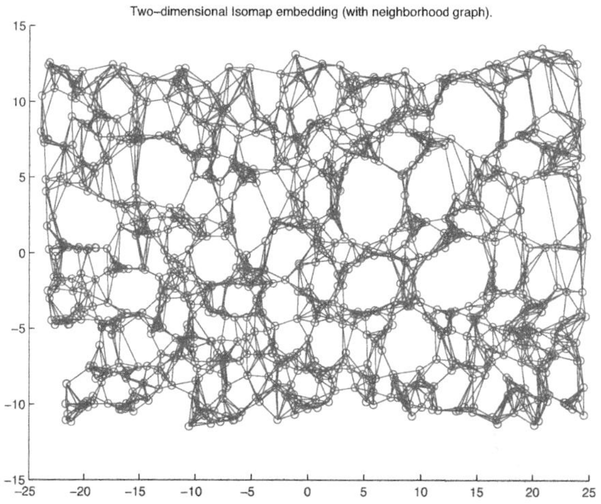

Figure 1.5: Two-dimensional LANDMARK ISOMAP embedding, with neighborhood graph, of the $n = 1{,}000$ Swiss roll data points. The number of neighborhood points is $K = 7$ and the number of landmark points is $m = 50$. Reproduced from Izenman (2008, Figure 16.10) with kind permission from Springer Science+Business Media.

graph solution is given in Figure 1.6.

1.5.2 Local Linear Embedding

The LOCAL LINEAR EMBEDDING (LLE) algorithm (Roweis and Saul, 2000; Saul and Roweis, 2003) for nonlinear dimensionality reduction is similar in spirit to the ISOMAP algorithm, but because it attempts to preserve local neighborhood information on the (Riemannian) manifold (without estimating the true geodesic distances), we view LLE as a *local* approach rather than the global approach exemplified by ISOMAP.

Like ISOMAP, the LLE algorithm also consists of three steps:

1. Nearest-neighbor search. Fix $K \ll r$ and let N_i^K denote the "neighborhood" of $\mathbf{x}_i$ that contains only its K nearest points, as measured by Euclidean distance (K could be different for each point $\mathbf{x}_i$). The success of LLE depends (as does ISOMAP) upon the choice of K: it must be sufficiently large so that the points can be well-reconstructed but also sufficiently small for the manifold to have little curvature. The LLE algorithm is best served if the graph formed by linking each point to its neighbors is connected. If the graph is not connected, the LLE algorithm can be applied separately to each of the disconnected subgraphs.

2. Constrained least-squares fits. Using the notion that every manifold is locally linear, we reconstruct $\mathbf{x}_i$ by a linear function of its K nearest neighbors,

$$\widehat{\mathbf{x}}_i = \sum_{j=1}^{n} w_{ij}\mathbf{x}_j, \tag{1.42}$$

where w_{ij} is a scalar weight for $\mathbf{x}_j$ with unit sum, $\sum_j w_{ij} = 1$, for translation invariance; if $\mathbf{x}_\ell \notin N_i^K$, then set $w_{i\ell} = 0$ in (1.42). Translation invariance implies that adding some constant $\mathbf{c}$ to each of the input vectors $\mathbf{x}_i$, $i = 1, 2, \ldots, n$, does not change the minimizing quantity: $\mathbf{x}_i - \sum_{j=1}^n w_{ij}\mathbf{x}_j \to (\mathbf{x}_i + \mathbf{c}) - \sum_{j=1}^n w_{ij}(\mathbf{x}_j + \mathbf{c}) = \mathbf{x}_i - \sum_{j=1}^n w_{ij}\mathbf{x}_j$. Set $\mathbf{W} = (w_{ij})$

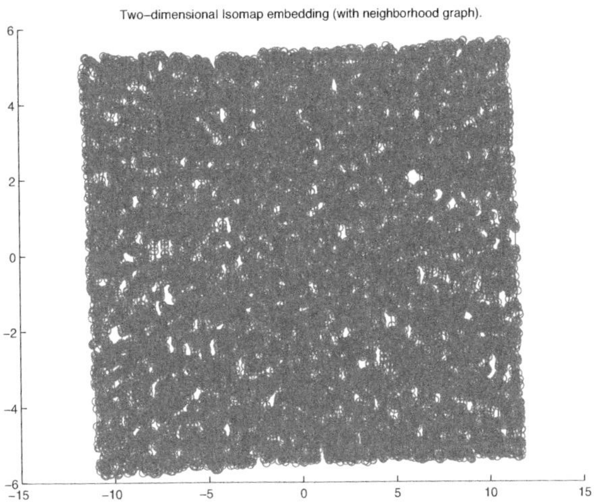

Figure 1.6: Two-dimensional LANDMARK ISOMAP embedding, with neighborhood graph, of the complete set of $n = 20{,}000$ Swiss roll data points. The number of neighborhood points is $K = 7$, and the number of landmark points is $m = 50$. Reproduced from Izenman (2008, Figure 16.11) with kind permission from Springer Science+Business Media.

to be a sparse $(n \times n)$-matrix of weights (there are only nK nonzero elements). Find optimal weights $\{\widehat{w}_{ij}\}$ by solving

$$\widehat{\mathbf{W}} = \arg\min_{\mathbf{W}} \sum_{i=1}^{n} \|\mathbf{x}_i - \sum_{j=1}^{n} w_{ij}\mathbf{x}_j\|^2, \tag{1.43}$$

subject to the *invariance constraint* $\sum_j w_{ij} = 1$, $i = 1, 2, \ldots, n$, and the *sparseness constraint* $w_{i\ell} = 0$ if $\mathbf{x}_\ell \notin N_i^K$. If we consider only convex combinations for (1.42) so that $w_{ij} \geq 0$ for all i, j, then the invariance constraint, $\sum_j w_{ij} = 1$, means that $\mathbf{W}$ could be viewed as a stochastic transition matrix.

The matrix $\widehat{\mathbf{W}}$ is obtained as follows. For a given point $\mathbf{x}_i$, we write the summand of (1.43) as

$$\|\sum_j w_{ij}(\mathbf{x}_i - \mathbf{x}_j)\|^2 = \mathbf{w}_i^\tau \mathbf{G}\mathbf{w}_i, \tag{1.44}$$

where $\mathbf{w}_i = (w_{i1}, \cdots, w_{in})^\tau$, only K of which are non-zero, and $\mathbf{G} = (G_{jk})$, where

$$G_{jk} = (\mathbf{x}_i - \mathbf{x}_j)^\tau(\mathbf{x}_i - \mathbf{x}_k), \quad j, k \in N_i^K,$$

is an $(n \times n)$ Gram matrix (i.e., symmetric and nonnegative-definite). Notice that the matrix $\mathbf{G}$ actually depends upon $\mathbf{x}_i$ and so we write $\mathbf{G}_i$. Using the Lagrangean multiplier μ, we minimize the function

$$f(\mathbf{w}_i) = \mathbf{w}_i^\tau \mathbf{G}_i \mathbf{w}_i - \mu(\mathbf{1}_n^\tau \mathbf{w}_i - 1).$$

Differentiating $f(\mathbf{w}_i)$ with respect to $\mathbf{w}_i$ and setting the result equal to zero yields $\widehat{\mathbf{w}}_i = \frac{\mu}{2}\mathbf{G}_i^{-1}\mathbf{1}_n$. Premultiplying this last result by $\mathbf{1}_n^\tau$ gives us the optimal weights

$$\widehat{\mathbf{w}}_i = \frac{\mathbf{G}_i^{-1}\mathbf{1}_n}{\mathbf{1}_n^\tau \mathbf{G}_i^{-1}\mathbf{1}_n},$$

where it is understood that for $\mathbf{x}_\ell \notin N_i^K$, the corresponding element, $\widehat{w}_{i\ell}$, of $\widehat{\mathbf{w}}_i$ is zero. Note that we can also write $\mathbf{G}_i(\frac{2}{\mu}\widehat{\mathbf{w}}_i) = \mathbf{1}_n$; so, the same result can be obtained by solving the linear system of n equations $\mathbf{G}_i\widehat{\mathbf{w}}_i = \mathbf{1}_n$, where any $\mathbf{x}_\ell \notin N_i^K$ has weight $\widehat{w}_{i\ell} = 0$, and then rescaling the weights to sum to one. The resulting optimal weights for each data point (and all other zero-weights) are collected into a sparse $(n \times n)$-matrix $\widehat{\mathbf{W}} = (\widehat{w}_{ij})$ having only nK nonzero elements.

3. Spectral embedding. Consider the optimal weight matrix $\widehat{\mathbf{W}}$ found at step 2 to be fixed. Now, we find the $(t \times n)$-matrix $\mathbf{Y} = (\mathbf{y}_1.\cdots,\mathbf{y}_n)$, $t \ll r$, of embedding coordinates that solves

$$\widehat{\mathbf{Y}} = \arg\min_{\mathbf{Y}} \sum_{i=1}^{n} \|\mathbf{y}_i - \sum_{j=1}^{n} \widehat{w}_{ij}\mathbf{y}_j\|^2, \tag{1.45}$$

subject to the constraints that the mean vector is zero (i.e., $\sum_i \mathbf{y}_i = \mathbf{Y}\mathbf{1}_n = \mathbf{0}$) and the covariance matrix is the identity (i.e., $n^{-1}\sum_i \mathbf{y}_i\mathbf{y}_i^\tau = n^{-1}\mathbf{Y}\mathbf{Y}^\tau = \mathbf{I}_t$). These constraints determine the translation, rotation, and scale of the embedding coordinates, and that helps ensure that the objective function will be invariant. The matrix of embedding coordinates (1.45) can be written as

$$\widehat{\mathbf{Y}} = \arg\min_{\mathbf{Y}} \mathrm{tr}\{\mathbf{Y}\mathbf{M}\mathbf{Y}^\tau\} \tag{1.46}$$

where $\mathbf{M}$ is the sparse, symmetric, and nonnegative-definite $(n \times n)$-matrix $\mathbf{M} = (\mathbf{I}_n - \widehat{\mathbf{W}})^\tau(\mathbf{I}_n - \widehat{\mathbf{W}})$.

The objective function $\mathrm{tr}\{\mathbf{Y}\mathbf{M}\mathbf{Y}^\tau\}$ in (1.46) has a unique global minimum given by the eigenvectors corresponding to the *smallest* $t+1$ eigenvalues of $\mathbf{M}$. The smallest eigenvalue of $\mathbf{M}$ is zero with corresponding eigenvector $\mathbf{v}_n = n^{-1/2}\mathbf{1}_n$. Because the sum of coefficients of each of the other eigenvectors, which are orthogonal to $n^{-1/2}\mathbf{1}_n$, is zero, if we ignore the smallest eigenvalue (and associated eigenvector), this will constrain the embeddings to have mean zero. The optimal solution then sets the rows of the $(t \times n)$-matrix $\widehat{\mathbf{Y}}$ to be the t remaining n-dimensional eigenvectors of $\mathbf{M}$,

$$\widehat{\mathbf{Y}} = (\widehat{\mathbf{y}}_1,\ldots,\widehat{\mathbf{y}}_n) = (\mathbf{v}_{n-1},\cdots,\mathbf{v}_{n-t})^\tau, \tag{1.47}$$

where $\mathbf{v}_{n-j}$ is the eigenvector corresponding to the $(j+1)$st smallest eigenvalue of $\mathbf{M}$. The sparseness of $\mathbf{M}$ enables eigencomputations to be carried out very efficiently.

Because LLE preserves local (rather than global) properties of the underlying manifold, it is less susceptible to introducing false connections in $\mathcal{G}$ and can successfully embed nonconvex manifolds. However, like ISOMAP, it has difficulty with manifolds that contain holes.

1.5.3 Laplacian Eigenmaps

The *Laplacian eigenmap* algorithm (Belkin and Niyogi, 2002) also consists of three steps. The first and third steps of the Laplacian eigenmap algorithm are very similar to the first and third steps, respectively, of the LLE algorithm.

1. Nearest-neighbor search. Fix an integer K or an $\epsilon > 0$. The neighborhoods of each data point are symmetrically defined: for a K-neighborhood N_i^K of the point $\mathbf{x}_i$, let $\mathbf{x}_j \in N_i^K$ iff $\mathbf{x}_i \in N_j^K$; similarly, for an ϵ-neighborhood N_i^ϵ, let $\mathbf{x}_j \in N_i^\epsilon$ iff $\|\mathbf{x}_i - \mathbf{x}_j\| < \epsilon$, where the norm is Euclidean norm. In general, let N_i denote the neighborhood of $\mathbf{x}_i$.

2. Weighted adjacency matrix. Let $\mathbf{W} = (w_{ij})$ be a symmetric $(n \times n)$ *weighted adjacency matrix* defined as follows:

$$w_{ij} = w(\mathbf{x}_i, \mathbf{x}_j) = \begin{cases} \exp\left\{-\frac{\|\mathbf{x}_i - \mathbf{x}_j\|^2}{2\sigma^2}\right\}, & \text{if } \mathbf{x}_j \in N_i; \\ 0, & \text{otherwise.} \end{cases} \tag{1.48}$$

These weights are determined by the isotropic Gaussian kernel (also known as the *heat kernel*), with scale parameter σ. Denote the resulting weighted graph by $\mathcal{G}$. If $\mathcal{G}$ is not connected, apply step 3 to each connected subgraph.

3. Spectral embedding. Let $\mathbf{D} = (d_{ij})$ be an $(n \times n)$ diagonal matrix with diagonal elements $d_{ii} = \sum_{j \in N_i} w_{ij} = (\mathbf{W}\mathbf{1}_n)_i$, $i = 1, 2, \ldots, n$. The $(n \times n)$ symmetric matrix $\mathbf{L} = \mathbf{D} - \mathbf{W}$ is known as the *graph Laplacian* for the graph $\mathcal{G}$. Let $\mathbf{y} = (y_i)$ be an n-vector. Then, $\mathbf{y}^\tau \mathbf{L}\mathbf{y} = \frac{1}{2}\sum_{i=1}^n \sum_{j=1}^n w_{ij}(y_i - y_j)^2$, so that $\mathbf{L}$ is nonnegative definite.

When data are uniformly sampled from a low-dimensional manifold $\mathcal{M}$ of $\Re^r$, the graph Laplacian $\mathbf{L} = \mathbf{L}_{n,\sigma}$ (considered as a function of n and σ) can be regarded as a discrete approximation to the continuous *Laplace–Beltrami operator* $\Delta_{\mathcal{M}}$ defined on the manifold $\mathcal{M}$, and converges to $\Delta_{\mathcal{M}}$ as $\sigma \to 0$ and $n \to \infty$. Furthermore, when the data are sampled from an arbitrary probability distribution P on the manifold $\mathcal{M}$, then, under certain conditions on $\mathcal{M}$ and P, the graph Laplacian converges to a weighted version of $\Delta_{\mathcal{M}}$ (Belkin and Niyogi, 2008).

The $(t \times n)$-matrix $\mathbf{Y} = (\mathbf{y}_1, \cdots, \mathbf{y}_n)$, which is used to embed the graph $\mathcal{G}$ into the low-dimensional space $\Re^t$, where $\mathbf{y}_i$ yields the embedding coordinates of the ith point, is determined by minimizing the objective function,

$$\sum_i \sum_j w_{ij} \|\mathbf{y}_i - \mathbf{y}_j\|^2 = \text{tr}\{\mathbf{Y}\mathbf{L}\mathbf{Y}^\tau\}. \tag{1.49}$$

In other words, we seek the solution,

$$\widehat{\mathbf{Y}} = \arg \min_{\mathbf{Y}:\mathbf{Y}\mathbf{D}\mathbf{Y}^\tau = \mathbf{I}_t} \text{tr}\{\mathbf{Y}\mathbf{L}\mathbf{Y}^\tau\}, \tag{1.50}$$

where we restrict $\mathbf{Y}$ such that $\mathbf{Y}\mathbf{D}\mathbf{Y}^\tau = \mathbf{I}_t$ to prevent a collapse onto a subspace of fewer than $t-1$ dimensions. The solution is given by the generalized eigenequation, $\mathbf{L}\mathbf{v} = \lambda\mathbf{D}\mathbf{v}$, or, equivalently, by finding the eigenvalues and eigenvectors of the matrix $\widehat{\mathbf{W}} = \mathbf{D}^{-1/2}\mathbf{W}\mathbf{D}^{-1/2}$. The smallest eigenvalue, λ_n, of $\widehat{\mathbf{W}}$ is zero. If we ignore the smallest eigenvalue (and its corresponding constant eigenvector $\mathbf{v}_n = \mathbf{1}_n$), then the best embedding solution in $\Re^t$ is similar to that given by LLE; that is, the rows of $\widehat{\mathbf{Y}}$ are the eigenvectors,

$$\widehat{\mathbf{Y}} = (\widehat{\mathbf{y}}_1, \cdots, \widehat{\mathbf{y}}_n) = (\mathbf{v}_{n-1}, \cdots, \mathbf{v}_{n-t})^\tau, \tag{1.51}$$

corresponding to the next t smallest eigenvalues, $\lambda_{n-1} \leq \cdots \leq \lambda_{n-t}$, of $\widehat{\mathbf{W}}$.

1.5.4 Diffusion Maps

The basic idea of DIFFUSION MAPS (Nadler, Lafon, Coifman, and Kevrekidis, 2005; Coifman and Lafon, 2006) uses a Markov chain constructed over a graph of the data points, followed by an eigenanalysis of the probability transition matrix of the Markov chain. As with the other algorithms in this Section, there are three steps in this algorithm, with the first and second steps the same as for Laplacian eigenmaps. Although a nearest-neighbor search (Step 1) was not explicitly considered in the above papers on diffusion maps as a means of constructing the graph (Step 2), a nearest-neighbor search is included in software packages for computing diffusion maps. For an example in astronomy of a diffusion map incorporating a nearest-neighbor search, see Freeman, Newman, Lee, Richards, and Schafer (2009).

1. Nearest-Neighbor Search. Fix an integer K or an $\epsilon > 0$. Define a K-neighborhood N_i^K or an ϵ-neighborhood N_i^ϵ of the point $\mathbf{x}_i$ as in Step 1 of Laplacian eigenmaps. In general, let N_i denote the neighborhood of $\mathbf{x}_i$.

2. Pairwise Adjacency Matrix. The n data points $\{\mathbf{x}_i\}$ in $\Re^r$ can be regarded as a graph $\mathcal{G} = \mathcal{G}(\mathcal{V}, \mathcal{E})$ with the data points playing the role of vertices $\mathcal{V} = \{\mathbf{x}_1, \ldots, \mathbf{x}_n\}$, and the set of edges $\mathcal{E}$ are the *connection strengths* (or *weights*), $w(\mathbf{x}_i, \mathbf{x}_j)$, between pairs of adjacent vertices,

$$w_{ij} = w(\mathbf{x}_i, \mathbf{x}_j) = \begin{cases} \exp\left\{-\frac{\|\mathbf{x}_i - \mathbf{x}_j\|^2}{2\sigma^2}\right\}, & \text{if } \mathbf{x}_j \in N_i; \\ 0, & \text{otherwise.} \end{cases} \tag{1.52}$$

This is a Gaussian kernel with width σ; however, other kernels may be used. Kernels such as (1.52) ensure that the closer two points are to each other, the larger the value of w. For convenience in exposition, we will suppress the fact that the elements of most of the matrices depend upon the value of σ. Then, $\mathbf{W} = (w_{ij})$ is a *pairwise adjacency matrix* between the n points. To make the matrix $\mathbf{W}$ even more sparse, values of its entries that are smaller than some given threshold (i.e., the points in question are far apart from each other) can be set to zero. The graph $\mathcal{G}$ with weight matrix $\mathbf{W}$ gives information on the local geometry of the data.

3. Spectral embedding. Define $\mathbf{D} = (d_{ij})$ to be a diagonal matrix formed from the matrix $\mathbf{W}$ by setting the diagonal elements, $d_{ii} = \sum_j w_{ij}$, to be the column sums of $\mathbf{W}$ and the off-diagonal elements to be zero. The $(n \times n)$ symmetric matrix $\mathbf{L} = \mathbf{D} - \mathbf{W}$ is the *graph Laplacian* for the graph $\mathcal{G}$. We are interested in the solutions of the generalized eigenequation, $\mathbf{L}\mathbf{v} = \lambda\mathbf{D}\mathbf{v}$, or, equivalently, of the matrix

$$\mathbf{P} = \mathbf{D}^{-1/2}\mathbf{L}\mathbf{D}^{-1/2} = \mathbf{I}_n - \mathbf{D}^{-1/2}\mathbf{W}\mathbf{D}^{-1/2}, \tag{1.53}$$

which is the *normalized graph Laplacian.* The matrix $\mathbf{H} = e^{t\mathbf{P}}$, $t \geq 0$, is usually referred to as the *heat kernel.* By construction, $\mathbf{P}$ is a stochastic matrix with all row sums equal to one, and, thus, can be interpreted as defining a random walk on the graph $\mathcal{G}$.

Let $\mathbf{X}(t)$ denote a Markov random walk over $\mathcal{G}$ using the weights $\mathbf{W}$ and starting at an arbitrary point at time $t = 0$. We choose the next point in our walk according to a given probability. The transition probability from point $\mathbf{x}_i$ to point $\mathbf{x}_j$ in one time step is obtained by normalizing the ith row of $\mathbf{W}$,

$$p(\mathbf{x}_j|\mathbf{x}_i) = \mathrm{P}\{\mathbf{X}(t+1) = \mathbf{x}_j|\mathbf{X}(t) = \mathbf{x}_i\} = \frac{w(\mathbf{x}_i, \mathbf{x}_j)}{\sum_{j=1}^{n} w(\mathbf{x}_i, \mathbf{x}_j)}. \tag{1.54}$$

Then, the matrix $\mathbf{P} = (p(\mathbf{x}_j|\mathbf{x}_i))$ is a *probability transition matrix* with all row sums equal to one, and this matrix defines the entire Markov chain on $\mathcal{G}$. The transition matrix $\mathbf{P}$ has a set of eigenvalues $\lambda_0 = 1 \geq \lambda_1 \geq \cdots \geq \lambda_{n-1} \geq 0$ and a set of left and right eigenvectors, which are defined by

$$\boldsymbol{\phi}_j^\tau \mathbf{P} = \lambda_j \boldsymbol{\phi}_j^\tau, \quad \mathbf{P}\boldsymbol{\psi}_j = \lambda_j \boldsymbol{\psi}_j, \tag{1.55}$$

respectively, where $\boldsymbol{\phi}_k$ and $\boldsymbol{\psi}_\ell$ are biorthogonal; i.e., $\boldsymbol{\phi}_k^\tau \boldsymbol{\psi}_\ell = 1$ if $k = \ell$ and zero otherwise. The largest eigenvalue, $\lambda_0 = 1$, has associated right eigenvector $\boldsymbol{\psi}_0 = \mathbf{1}_n = (1, 1, \cdots, 1)^\tau$ and left eigenvector $\boldsymbol{\phi}_0$. Thus, $\mathbf{P}$ is diagonalizable as the product,

$$\mathbf{P} = \boldsymbol{\Psi}\boldsymbol{\Lambda}\boldsymbol{\Phi}^\tau, \tag{1.56}$$

where $\boldsymbol{\Phi} = (\boldsymbol{\phi}_0, \cdots, \boldsymbol{\phi}_{n-1})$, $\boldsymbol{\Psi} = (\boldsymbol{\psi}_0, \cdots, \boldsymbol{\psi}_{n-1})$, and $\boldsymbol{\Lambda} = \mathrm{diag}\{\lambda_0, \cdots, \lambda_{n-1}\}$.

The matrix $\mathbf{P}^m = (p_m(\mathbf{x}_j|\mathbf{x}_i))$ is the matrix $\mathbf{P}$ multiplied by itself m times, and its ijth element,

$$p_m(\mathbf{x}_j|\mathbf{x}_i) = \mathrm{P}\{\mathbf{X}(t+m) = \mathbf{x}_j|\mathbf{X}(t) = \mathbf{x}_i\}, \tag{1.57}$$

gives the transition probability of going from point $\mathbf{x}_i$ to point $\mathbf{x}_j$ in m time steps. As m increases, the local influence of each vertex in $\mathcal{G}$ spreads out to its neighboring points. Fix

$0 < m < \infty$. We wish to construct a distance measure on $\mathcal{G}$ so that two points, $\mathbf{x}_i$ and $\mathbf{x}_j$, say, will be close if the corresponding conditional probabilities, $p_m(\cdot|\mathbf{x}_i)$ and $p_m(\cdot|\mathbf{x}_j)$, are close. We define the *diffusion distance*

$$d_m^2(\mathbf{x}_i, \mathbf{x}_j) = \| p_m(\cdot|\mathbf{x}_i) - p_m(\cdot|\mathbf{x}_j) \|^2 \tag{1.58}$$

$$= \sum_{\mathbf{z}} (p_m(\mathbf{z}|\mathbf{x}_i) - p_m(\mathbf{z}|\mathbf{x}_j))^2 w(\mathbf{z}), \tag{1.59}$$

with weight function $w(\mathbf{z}) = 1/\phi_0(\mathbf{z})$, where $\phi_0(\mathbf{z})$ is the unique stationary probability distribution for the Markov chain on the graph $\mathcal{G}$ (i.e., $\phi_0(\mathbf{z})$ is the probability of reaching point $\mathbf{z}$ after taking an infinite number of steps — independent of the starting point) and also measures the density of the data points. The diffusion distance gives us information concerning how many paths exist between the points $\mathbf{x}_i$ and $\mathbf{x}_j$; the distance will be small if they are connected by many paths in the graph. From (1.56), the matrix $\mathbf{P}^m$ can be written as

$$\mathbf{P}^m = \mathbf{\Psi}\mathbf{\Lambda}^m\mathbf{\Phi}^\tau, \tag{1.60}$$

with ijth element,

$$p_m(\mathbf{x}_j|\mathbf{x}_i) = \phi_0(\mathbf{x}_j) + \sum_{k=1}^{n-1} \lambda_k^m \psi_k(\mathbf{x}_i)\phi_k(\mathbf{x}_j). \tag{1.61}$$

Substituting (1.61) into (1.59) yields

$$d_m^2(\mathbf{x}_i, \mathbf{x}_j) = \sum_{k=1}^{n-1} \lambda_k^{2m} (\psi_k(\mathbf{x}_i) - \psi_k(\mathbf{x}_j))^2. \tag{1.62}$$

Because the eigenvalues of $\mathbf{P}$ decay relatively fast, we only need to retain the first t terms in the sum (1.62). This gives us a rank-t approximation of $\mathbf{P}^m$. So, we can approximate closely the diffusion distance using only the first t eigenvalues and corresponding eigenvectors,

$$d_m^2(\mathbf{x}_i, \mathbf{x}_j) \approx \sum_{k=1}^{t} \lambda_k^{2m} (\psi_k(\mathbf{x}_i) - \psi_k(\mathbf{x}_j))^2 \tag{1.63}$$

$$= \| \mathbf{\Psi}_m(\mathbf{x}_i) - \mathbf{\Psi}_m(\mathbf{x}_j) \|^2, \tag{1.64}$$

where the *diffusion map* is

$$\mathbf{\Psi}_m(\mathbf{x}) = (\lambda_1^m \psi_1(\mathbf{x}), \cdots, \lambda_t^m \psi_t(\mathbf{x}))^\tau. \tag{1.65}$$

Thus, $\mathbf{\Psi}_m$ embeds the r-vectors $\mathbf{x}_1, \ldots, \mathbf{x}_n$ in $\Re^t$ so that the diffusion distance between the two points $\mathbf{x}_i$ and $\mathbf{x}_j$ is approximated by Euclidean distance in $\Re^t$ between the right eigenvectors (corresponding to the two points) each weighted by λ_k^m. The coordinates of the n t-vectors are given by

$$\widehat{\mathbf{Y}} = (\widehat{\mathbf{y}}_1, \cdots, \widehat{\mathbf{y}}_n) = (\mathbf{\Psi}_m(\mathbf{x}_1), \cdots, \mathbf{\Psi}_m(\mathbf{x}_n)). \tag{1.66}$$

Thus, we see that nonlinear manifold learning using diffusion maps depends upon K or ϵ for the neighborhood definition, the number m of steps taken by the random walk, the scale parameter σ in the Gaussian kernel, and the spectral decay in the eigenvalues of $\mathbf{P}^m$.

1.5.5 Hessian Eigenmaps

Recall that, in certain situations, the convexity assumption for ISOMAP may be too restrictive. Instead, we may require that the manifold $\mathcal{M}$ be locally isometric to an open, connected subset of $\Re^t$. Popular examples include families of "articulated" images (i.e., translated or rotated images of the same object, possibly through time) that are found in a high-dimensional, digitized-image library (e.g., faces, pictures, handwritten numbers or letters). However, if the pixel elements of each 64-pixel-by-64-pixel digitized image are represented as a 4,096-dimensional vector in "pixel space," it would be very difficult to show that the images really live on a low-dimensional manifold, especially if that *image manifold* is unknown.

We can model such images using a vector of smoothly varying *articulation parameters* $\boldsymbol{\theta} \in \Theta$. For example, digitized images of a person's face that are varied by pose and illumination can be parameterized by two pose parameters (expression [happy, sad, sleepy, surprised, wink] and glasses–no glasses) and a lighting direction (centerlight, leftlight, rightlight, normal); similarly, handwritten "2"s appear to be parameterized essentially by two features, bottom loop and top arch (Tenenbaum, de Silva, and Langford, 2000; Roweis and Saul, 2000). To some extent, learning about an underlying image manifold depends upon whether the images are sufficiently scattered around the manifold and how good is the quality of digitization of each image?

HESSIAN EIGENMAPS (Donoho and Grimes, 2003b) were proposed for recovering manifolds of high-dimensional libraries of articulated images where the convexity assumption is often violated. Let $\Theta \subset \Re^t$ be the parameter space and suppose that $\phi : \Theta \to \Re^r$, where $t < r$. Assume $\mathcal{M} = \phi(\Theta)$ is a smooth manifold of articulated images. The isometry and convexity requirements of ISOMAP are replaced by the following weaker requirements:

- *Local Isometry:* ϕ is a locally isometric embedding of Θ into $\Re^r$. For any point $\mathbf{x}'$ in a sufficiently small neighborhood around each point $\mathbf{x}$ on the manifold $\mathcal{M}$, the geodesic distance equals the Euclidean distance between their corresponding parameter points $\boldsymbol{\theta}, \boldsymbol{\theta}' \in \Theta$; that is,
$$d^{\mathcal{M}}(\mathbf{x}, \mathbf{x}') = \|\boldsymbol{\theta} - \boldsymbol{\theta}'\|_{\Theta}, \tag{1.67}$$
where $\mathbf{x} = \phi(\boldsymbol{\theta})$ and $\mathbf{x}' = \phi(\boldsymbol{\theta}')$.

- *Connectedness:* The parameter space Θ is an open, connected subset of $\Re^t$.

The goal is to recover the parameter vector $\boldsymbol{\theta}$ (up to a rigid motion).

First, consider the differentiable manifold $\mathcal{M} \subset \Re^r$. Let $\mathcal{T}_{\mathbf{x}}(\mathcal{M})$ be a tangent space of the point $\mathbf{x} \in \mathcal{M}$, where $\mathcal{T}_{\mathbf{x}}(\mathcal{M})$ has the same number of dimensions as $\mathcal{M}$ itself. We endow $\mathcal{T}_{\mathbf{x}}(\mathcal{M})$ with a (non-unique) system of orthonormal coordinates having the same inner product as $\Re^r$. Think of $\mathcal{T}_{\mathbf{x}}(\mathcal{M})$ as an affine subspace of $\Re^r$ that is spanned by vectors tangent to $\mathcal{M}$ and passes through the point $\mathbf{x}$, with the origin $0 \in \mathcal{T}_{\mathbf{x}}(\mathcal{M})$ identified with $\mathbf{x} \in \mathcal{M}$. Let $N_{\mathbf{x}}$ be a neighborhood of $\mathbf{x}$ such that each point $\mathbf{x}' \in N_{\mathbf{x}}$ has a unique closest point $\boldsymbol{\xi}' \in \mathcal{T}_{\mathbf{x}}(\mathcal{M})$; a point in $N_{\mathbf{x}}$ has local coordinates, $\boldsymbol{\xi} = \boldsymbol{\xi}(\mathbf{x}) = (\xi_1(\mathbf{x}), \ldots, \xi_t(\mathbf{x}))^{\tau}$, say, and these coordinates are referred to as *tangent coordinates*.

Suppose $f : \mathcal{M} \to \Re$ is a C^2-function (i.e., a function with two continuous derivatives) near $\mathbf{x}$. If the point $\mathbf{x}' \in N_{\mathbf{x}}$ has local coordinates $\boldsymbol{\xi} = \boldsymbol{\xi}(\mathbf{x}) \in \Re^t$, then the rule $g(\boldsymbol{\xi}) = f(\mathbf{x}')$ defines a C^2-function $g : U \to \Re$, where U is a neighborhood of $0 \in \Re^r$. The *tangent Hessian* matrix, which measures the "curviness" of f at the point $\mathbf{x} \in \mathcal{M}$, is defined as the ordinary $(t \times t)$ Hessian matrix of g,

$$\mathbf{H}_f^{\tan}(\mathbf{x}) = \left(\frac{\partial^2 g(\boldsymbol{\xi})}{\partial \xi_i \partial \xi_j} \bigg|_{\boldsymbol{\xi}=\mathbf{0}} \right). \tag{1.68}$$

The average "curviness" of f over $\mathcal{M}$ is then the quadratic form,

$$\mathcal{H}(f) = \int_{\mathcal{M}} \| \mathbf{H}_f^{\tan}(\mathbf{x}) \|_F^2 \, d\mathbf{x}, \tag{1.69}$$

where $\| \mathbf{H} \|_F^2 = \sum_i \sum_j H_{ij}^2$ is the squared Frobenius norm of a square matrix $\mathbf{H} = (H_{ij})$. Note that even if we define two different orthonormal coordinate systems for $\mathcal{T}_{\mathbf{x}}(\mathcal{M})$, and hence two different tangent Hessian matrices, $\mathbf{H}_f$ and $\mathbf{H}'_f$, at $\mathbf{x}$, they are related by $\mathbf{H}'_f = \mathbf{U}\mathbf{H}_f\mathbf{U}^\tau$, where $\mathbf{U}$ is orthogonal, so that their Frobenius norms are equal and $\mathcal{H}(f)$ is well-defined.

Donoho and Grimes showed that $\mathcal{H}(f)$ has a $(t+1)$-dimensional nullspace consisting of the constant function and a t-dimensional space of functions spanned by the original isometric coordinates, $\theta_1, \ldots, \theta_t$, which can be recovered (up to a rigid motion) from the null space of $\mathcal{H}(f)$.

The Hessian Locally Linear Embedding (HLLE) algorithm computes a discrete approximation to the functional $\mathcal{H}$ using the data lying on $\mathcal{M}$. There are, again, three steps to this algorithm, which essentially substitutes a quadratic form based upon the Hessian instead of one based upon the Laplacian.

1. Nearest-Neighbor Search. We begin by identifying a neighborhood of each point as in Step 1 of the LLE algorithm. Fix an integer K and let N_i^K denote the K nearest neighbors of the data point $\mathbf{x}_i$ using Euclidean distance.

2. Estimate Tangent Hessian Matrices. Assuming local linearity of the manifold $\mathcal{M}$ in the region of the neighborhood N_i^K, form the $(r \times r)$ covariance matrix $\mathbf{M}_i$ of the K neighborhood-centered points $\mathbf{x}_j - \bar{\mathbf{x}}_i$, $j \in N_i^K$, where $\bar{\mathbf{x}}_i = n^{-1} \sum_{j \in N_i^K} \mathbf{x}_j$, and compute a PCA of the matrix $\mathbf{M}_i$. Assuming $K \geq t$, the first t eigenvectors of $\mathbf{M}_i$ give the tangent coordinates of the K points in N_i^K and provide the best-fitting t-dimensional linear subspace corresponding to $\mathbf{x}_i$. Next, construct an LS estimate, $\widehat{\mathbf{H}}_i$, of the local Hessian matrix $\mathbf{H}_i$ as follows: build a matrix $\mathbf{Z}_i$ by putting all squares and cross-products of the columns of $\mathbf{M}_i$ up to the tth order in its columns, including a column of 1s; so, $\mathbf{Z}_i$ has $1+t+t(t+1)/2$ columns and K rows. Then, apply a Gram–Schmidt orthonormalization to $\mathbf{Z}_i$. The estimated $(t(t+1)/2 \times K)$ tangent Hessian matrix $\widehat{\mathbf{H}}_i$ is given by the transpose of the last $t(t+1)/2$ orthonormal columns of $\mathbf{Z}_i$.

3. Spectral embedding. The estimated local Hessian matrices, $\widehat{\mathbf{H}}_i, i = 1, 2, \ldots, n$, are used to construct a sparse, symmetric, $(r \times r)$-matrix $\widehat{\mathcal{H}} = (\widehat{\mathcal{H}}_{k\ell})$, where

$$\widehat{\mathcal{H}}_{k\ell} = \sum_i \sum_j ((\widehat{\mathbf{H}}_i)_{jk} (\widehat{\mathbf{H}}_i)_{j\ell}). \tag{1.70}$$

$\widehat{\mathcal{H}}$ is a discrete approximation to the functional $\mathcal{H}$. We now follow Step 3 of the LLE algorithm, this time performing an eigenanalysis of $\widehat{\mathcal{H}}$. To obtain the low-dimensional representation that will minimize the curviness of the manifold, we find the *smallest* $t+1$ eigenvectors of $\widehat{\mathcal{H}}$; the smallest eigenvalue will be zero, and its associated eigenvector will consist of constant functions; the remaining t eigenvectors provide the embedding coordinates for $\widehat{\boldsymbol{\theta}}$.

1.5.6 Nonlinear PCA

Another way of dealing with nonlinear manifold learning is to construct nonlinear versions of linear manifold learning techniques. We have already seen how Isomap provides a nonlinear generalization of MDS. How can we generalize PCA to the nonlinear case? In this Section, we briefly describe the basic ideas behind POLYNOMIAL PCA, PRINCIPAL CURVES AND SURFACES, MULTILAYER AUTOASSOCIATIVE NEURAL NETWORKS, and KERNEL PCA.

Polynomial PCA

There have been several different attempts to generalize PCA to data living on or near nonlinear manifolds of a lower-dimensional space than input space. The first such idea was to add to the set of r input variables quadratic, cubic, or higher-degree polynomial transformations of those input variables, and then apply linear PCA. The result is POLYNOMIAL PCA (Gnanadesikan and Wilk, 1969), whose embedding coordinates are the eigenvectors corresponding to the smallest few eigenvalues of the expanded covariance matrix.

In the original study of polynomial PCA, the method was illustrated with a quadratic transformation of bivariate input variables. In this scenario, (X_1, X_2) expands to become $(X_1, X_2, X_1^2, X_2^2, X_1X_2)$. This formulation is feasible, but for larger problems, the possibilities become more complicated. First, the variables in the expanded set will not be scaled in a uniform manner, so that standardization will be necessary, and second, the number of variables in the expanded set will increase rapidly with large r, which will lead to bigger computational problems. Gnanadesikan and Wilk's article, however, gave rise to a variety of attempts to define a more general nonlinear version of PCA.

Principal Curves and Surfaces

The next attempt at creating a nonlinear PCA was PRINCIPAL CURVES AND SURFACES (Hastie, 1984; Hastie and Stuetzle, 1989). A principal curve is a smooth one-dimensional curve that passes through the "middle" of the data, and a principal surface (or *principal manifold*) is a generalization of a principal curve to a smooth two- or higher-dimensional manifold. So, we can visualize principal curves and surfaces as defining a nonlinear manifold in higher-dimensional input space.

Let $\mathbf{x} \in \Re^r$ be a data point and let $\mathbf{f}(\lambda)$ be a curve, $\lambda \in \Lambda$; see Section 1.2.4 for definitions. Project $\mathbf{x}$ to a point on $\mathbf{f}(\lambda)$ that is closest in Euclidean distance to $\mathbf{x}$. Define the *projection index*

$$\lambda_{\mathbf{f}}(\mathbf{x}) = \sup_{\lambda} \left\{ \lambda : \| \mathbf{x} - \mathbf{f}(\lambda) \| = \inf_{\mu} \| \mathbf{x} - \mathbf{f}(\mu) \| \right\}, \tag{1.71}$$

which produces a value of λ for which $\mathbf{f}(\lambda)$ is closest to $\mathbf{x}$. In the case of ties (termed *ambiguous points*), we choose that λ for which the projection index is largest.

We would like to find the $\mathbf{f}$ that minimizes the *reconstruction error*,

$$D^2(\mathbf{X}, \mathbf{f}) = \mathrm{E}\{\| \mathbf{X} - \mathbf{f}(\lambda_{\mathbf{f}}(\mathbf{X})) \|^2\}. \tag{1.72}$$

The quantity $\| \mathbf{x} - \mathbf{f}(\lambda_{\mathbf{f}}(\mathbf{x})) \|^2$, which is the *projection distance* from the point $\mathbf{x}$ to its projected point $\mathbf{f}(\lambda_{\mathbf{f}}(\mathbf{x}))$ on the curve, is an orthogonal distance, not the vertical distance that we use in least-squares regression. If $\mathbf{f}(\lambda)$ satisfies

$$\mathbf{f}(\lambda) = \mathrm{E}\{\mathbf{X} | \lambda_{\mathbf{f}}(\mathbf{X}) = \lambda\}, \quad \text{for almost every } \lambda \in \Lambda, \tag{1.73}$$

then $\mathbf{f}(\lambda)$ is said to be *self-consistent* for $\mathbf{X}$; this property implies that $\mathbf{f}(\lambda)$ is the average of all those data values that project to that point. If $\mathbf{f}(\lambda)$ does not intersect itself and is self-consistent, then it is a principal curve for $\mathbf{X}$. In a variational sense, it can be shown that the principal curve $\mathbf{f}$ is a stationary (or critical) value of the reconstruction error (Hastie and Steutzle, 1989); unfortunately, *all* principal curves are saddle points and so cannot be local minima of the reconstruction error. This implies that cross-validation cannot be used to aid in determining principal curves. So, we look in a different direction for a method to estimate principal curves.

Suppose we are given n observations, $\mathbf{X}_1, \ldots, \mathbf{X}_n$, on $\mathbf{X}$. Estimate the reconstruction error (1.72) by

$$D^2(\{\mathbf{x}_i\}, \mathbf{f}) = \sum_{i=1}^{n} \| \mathbf{x}_i - \mathbf{f}(\lambda_{\mathbf{f}}(\mathbf{x}_i)) \|^2, \tag{1.74}$$

and estimate $\mathbf{f}$ by minimizing (1.74); that is,

$$\widehat{\mathbf{f}} = \arg\min_{\mathbf{f}} D^2(\{\mathbf{x}_i\}, \mathbf{f}). \tag{1.75}$$

This minimization is carried out using an algorithm that alternates between a projection step (estimating λ assuming a fixed $\mathbf{f}$) and an expectation step (estimating $\mathbf{f}$ assuming a fixed λ); see Izenman (2008, Section 16.3.3) for details. This algorithm, however, can yield biased estimates of $\mathbf{f}$. A modification of the algorithm (Banfield and Raftery, 1992), which reduced the bias, was applied to the problem of charting the outlines of ice floes above a certain size from satellite images of the polar regions.

These basic ideas were extended to *principal surfaces* for two (or higher) dimensions (Hastie, 1984; LeBlanc and Tibshirani, 1994). In the two-dimensional case, for example, $\boldsymbol{\lambda} = (\lambda_1, \lambda_2) \in \Lambda \subseteq \Re^2$ and $\mathbf{f} : \Lambda \rightarrow \Re^r$. A continuous two-dimensional surface in $\Re^r$ is given by

$$\mathbf{f}(\boldsymbol{\lambda}) = (f_1(\boldsymbol{\lambda}), \cdots, f_r(\boldsymbol{\lambda}))^\tau = (f_1(\lambda_1, \lambda_2), \cdots, f_r(\lambda_1, \lambda_2))^\tau, \tag{1.76}$$

which is an r-vector of smooth, continuous, coordinate functions parametrized by $\boldsymbol{\lambda} = (\lambda_1, \lambda_2)$. The generalization of the projection index (1.71) is given by

$$\boldsymbol{\lambda}_{\mathbf{f}}(\mathbf{x}) = \sup_{\boldsymbol{\lambda}} \left\{ \boldsymbol{\lambda} : \| \mathbf{x} - \mathbf{f}(\boldsymbol{\lambda}) \| = \inf_{\boldsymbol{\mu}} \| \mathbf{x} - \mathbf{f}(\boldsymbol{\mu}) \| \right\}, \tag{1.77}$$

which yields the value of $\boldsymbol{\lambda}$ corresponding to the point on the surface closest to $\mathbf{x}$. A principal surface satisfies the self-consistency property,

$$\mathbf{f}(\boldsymbol{\lambda}) = \mathrm{E}\{\mathbf{X} | \boldsymbol{\lambda}_{\mathbf{f}}(\mathbf{X}) = \boldsymbol{\lambda}\}, \quad \text{for almost every } \boldsymbol{\lambda} \in \Lambda. \tag{1.78}$$

The estimated reconstruction error corresponding to (1.74) is given by

$$D^2(\{\mathbf{x}_i\}, \mathbf{f}) = \sum_{i=1}^{n} \| \mathbf{x}_i - \mathbf{f}(\boldsymbol{\lambda}_{\mathbf{f}}(\mathbf{x}_i)) \|^2, \tag{1.79}$$

and $\mathbf{f}$ is estimated by minimizing (1.79). There are difficulties, however, in generalizing the projection-expectation algorithm for principal curves to principal surfaces, and an alternative approach is necessary. LeBlanc and Tibshirani (1994) propose an adaptive algorithm for obtaining $\mathbf{f}$ and they give some examples. See also Malthouse (1998).

Multilayer Autoassociative Neural Networks

Principal curves are closely related to MULTILAYER AUTOASSOCIATIVE NEURAL NETWORKS (Kramer, 1991), which is another proposed formulation of nonlinear PCA. This special type of artificial neural network consists of (at least) a five-layer model in which the middle three hidden layers of nodes are the mapping layer, the bottleneck layer, and the demapping layer, respectively, and each is defined by a nonlinear activation function. There may be more than one mapping and demapping layers. The bottleneck layer has fewer nodes than either the mapping or demapping layers, and is the most important feature of the network because it reduces the dimensionality of the inputs through data compression. The network is run using feedforward connections trained by backpropagation. Although the projection index $\lambda_{\mathbf{f}}$ used in the definition of principal curves can be a discontinuous function, the neural network version of $\lambda_{\mathbf{f}}$ is a continuous function, and this difference causes severe problems with the latter's application as a version of nonlinear PCA; see Malthouse (1998) for details.

Kernel PCA

The most popular nonlinear PCA technique is that of KERNEL PCA (Scholkopf, Smola, and Muller, 1998), which builds upon the theory of kernel methods used to define support vector machines.

Suppose we have a set of n input data points $\mathbf{x}_i \in \Re^r$, $i = 1, 2, \ldots, n$. Kernel PCA is the following two-step process:

1. Make a nonlinear transformation of the ith input data point $\mathbf{x}_i \in \Re^r$ into the point
$$\mathbf{\Phi}(\mathbf{x}_i) = (\phi_1(\mathbf{x}_i), \cdots, \phi_{N_{\mathcal{H}}}(\mathbf{x}_i))^\tau \in \mathcal{H}, \tag{1.80}$$
where $\mathcal{H}$ is an $N_{\mathcal{H}}$-dimensional *feature space*, $i = 1, 2, \ldots, n$. The map $\mathbf{\Phi} : \Re^r \to \mathcal{H}$ is called a *feature map*, and each of the $\{\phi_j\}$ is a nonlinear map. Note that $\mathcal{H}$ could be an infinite-dimensional space.

2. Given the points $\mathbf{\Phi}(\mathbf{x}_1), \ldots, \mathbf{\Phi}(\mathbf{x}_n) \in \mathcal{H}$, with $\sum_{i=1}^n \mathbf{\Phi}(\mathbf{x}_i) = \mathbf{0}$, solve a linear PCA problem in feature space $\mathcal{H}$, where $N_{\mathcal{H}} > r$.

The basic idea here is that low-dimensional structure may be more easily discovered if it is embedded in a larger space $\mathcal{H}$.

In the following, we assume that $\mathcal{H}$ is an $N_{\mathcal{H}}$-dimensional Hilbert space with inner product $\langle \cdot, \cdot \rangle$ and norm $\| \cdot \|_{\mathcal{H}}$. For example, suppose $\boldsymbol{\xi}_j = (\xi_{j1}, \cdots, \xi_{jN_{\mathcal{H}}})^\tau \in \mathcal{H}$, $j = 1, 2$. Then $\langle \boldsymbol{\xi}_1, \boldsymbol{\xi}_2 \rangle = \boldsymbol{\xi}_1^\tau \boldsymbol{\xi}_2 = \sum_{i=1}^{N_{\mathcal{H}}} \xi_{1i}\xi_{2i}$, and if $\boldsymbol{\xi} = (\xi_1, \cdots, \xi_{N_{\mathcal{H}}})^\tau \in \mathcal{H}$, then $\| \boldsymbol{\xi} \|_{\mathcal{H}}^2 = \sum_{i=1}^{N_{\mathcal{H}}} \xi_i^2$.

Following Step 1, suppose we have n $N_{\mathcal{H}}$-vectors $\mathbf{\Phi}(\mathbf{x}_1), \ldots, \mathbf{\Phi}(\mathbf{x}_n)$, which we have standardized so that $\sum_{i=1}^n \mathbf{\Phi}(\mathbf{x}_i) = \mathbf{0}$. Step 2 says that we carry out a linear PCA of these transformed points. This is accomplished by computing the sample covariance matrix,

$$\mathbf{C} = n^{-1} \sum_{i=1}^{n} \mathbf{\Phi}(\mathbf{x}_i)\mathbf{\Phi}(\mathbf{x}_i)^\tau, \tag{1.81}$$

and then computing the eigenvalues and associated eigenvectors of $\mathbf{C}$. The eigenequation is $\mathbf{C}\mathbf{v} = \lambda\mathbf{v}$, where $\mathbf{v} \in \mathcal{H}$ is the eigenvector corresponding to the eigenvalue $\lambda \geq 0$ of $\mathbf{C}$. We can rewrite this eigenequation in an equivalent form as

$$\langle \mathbf{\Phi}(\mathbf{x}_i), \mathbf{C}\mathbf{v} \rangle = \lambda \langle \mathbf{\Phi}(\mathbf{x}_i), \mathbf{v} \rangle, \quad i = 1, 2, \ldots, n. \tag{1.82}$$

Now, from (1.81),

$$\mathbf{C}\mathbf{v} = n^{-1} \sum_{i=1}^{n} \mathbf{\Phi}(\mathbf{x}_i) \langle \mathbf{\Phi}(\mathbf{x}_i), \mathbf{v} \rangle. \tag{1.83}$$

So, all solutions $\mathbf{v}$ with nonzero eigenvalue λ are contained in the span of $\mathbf{\Phi}(\mathbf{x}_1), \ldots, \mathbf{\Phi}(\mathbf{x}_n)$. Thus, there exist coefficients, $\alpha_1, \ldots, \alpha_n$, such that

$$\mathbf{v} = \sum_{i=1}^{n} \alpha_i \mathbf{\Phi}(\mathbf{x}_i). \tag{1.84}$$

Substituting (1.84) for $\mathbf{v}$ into (1.82), we get that

$$n^{-1}\sum_{j=1}^{n}\alpha_j\left\langle \mathbf{\Phi}(\mathbf{x}_i), \sum_{k=1}^{n}\mathbf{\Phi}(\mathbf{x}_k)\right\rangle \langle\mathbf{\Phi}(\mathbf{x}_k), \mathbf{\Phi}(\mathbf{x}_j)\rangle = \lambda\sum_{k=1}^{n}\alpha_k\langle\mathbf{\Phi}(\mathbf{x}_i), \mathbf{\Phi}(\mathbf{x}_k)\rangle, \tag{1.85}$$

for all $i = 1, 2, \ldots, n$. Solving this eigenequation depends upon being able to compute inner products of the form $\langle\mathbf{\Phi}(\mathbf{x}_i), \mathbf{\Phi}(\mathbf{x}_j)\rangle$ in feature space $\mathcal{H}$. Computing these inner products in $\mathcal{H}$ would be computationally intensive and expensive bcause of the high dimensionality involved. This is where we apply the so-called *kernel trick*. The trick is to use a nonlinear kernel function,

$$K_{ij} = K(\mathbf{x}_i, \mathbf{x}_j) = \langle\mathbf{\Phi}(\mathbf{x}_i), \mathbf{\Phi}(\mathbf{x}_j)\rangle, \tag{1.86}$$

in input space. There are several types of kernel functions that are used in contexts such as this. Examples of kernel functions include a polynomial of degree d ($K(\mathbf{x}, \mathbf{y}) = (\langle\mathbf{x}, \mathbf{y}\rangle + c)^d$) and a Gaussian radial-basis function ($K(\mathbf{x}, \mathbf{y}) = \exp\{-\parallel \mathbf{x} - \mathbf{y} \parallel^2 /2\sigma^2\}$). For further details on kernel functions, see Izenman (2008, Sections 11.3.2–11.3.4) or Shawe-Taylor and Cristianini (2004).

Define the $(n \times n)$-matrix $\mathbf{K} = (K_{ij})$. Then, we can rewrite the eigenequation (1.85) as

$$\mathbf{K}^2 = n\lambda\mathbf{K}\boldsymbol{\alpha}, \tag{1.87}$$

or

$$\mathbf{K}\boldsymbol{\alpha} = \widetilde{\lambda}\boldsymbol{\alpha}, \tag{1.88}$$

where $\boldsymbol{\alpha} = (\alpha_1, \cdots, \alpha_n)^\tau$ and $\widetilde{\lambda} = n\lambda$. Denote the ordered eigenvalues of $\mathbf{K}$ by $\widetilde{\lambda}_1 \geq \widetilde{\lambda}_2 \geq \cdots \geq \widetilde{\lambda}_n \geq 0$, with associated eigenvectors $\boldsymbol{\alpha}_1, \ldots, \boldsymbol{\alpha}_n$, where $\boldsymbol{\alpha}_i = (\alpha_{i1}, \cdots, \alpha_{in})^\tau$. If we require that $\langle\mathbf{v}_i, \mathbf{v}_i\rangle = 1$, $i = 1, 2, \ldots, n$, then, using the expansion (1.84) for $\mathbf{v}_i$ and the eigenequation (1.85), we have that

$$\begin{aligned} 1 &= \sum_{j=1}^{n}\sum_{k=1}^{n}\alpha_{ij}\alpha_{ik}\langle\mathbf{\Phi}(\mathbf{x}_j), \mathbf{\Phi}(\mathbf{x}_k)\rangle \\ &= \sum_{j=1}^{n}\sum_{k=1}^{n}\alpha_{ij}\alpha_{ik}K_{jk} \\ &= \langle\boldsymbol{\alpha}_i, \mathbf{K}\boldsymbol{\alpha}_i\rangle = \widetilde{\lambda}_i\langle\boldsymbol{\alpha}_i, \boldsymbol{\alpha}_i\rangle, \end{aligned} \tag{1.89}$$

which determines the normalization for the vectors $\boldsymbol{\alpha}_1, \ldots, \boldsymbol{\alpha}_n$.

The *nonlinear principal component scores* of a test point $\mathbf{x}$ corresponding to $\mathbf{\Phi}$ are given by the projection of $\mathbf{\Phi}(\mathbf{x}) \in \mathcal{H}$ onto the eigenvectors $\mathbf{v}_k \in \mathcal{H}$,

$$\langle\mathbf{v}_k, \mathbf{\Phi}(\mathbf{x})\rangle = \lambda_k^{-1/2}\sum_{i=1}^{n}\alpha_{ki}K(\mathbf{x}_i, \mathbf{x}), \quad k = 1, 2, \ldots, n, \tag{1.90}$$

where we used (1.86) and the $\lambda_k^{-1/2}$ term is included so that $\langle\mathbf{v}_k, \mathbf{v}_k\rangle = 1$. Suppose we set $\mathbf{x} = \mathbf{x}_m$ in (1.90). Then, $\langle\mathbf{v}_k, \mathbf{\Phi}(\mathbf{x}_m)\rangle = \lambda_k^{-1/2}\sum_i \alpha_{ki}K_{im} = \lambda_k^{-1/2}(\mathbf{K}\boldsymbol{\alpha}_k)_m = \lambda_k^{-1/2}(\lambda_k\boldsymbol{\alpha}_k)_m \propto \alpha_{km}$, where $(\mathbf{A})_m$ stands for the mth row of the matrix $\mathbf{A}$.

We assumed in Step 2 that the $\mathbf{\Phi}$-images in feature space have been centered; i.e., $\sum_{i=1}^{n}\mathbf{\Phi}(\mathbf{x}_i) = \mathbf{0}$. How can we do this if we do not know $\mathbf{\Phi}$? It turns out that knowing $\mathbf{\Phi}$ is not necessary because all we need to know is $\mathbf{K}$. Following our discussion on multidimensional scaling (see Section 1.3.2), let $\mathbf{H} = \mathbf{I}_n - n^{-1}\mathbf{J}_n$ be a "centering" matrix, where

$\mathbf{J}_n = \mathbf{1}_n\mathbf{1}_n^\tau$ is an $(n \times n)$-matrix of all ones, and $\mathbf{1}_n$ is an n-vector of all ones. Set

$$\begin{aligned} \widetilde{\mathbf{K}} &= \mathbf{HKH} \\ &= \mathbf{K} - \mathbf{K}(n^{-1}\mathbf{J}_n) - (n^{-1}\mathbf{J}_n)\mathbf{K} + (n^{-1}\mathbf{J}_n)\mathbf{K}(n^{-1}\mathbf{J}_n), \end{aligned} \tag{1.91}$$

which corresponds to starting with a centered $\mathbf{\Phi}$.

It can be shown that by defining the kernel in an appropriate way, kernel PCA is closely related to the ISOMAP, LLE, and LAPLACIAN EIGENMAPS algorithms; see Ham, Lee, Mika, and Scholkopf (2003).

1.6 Summary

When high-dimensional data, such as those obtained from images or videos, lie on or near a manifold of a lower-dimensional space, it is important to learn the structure of that manifold. This chapter presents an overview of various methods proposed for manifold learning. We first reviewed the notion of a smooth manifold using basic concepts from topology and differential geometry. To learn a linear manifold, we described the global embedding algorithms of principal component analysis and multidimensional scaling. In some situations, however, linear methods fail to discover the structure of curved or nonlinear manifolds. Methods for learning nonlinear manifolds attempt to preserve either the local or global structure of the manifold. This led to the development of spectral embedding algorithms such as Isomap, local linear embedding, Laplacian eigenmaps, Hessian eigenmaps, and diffusion maps. We showed that such algorithms consist of three steps: a nearest-neighbor search in high-dimensional input space, a computation of distances between points based upon the neighborhood graph obtained from the previous step, and an eigenproblem for embedding the points into a lower-dimensional space. We also described various nonlinear versions of principal component analysis.

1.7 Acknowledgment

The author thanks Boaz Nadler for helpful correspondence on diffusion maps.

Bibliography

[1] Anderson, T.W. (1963). Asymptotic theory for principal component analysis, *Annals of Mathematical Statistics*, 36, 413–432.

[2] Aswani, A., Bickel, P., and Tomlin, C. (2011). Regression on manifolds: estimation of the exterior derivative, *The Annals of Statistics*, 39, 48–81.

[3] Baik, J. and Silverstein, J.W. (2006). Eigenvalues of large sample covariance matrices of spiked population models, *Journal of Multivariate Analysis*, 97, 1382–1408.

[4] Bai, Z.D. and Silverstein, J.W. (2009). *Spectral Analysis of Large Dimensional Random matrices, 2nd Edition*, New York: Springer.

[5] Banfield, J.D. and Raftery, A.E. (1992). Ice floe identification in satellite images using mathematical morphology and clustering about principal curves, *Journal of the American Statistical Association*, 87, 7–16.

[6] Belkin, M. and Niyogi, P. (2002). Laplacian eigenmaps and spectral techniques for embedding and clustering, *Advances in Neural Information Processing Systems 14* (T.G. Dietterich, S. Becker, and Z. Ghahramani, eds.), Cambridge, MA: MIT Press, pp. 585–591.

[7] Belkin, M. and Niyogi, P. (2008). Towards a theoretical foundation for Laplacian-based manifold methods, *Journal of Computer and System Sciences*, 74, 1289–1308.

[8] Belkin, M., Niyogi, P., and Sindhwani, V. (2006). Manifold regularization: a geometric framework for learning from labeled and unlabeled examples, *Journal of Machine Learning Research*, 7, 2399–2434.

[9] Bernstein, M., de Silva, V., Langford, J.C., and Tenenbaum, J.B. (2001). Graph approximations to geodesics on embedded manifolds, Unpublished Technical Report, Stanford University.

[10] Bickel, P.J. and Li, B. (2007). Local polynomial regression on unknown manifolds, in *Complex Datasets and Inverse Problems: Tomography, Networks, and Beyond, Institute of Mathematical Statistics Lecture Notes – Monograph Series*, 54, 177–186, Beachwood, OH: IMS.

[11] Bourbaki, N. (1989). *General Topology* (2 volumes), New York: Springer.

[12] Brillinger, D.R. (1969). The canonical analysis of stationary time series, in *Multivariate Analysis II* (ed. P.R. Krishnaiah), pp. 331–350, New York: Academic Press.

[13] De Silva, V. and Tenenbaum, J.B. (2003). Unsupervised learning of curved manifolds, In *Nonlinear Estimation and Classification* (D.D. Denison, M.H. Hansen, C.C. Holmes, B. Mallick, and B. Yu, eds.), *Lecture Notes in Statistics*, 171, pp. 453–466, New York: Springer.

[14] De la Torres, F. and Black, M.J. (2001). Robust principal component analysis for computer vision, *Proceedings of the International Conference on Computer Vision*, Vancouver, Canada.

[15] Diaconis, P., Goel, S., and Holmes, S. (2008). Horseshoes in multidimensional scaling and local kernel methods, *The Annals of Applied Statistics*, 2, 777–807.

[16] Dieudonné, J. (2009). *A History of Algebraic and Differential Topology, 1900–1960*, Boston, MA: Birkhäuser.

[17] Dijkstra, E.W. (1959). A note on two problems in connection with graphs, *Numerische Mathematik*, 1, 269–271.

[18] Donoho, D. and Grimes, C. (2003a). Local ISOMAP perfectly recovers the underlying parametrization of occluded/lacunary libraries of articulated images, unpublished technical report, Department of Statistics, Stanford University.

[19] Donoho, D. and Grimes, C. (2003b). Hessian eigenmaps: locally linear embedding techniques for high-dimensional data, *Proceedings of the National Academy of Sciences*, 100, 5591–5596.

[20] Floyd, R.W. (1962). Algorithm 97, *Communications of the ACM*, 5, 345.

[21] Fréchet, M. (1906). *Sur Quelques Points du Calcul Fontionnel*, doctorol dissertation, École normale Supérieure, Paris, France.

[22] Freeman, P.E., Newman, J.A., Lee, A.B., Richards, J.W., and Schafer, C.M. (2009). Photometric redshift estimation using spectral connectivity analysis, *Monthly Notices of the Royal Astronomical Society*, 398, 2012–2021.

[23] Gnanadesikan, R. and Wilk, M.B. (1969). Data analytic methods in multivariate statistical analysis, In *Multivariate Analysis II* (P.R. Krishnaiah, ed.), New York: Academic Press.

[24] Goldberg, Y., Zakai, A., Kushnir, D., and Ritov, Y. (2008). Manifold learning: the price of normalization, *Journal of Machine Learning Research*, 9, 1909–1939.

[25] Ham, J., Lee, D.D., Mika, S., and Schölkopf, B. (2003). A kernel view of the dimensionality reduction of manifolds, Technical Report TR–110, Max Planck Institut für biologische Kybernetik, Germany.

[26] Hastie, T. (1984). Principal curves and surfaces, Technical Report, Department of Statistics, Stanford University.

[27] Hastie, T. and Steutzle, W. (1989). Principal curves, *Journal of the American Statistical Association*, 84, 502–516.

[28] Holm, L. and Sander, C. (1996). Mapping the protein universe, *Science*, 273, 595–603.

[29] Hotelling, H. (1933). Analysis of a complex of statistical variables into principal components, *Journal of Educational Psychology*, 24, 417–441, 498–520.

[30] Hou, J., Sims, G.E., Zhang, C., and Kim, S.-H. (2003). A global representation of the protein fold space, *Proceedings of the National Academy of Sciences*, 100, 2386–2390.

[31] Hou, J., Jun, S.-R., Zhang, C., and Kim, S.-H. (2005). Global mapping of the protein structure space and application in structure-based inference of protein function, *Proceedings of the National Academy of Sciences*, 102, 3651–3656.

[32] Izenman, A.J. (2008). *Modern Multivariate Statistical Techniques: Regression, Classification, and Manifold Learning*, New York: Springer.

[33] James, I.M. (ed.) (1999). *History of Topology*, Amsterdam, Netherlands: Elsevier B.V.

[34] Johnstone, I.M. (2001). On the distribution of the largest eigenvalue in principal components analysis, *The Annals of Statistics*, 29, 295–327.

[35] Johnstone, I.M. (2006). High dimensional statistical inference and random matrices, *Proceedings of the International Congress of Mathematicians*, Madrid, Spain, 307–333.

[36] Kelley, J.L. (1955). *General Topology*, Princeton, NJ: Van Nostrand. Reprinted in 1975 by Springer.

[37] Kim, J., Ahn, Y., Lee, K., Park, S.H., and Kim, S. (2010). A classification approach for genotyping viral sequences based on multidimensional scaling and linear discriminant analysis, *BMC Bioinformatics*, 11, 434.

[38] Kramer, M.A. (1991). Nonlinear principal component analysis using autoassociative neural networks, *AIChE Journal*, 37, 233–243.

[39] Kreyszig, E. (1991). *Differential Geometry*, Dover Publications.

[40] Kühnel, W. (2000). *Differential Geometry: Curves–Surfaces–Manifolds, 2nd Edition*, Providence, RI: American Mathematical Society.

[41] Lafferty, J. and Wasserman, L. (2007). Statistical analysis of semisupervised regression, in *Advances in Neural Information Processing Systems (NIPS)*, 20, 8 pages.

[42] LeBlanc, M. and Tibshirani, R. (1994). Adaptive principal surfaces, *Journal of the American Statistical Association*, 89, 53–64.

[43] Lee, J.A. and Verleysen, M. (2007). *Nonlinear Dimensionality Reduction*, New York: Springer.

[44] Lee, J.M. (2002). *Introduction to Smooth Manifolds*, New York: Springer.

[45] Lin, Z. and Altman, R.B. (2004). Finding haplotype-tagging SNPs by use of principal components analysis, *American Journal of Human Genetics*, 75, 850–861.

[46] Lu, F., Keles, S., Wright, S.J., and Wahba, G. (2005). Framework for kernel regularization with application to protein clustering, *Proceedings of the National Academy of Sciences*, 102, 12332–12337.

[47] Malthouse, E.C. (1998). Limitations on nonlinear PCA as performed with generic neural networks, *IEEE Transactions on Neural Networks*, 9, 165–173.

[48] Mardia, K.V. (1978). Some properties of classical multidimensional scaling, *Communications in Statistical Theory and Methods, Series A*, 7, 1233–1241.

[49] Mehta, M.L. (2004). *Random Matrices, 3rd Edition*, Pure and Applied Mathematics (Amsterdam), 142, Amsterdam, Netherlands: Elsevier/Academic Press.

[50] Mendelson, B. (1990). *Introduction to Topology, 3rd Edition*, New York: Dover Publications.

[51] Nadler, B., Lafon, S., Coifman, R.R., and Kevrekidis, I.G. (2005). Diffusion maps, spectral clustering, and eigenfunctions of Fokker–Planck operators, *Neural Information Processing Systems (NIPS)*, 18, 8 pages.

[52] Niyogi, P. (2008). Manifold regularization and semi-supervised learning: some theoretical analyses, Technical Report TR-2008-01, Department of Computer Science, The University of Chicago.

[53] Pressley, A. (2010). *Elementary Differential Geometry, 2nd Edition*, New York: Springer.

[54] Riemann, G.F.B. (1851). *Grundlagen für eine allgemeine Theorie der Functionen einer veränderlichen complexen Grösse*, doctoral dissertation, University of Göttingen, Germany.

[55] Roweis, S.T. and Saul, L.K. (2000). Nonlinear dimensionality reduction by locally linear embedding, *Science*, 290, 2323–2326.

[56] Saul, L.K. and Roweis, S.T. (2003). Think globally, fit locally: unsupervised learning of low dimensional manifolds, *Journal of Machine Learning Research*, 4, 119–155.

[57] Schölkopf, B., Smola, A.J., and Muller, K.-R. (1998). Nonlinear component analysis as a kernel eigenvalue problem, *Neural Computation*, 10, 1299–1319.

[58] Shawe-Tayor, J. and Cristianini, N. (2004). *Kernel Methods for Pattern Analysis*, Cambridge, U.K.: Cambridge University Press.

[59] Spivak, M. (1965). *Calculus on Manifolds*, New York: Benjamin.

[60] Steen, L.A. (1995). *Counterexamples in Topology*, New York: Dover Publications.

[61] Steiner, J.E., Menezes, R.B., Ricci, T.V., and Oliveira, A.S. (2009). PCA tomography: how to extract information from datacubes, *Monthly Notices of the Royal Astronomical Society*, 395, 64–75.

[62] Tenenbaum, J.B., de Silva, V., and Langford, J.C. (2000). A global geometric framework for nonlinear dimensionality reduction, *Science*, 290, 2319–2323.

[63] Torgerson, W.S. (1952). Multidimensional scaling: I. Theory and method, *Psychometrika*, 17, 401–419.

[64] Torgerson, W.S. (1958). *Theory and Methods of Scaling*, New York: Wiley.

[65] Turk, M.A. and Pentland, A.P. (1991). Face recognition using eigenfaces, *Proceedings of the IEEE Conference on Computer Vision and Pattern Recognition*, Maui, Hawaii, pp. 586–591.

[66] Whitney, H. (1936). Differentiable manifolds, *Annals of Mathematics*, 37, 645–680.

[67] Willard, S. (1970). *General Topology*, Reading, MA: Addison-Wesley. Republished in 2004 by Dover Publications.

[68] Yeung, K.Y. and Ruzzo, W.L. (2001). Principal component analysis for clustering gene expression data, *Human Molecular Genetics*, 17, 763–774.

[69] Zheng-Bradley, X., Rung, J., Parkinson, H., and Brazma, A. (2010). Large-scale comparison of global gene expression patterns in human and mouse, *Genome Biology*, 11, R124 (11 pages).

Chapter 2

Robust Laplacian Eigenmaps Using Global Information

Shounak Roychowdhury and Joydeep Ghosh

2.1 Introduction

Dimensionality reduction is an important process that is often required to understand the data in a more tractable and humanly comprehensible way. This process has been extensively studied in terms of linear methods such as Principal Component Analysis (PCA), Independent Component Analysis (ICA), Factor Analysis etc. [8]. However, it has been noticed that many high dimensional data, such as a series of related images, lie on a manifold [12] and are not scattered throughout the feature space.

Belkin and Niyogi in [2] proposed Laplacian Eigenmaps (LEM), a method that approximates the Laplace–Beltrami Operator which is able to capture the properties of any Riemaniann manifold. The motivation of our work derives from our experimental observations that when the graph that used Laplacian Eigenmaps (LEM) [2] is not well-constructed (either it has lot of isolated vertices or there are islands of subgraphs) the data is difficult to interpret after a dimension reduction. This paper discusses how global information can be used in addition to local information in the framework of Laplacian Eigenmaps to address such situations. We make use of an interesting result by Costa and Hero that shows that Minimum Spanning Tree on a manifold can reveal its intrinsic dimension and entropy [4]. In other words, it implies that MSTs can capture the underlying global structure of the manifold if it exists. We use this finding to extend the dimension reduction technique using LEM to exploit both local and global information.

LEM depends on the Graph Laplacian matrix and so does our work. Fiedler initially proposed the Graph Laplacian matrix as a means to comprehend the notion of algebraic connectivity of a graph [6]. Merris has extensively discussed the wide variety of properties of the Laplacian matrix of a graph such as invariance, on various bounds and inequalities, extremal examples and constructions, etc., in his survey [10]. A broader role of the Laplacian matrix can be seen in Chung's book on Spectral Graph Theory [3].

The second section touches on the Graph Laplacian matrix. The role of global information in manifold learning is then presented, followed by our proposed approach of augmenting LEM by including global information about the data. Experimental results confirm that global information can indeed help when the local information is limited for manifold learning.

2.2 Graph Laplacian

In this section we will briefly review the definitions of a Graph Laplacian matrix and Laplacian of Graph Sum.

2.2.1 Definitions

Let us consider a weighted graph $G = (V, E)$, where $V = V(G) = \{v_1, v_2, ..., v_n\}$ is the set of vertices (also called vertex set) and $E = E(G) = \{e_1, e_2, ..., e_n\}$ is the set of edges (also called edge set). The weight w function is defined as $w : V \times V \to \Re$ such that $w(v_i, v_j) = w(v_j, v_i) = w_{ij}$.

Definition 1: The Laplacian [6] of a graph without loops of multiple edges is defined as the following:

$$L(G) = \begin{cases} d_{v_i} & \text{if } v_i = v_j, \\ -1 & \text{if } v_i \text{ are } v_j \text{ adjacent,} \\ 0 & \text{Otherwise.} \end{cases} \tag{2.1}$$

Fiedler [6] defined the Laplacian of a graph as a symmetric matrix for a regular graph, where A is an adjacency matrix (A^T is the transpose of adjacency matrix), I is the identity matrix, and n is the degree of the regular graph:

$$L(G) = nI - A. \tag{2.2}$$

A definition by Chung (see [3]) — which is given below — generalizes the Laplacian by adding the weights on the edges of the graph. It can be viewed as Weighed Graph Laplacian. Simply, it is a difference between the diagonal matrix D and W, the weighted adjacency matrix.

$$L_W(G) = D - W, \tag{2.3}$$

where the diagonal element in D is defined as $d_{v_i} = \sum_{j=1}^{n} w(v_i, v_j)$.

Definition 2: The Laplacian of weighted graph (operator) is defined as the following:

$$L_w(G) = \begin{cases} d_{v_i} - w(v_i, v_j) & \text{if } v_i = v_j \\ -w(v_i, v_j) & \text{if } v_i \text{ are } v_j \text{ connected} \\ 0 & \text{otherwise.} \end{cases} \tag{2.4}$$

$L_w(G)$ reduces to $L(G)$ when the edges have unit weights.

2.2.2 Laplacian of Graph Sum

Here we are primarily interested in knowing how to derive Laplacian of a resultant graph derived from two different graphs of the same order (on a given data set). In fact we are superimposing two graphs having the same set of vertices together to form a new graph. We do graph fusion as we are interested in combining a local-neighborhood graph and a global graph which we will describe later on.

Harary in [7] introduced a graph operator called Graph Sum; the operator is denoted by $\oplus : G \times H \to J$, to sum up two graphs of the same order ($|V(G)| = |V(H)| = |V(J)|$). The operator is quite simple — adjacency matrices of each graph are numerically added to form the adjacency matrix of the summed graph. Let G and H be two graphs of order n, and the new summed graph be denoted by J as shown in Figure 2.1. Furthermore, let A_G, A_H, and A_J be the adjacency of each graph respectively. Then

$$J = G \oplus H,$$

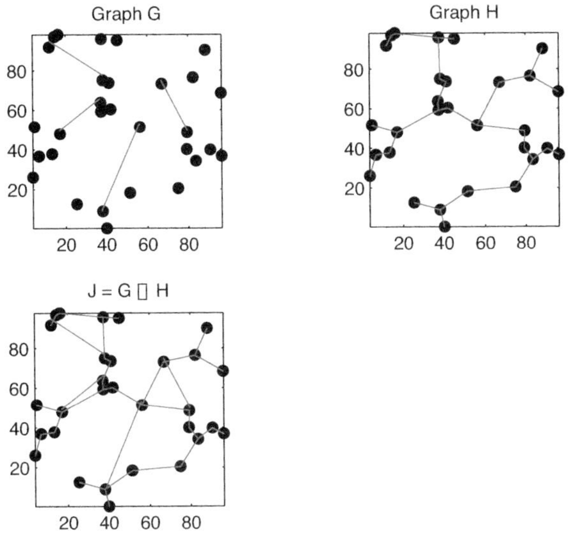

Figure 2.1: The top-left figure shows a graph G; top-right figure shows an MST graph H; and the bottom-left figure shows the graph sum $J = G \oplus H$. Note how the graphs superimpose on each other to form a new graph.

and

$$A_J = A_G + A_H.$$

From Definition 2, it is obvious that

$$L_w(J) = L_w(G) + L_w(H). \tag{2.5}$$

2.3 Global Information of Manifold

Global information has not been used in manifold learning since it is widely believed that global information may capture unnecessary data (like ambient data points) that should be avoided when dealing with manifolds.

However, some recent research results show that that it might be useful to to explore global information in a more constrained manner for manifold learning. Costa and Hero show that it is possible to use a Geodesic Minimum Spanning Tree (GMST) on the manifold to estimate the intrinsic dimension and intrinsic entropy of the manifold [4].

Costa and Hero showed in the following theorem that is possible to learn the intrinsic entropy and intrinsic dimension of a non-linear manifold by extending the BHH theorem [1], a well-known result in Geometric Probability.

Theorem: *[Generalization of BHH Theorem to Embedded manifolds: [4]] Let $\mathcal{M}$ be a smooth compact m-dimensional manifold embedded in $\mathbb{R}^d$ through the diffeomorphism $\phi : \Omega \to \mathcal{M}$, and $\Omega \in \mathbb{R}^d$. Assume $2 \leq m \leq d$ and $0 < \gamma < m$. Suppose that Y_1, Y_2, ... are iid random vectors on $\mathcal{M}$ having a common density function f with respect to a Lebesgue measure $\mu_{\mathcal{M}}$ on $\mathcal{M}$. Then the length functional $T_\gamma^{\mathbb{R}^m} \phi_{-1}(Y_n)$ of the MST spanning $\phi^{-1}(Y_n)$ satisfies the equation shown below in an almost sure sense:*

$$\lim_{n\to\infty} \frac{T_\gamma^{\mathbb{R}^m} \phi_{-1}(Y_n)}{n^{\frac{(d-1)}{d}}} =$$

$$\begin{cases} \infty & d' < m \\ \beta_m \int_{\mathcal{M}} [det(J_\phi^T J_\phi)] f(x)^\alpha \mu_{\mathcal{M}} d(y) \quad a.s., & d' = m \\ 0 & d' > m \end{cases} \tag{2.6}$$

where $\alpha = (m - \gamma)/m$, and is always between $0 < \alpha < 1$, J is the Jacobian, and β_m is a constant which depends on m.

Based on the above theorem we use MST on the entire data set as a source of global information. For more details see [4], and more background information see [15] and [13].

The basic principle of GLEM is quite straightforward. The objective function that is to be minimized is given by the following (it is has the same flavor and notation used in [2]):

$$\sum_{i,j} ||\mathbf{y}^{(\mathbf{i})} - \mathbf{y}^{(\mathbf{j})}||_2^2 (W_{ij}^{NN} + W_{ij}^{MST})$$

$$\begin{aligned} &= tr(\mathbf{Y}^T L(G_{NN})\mathbf{Y} + \mathbf{Y}^T L(G_{MST})\mathbf{Y}) \qquad (2.7) \\ &= tr(\mathbf{Y}^T (L(G_{NN}) + L(G_{MST}))\mathbf{Y}) \\ &= tr(\mathbf{Y}^T L(J)\mathbf{Y}). \end{aligned}$$

where $\mathbf{y}^{(\mathbf{i})} = [y_1(i), ..., y_m(i)]^T$, and m is the dimension of embedding. W_{ij}^{NN} and W_{ij}^{MST} are weighted matrices of k-Nearest Neighbor graph and the MST graph respectively. In other words, we have

$$\underset{\mathbf{Y^T D Y = I}}{argmin_{\;Y}} \; \mathbf{Y}^T L \mathbf{Y} \tag{2.8}$$

such that $Y = [\mathbf{y_1}, \mathbf{y_2}, ..., \mathbf{y_m}]$ and $\mathbf{y}^{(\mathbf{i})}$ is the m-dimensional representation of ith vertex. The solutions to this optimization problem are the eigenvectors of the generalized eigenvalue problem

$$L\mathbf{Y} = \Lambda D\mathbf{Y}.$$

The GLEM algorithm is described in Algorithm 1.

2.4 Laplacian Eigenmaps with Global Information

In this section we describe our approach of using the global information of the manifold which is modeled as an MST. Similar to the LEM approach our method (GLEM) also actually builds an adjacency graph G_{NN} by using neighborhood information. We compute Laplacian of the adjacency graph ($L(G_{NN})$). The weights of the edges of the G_{NN} are determined by the Heat Kernel $H(x_i, x_j) = exp(||x_i - x_j||^2)/t$ where $t > 0$. Thereafter compute the MST graph G_{MST} on the data set, and its Laplacian $L(G_{MST})$. Based on Laplacian Graph summation as described earlier, we now combine two graphs G_{NN} and G_{MST} by effectively adding their Laplacians $L(G_{NN})$ and $L(G_{MST})$.

2.5 Experiments

Here we show the results of our experiments conducted on two well-known manifold data sets: 1) S-Curve and 2) ISOMAP face data set [9] using LEM, which uses the local neighborhood information, and GLEM, which exploits local as well as global information of the

Algorithm 1 Global Laplacian Eigenmaps (GLEM)

Data: Data: $X_{n \times p}$ where n is the number of data points and p is the number of dimensions. k: number of fixed neighbors or $\epsilon - ball$: using neighbors falling within a ball of radius ϵ, and $0 \leq \lambda \leq 1$

Result: Low-dimensional subspace; $X_{n \times m}$ where n is the number of data points and m is the number of selected eigen-dimensions such that $m \ll p$.

begin

Construct the graph G_{NN} either by using local neighbor k or $\epsilon - ball$.

  Construct the Adjacency Graph $A(G_{NN})$ of graph G_{NN}.

  Compute the weight matrix (W_{NN}) from the weights of the edges of graph G_{NN} using the Heat Kernel function.

  Compute Laplacian matrix $L(G_{NN}) = D_{NN}$ - W_{NN}

  /* D_{NN} is the Diagonal Matrix of the NN Graph. */

Construct the graph G_{MST}.

  Construct the Adjacency Graph $A(G_{MST})$ of graph G_{MST}.

  Compute the weight matrix (W_{MST}) from the weights of the edges of graph G_{MST} using the Heat Kernel function.

  Compute Laplacian matrix $L(G_{MST}) = D_{MST}$ - W_{MST}

  /* D_{MST} is the Diagonal Matrix of the MST Graph. */

$L(G) = L(G_{NN}) + \lambda L(G_{MST})$

return the subspace $X_{n \times m}$ by selecting first m eigenvectors of $L(G)$

end

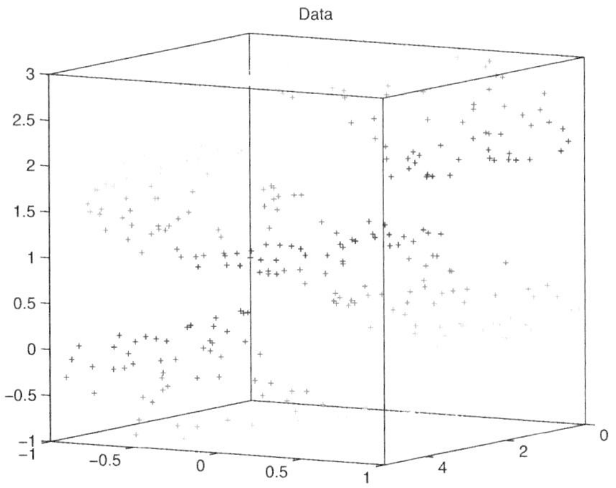

Figure 2.2: (See Color Insert.) S-Curve manifold data. The graph is easier to understand in color.

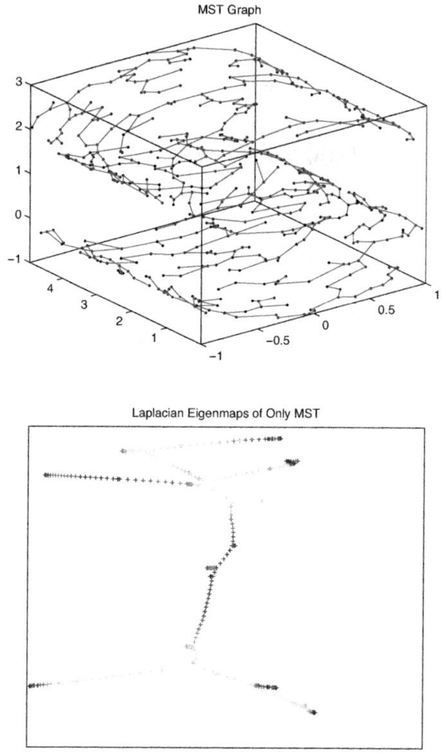

Figure 2.3: (See Color Insert.) The MST graph and the embedded representation.

manifold. For calculation of local neighborhood we use the kNN method. The S-Curve data is shown in Figure 2.2 for reference.

The MST on the S-Curve data is shown in Figure 2.3. The top figure shows the MST of data, while the bottom figure shows the embedding of the graph. Notice how the data is embedded in a tree-like structure, yet the local information of the data is completely preserved. Figure 2.4 shows embedding of the ISOMAP face data set using the MST graph. We use a limited number of face images to clearly show the embedded structure; the data points are shown by '+' in embedded space.

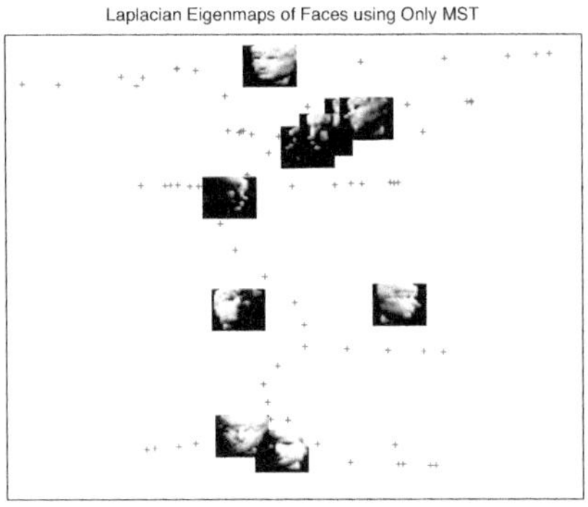

Figure 2.4: Embedded representation for face images using the MST graph. The sign '+' denotes a data point.

2.5.1 LEM Results

Figures 2.5–2.7 show the results after using LEM for different values of k. As the value of k increases from 1 to higher values we notice the spreading of the embedded data. The bottom subplot shows the nearest neighbor graph with $k = 1$ as shown in Figure 2.7. The right plot shows the embedding of the graph. It is interesting to observe how the embedded data loses its local neighborhood information. The embedding practically happens along the second principal eigenvector (the first being Zero Vector). As the value of k is increased to 2, we observe that embedding happens along the second and third principal axes. See Figure 2.7. For $k = 1$ the graph is highly disconnected and for $k = 2$ the graphs has much less isolated pieces of graphs. One interesting thing to observe is that as the connectivity of the graph increases the low-dimensional representation begins to preserve the local information.

The graph with $k = 2$ and its embedding is shown in Figure 2.8. Increasing the neighborhood information to 2 neighbors is still not able to represent the continuity of the original manifold. Figure 2.7 shows the graph with $k = 3$ and its embedding. Increasing the neighborhood information to 3 neighbors better represents the continuity of the original manifold. Figure 2.5 shows the graph with $k = 5$ and its embedding. Increasing the neighborhood information to 5 neighbors better represents the continuity of the original manifold. Similar results are obtained by increasing the the number of neighbors, however, it should be noted that when the number of neighbors is very high then the graph starts to get influenced by ambient neigbhors.

We see similar results for the face images. The three plots in Figure 2.6 show the embedding results obtained using LEM when the neighborhood graphs are created using $k = 1$, $k = 2$, and $k = 5$. The top and the middle plot validate the limitation of LEM

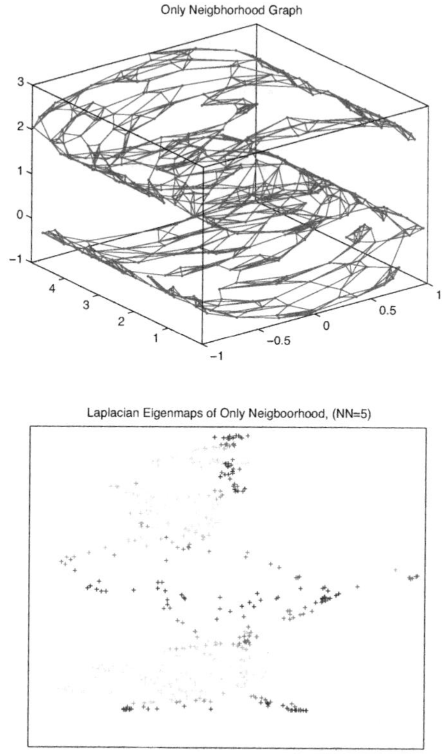

Figure 2.5: (See Color Insert.) The graph with $k = 5$ and its embedding using LEM. Increasing the neighborhood information to 5 neighbors better represents the continuity of the original manifold.

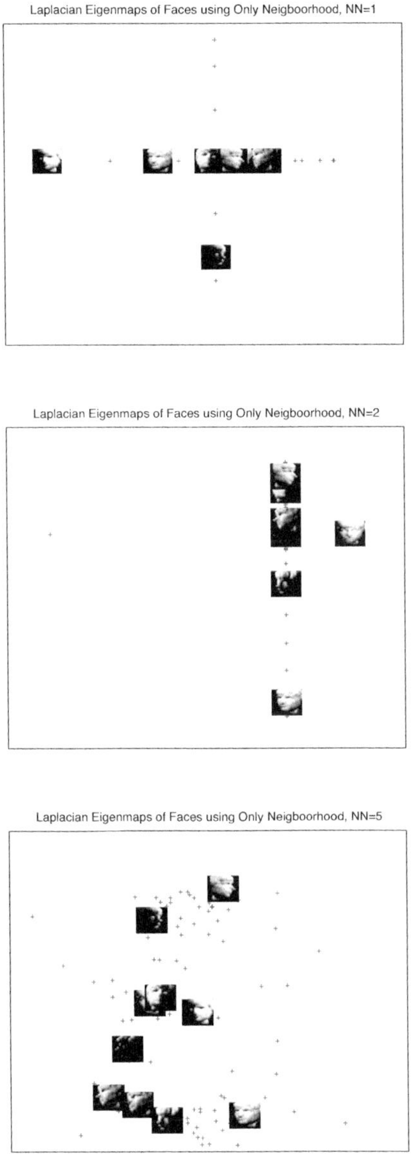

Figure 2.6: The embedding of the face images using LEM. The top and middle plots show embedding using $k = 1$ and $k = 2$, respectively. The bottom plot shows embedding for $k = 5$. Few faces have been shown to maintain clarity of the embeddings.

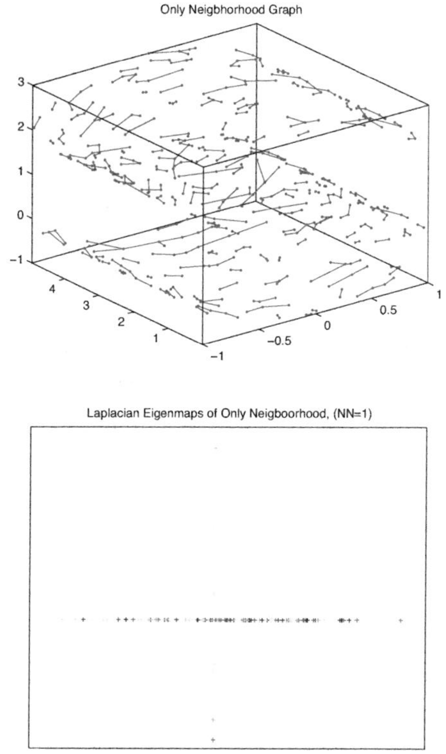

Figure 2.7: (See Color Insert.) The graph with $k = 1$ and its embedding using LEM. Because of very limited neighborhood information the embedded representation cannot capture the continuity of the original manifold.

for $k = 1$ and $k = 2$. As expected, for $k = 5$ there is continuity of facial images in the embedded space.

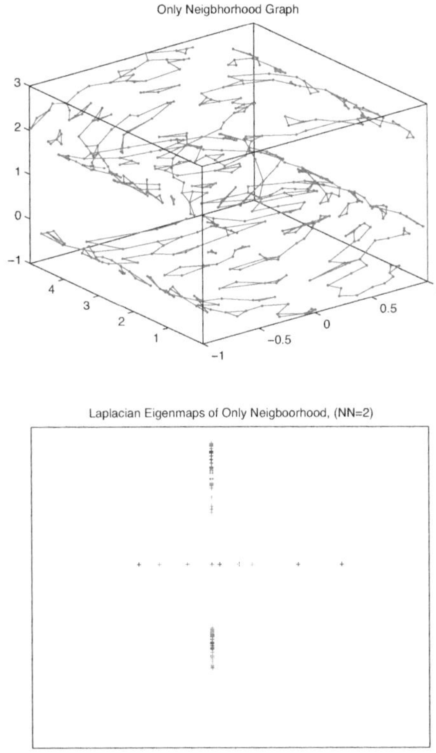

Figure 2.8: (See Color Insert.) The graph with $k = 2$ and its embedding using LEM. Increasing the neighborhood information to 2 neighbors is still not able to represent the continuity of the original manifold.

2.5.2 GLEM Results

Figures 2.9–2.11 show the GLEM results for $k = 1$, $k = 2$, and $k = 5$ respectively. Figure 2.9 shows a graph sum of a graph with neighborhood of $k = 1$ and MST and its embedding. In spite of very limited neighborhood information GLEM preserves continuity of the original manifold in the embedded representation and which is due to the MST's contribution. On comparing Figure 2.7 and Figure 2.9 it becomes clear that addition of some amount of global information can help to preserve the manifold structure. Similarly, in Figure 2.10, MST dominates the embedding's continuity. However, on increasing $k = 5$ (Figure. 2.11) the dominance of MST starts to decrease as the local neighborhood graph starts dominating. The λ in GLEM plays the role of a simple regularizer. Figure 2.12 shows the effect of different values of $\lambda \in \{0, 0.2, 0.5, 0.8, 1.0\}$.

The results of GLEM on ISOMAP face images are shown in Figure 2.13 where the neighborhood graphs are created using $k = 1$, $k = 2$, and $k = 5$. The top and the middle plots of Figure 2.13 reveal the contribution of MST for $k = 1$ and $k = 2$, similar to Figures 2.9–2.10. For $k = 5$ we clearly see the alignment of faces' images in the bottom figure.

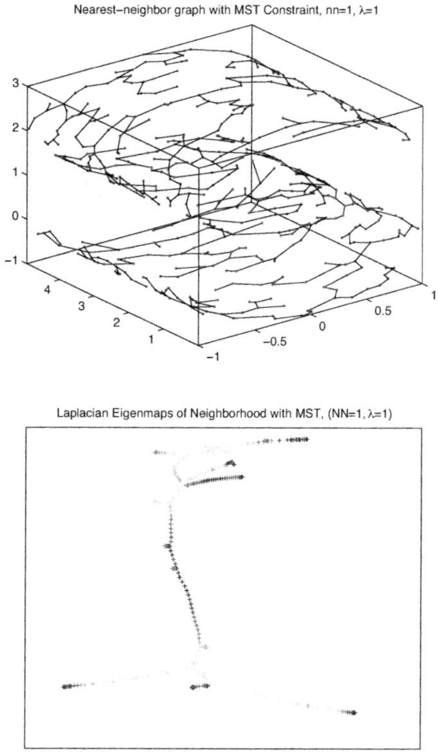

Figure 2.9: (See Color Insert.) The graph sum of a graph with neighborhood of $k = 1$ and MST, and its embedding. In spite of very limited neighborhood information the GLEM is able to preserve the continuity of the original manifold and is primarily due to MST's contribution.

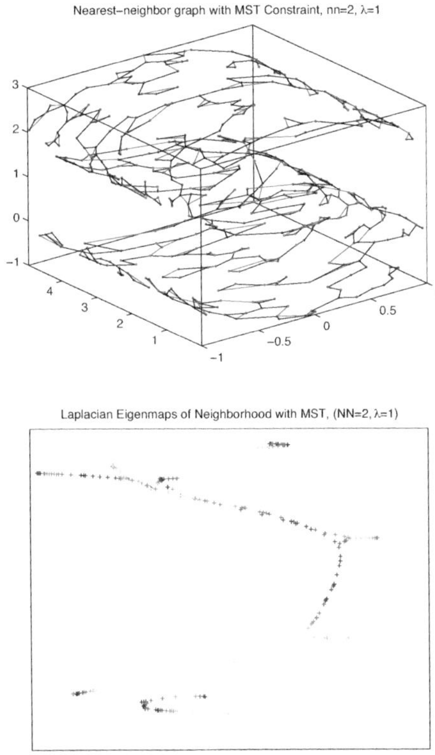

Figure 2.10: (See Color Insert.) GLEM results for $k = 2$ and MST, and its embedding GLEM. In this case also, embedding's continuity is dominated by the MST.

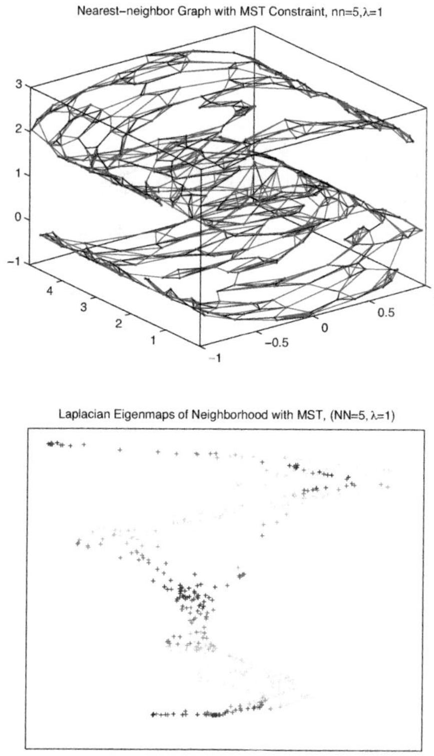

Figure 2.11: Increase the neighbors to $k = 5$ and the neighborhood graph starts dominating, and the embedded representation is similar to Figure 2.5.

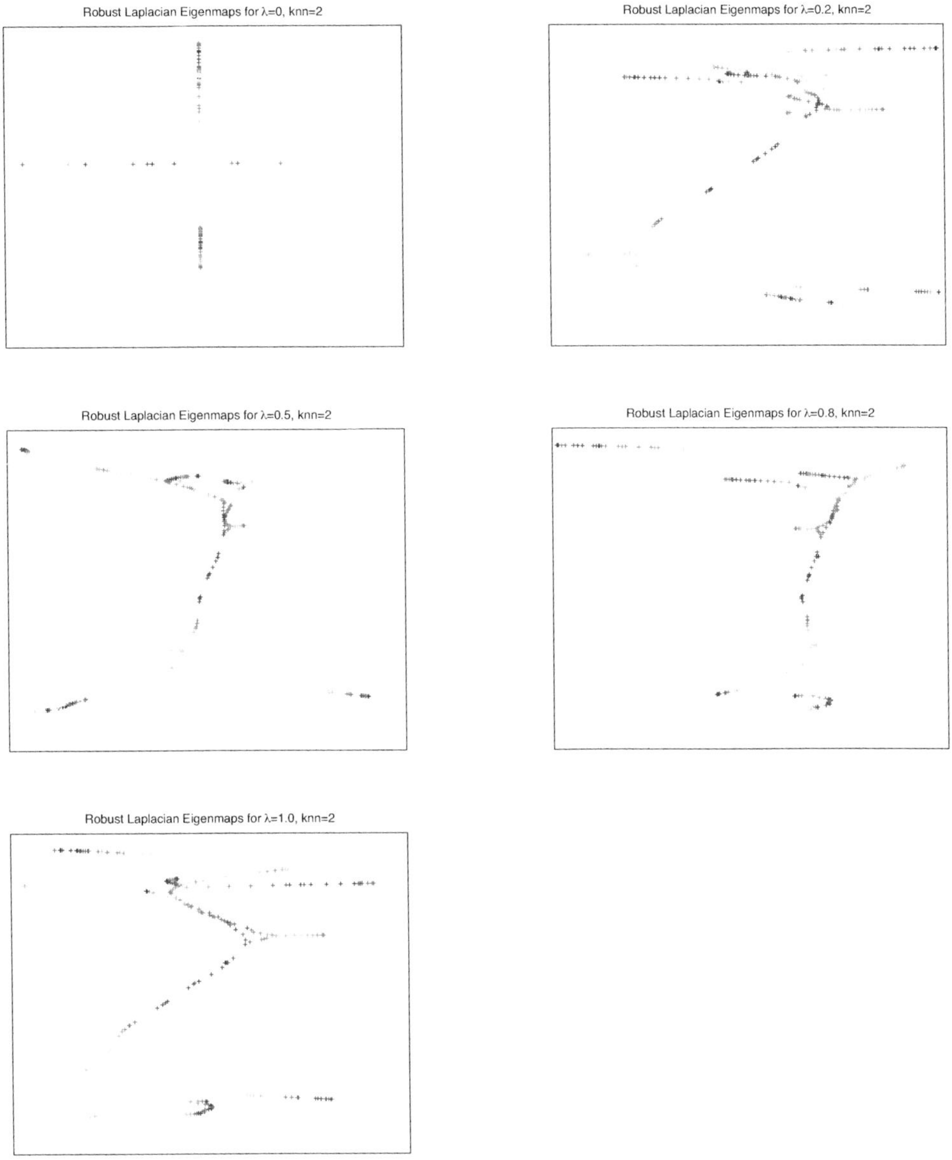

Figure 2.12: (See Color Insert.) Change in regularization parameter $\lambda \in \{0, 0.2, 0.5, 0.8, 1.0\}$ for $k = 2$. In fact the results here show that the embedded representation is controlled by the MST.

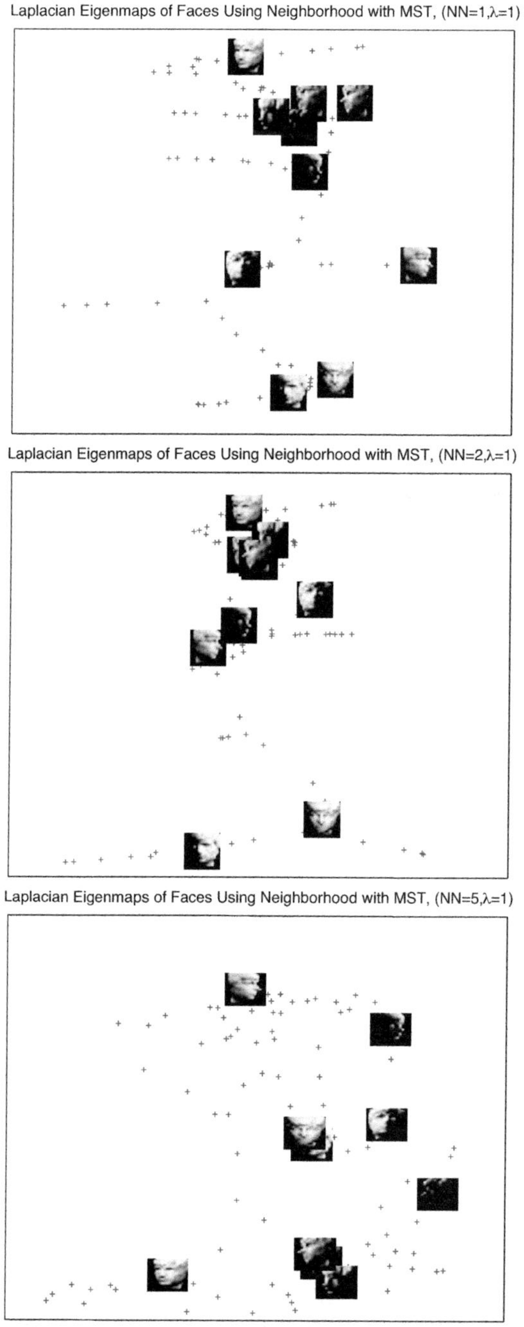

Figure 2.13: (See Color Insert.) The embedding of face images using LEM. The top and middle plots show embedding using $k = 1$ and $k = 2$, respectively. The bottom plot shows embedding for $k = 5$. Few faces have been shown to maintain clarity of the embeddings. In this figure we see how the MST preserves the embedding.

2.6 Summary

In this paper we show that when the neighborhood information of the manifold graph is limited, then the use of global information of the data can be very helpful. In this short study we proposed the use of local neighborhood graphs along with Minimal Spanning trees for the Laplacian Eigenmaps by leveraging the theorem proposed by Costa and Hero regarding MSTs and manifolds. This work also indicates the potential for using different geometric sub-additive graphical structures [15] in non-linear dimension reduction and manifold learning.

2.7 Bibliographical and Historical Remarks

Dimensionality reduction is an important research area in data analysis with an extensive research literature. Both linear and non-linear methods exist, and each category has both supervised and unsupervised versions. In this section we will briefly mention some of the salient works that have been proposed in the area of locally preserving manifold learning; see [8] for a broader survey.

Lee and Seung [12] showed that many high dimensional data such as a series of related images, video frames, etc. lie on a much lower-dimensional manifold instead of being scattered throughout the feature space. This particular observation has motivated researchers to develop dimension reduction algorithms that try to learn an embedded manifold in a high-dimensional space.

ISOMAP [14] learns the manifold by exploring geodesic distances. In fact the algorithm tries to preserve the geometry of the data on the manifold by noting the points in the neighborhood of each point. The algorithm is defined as such:

1. Form a neighborhood graph G for the dataset, based, for instance, on the K nearest neighbors of each point x_i.

2. For every pair of nodes in the graph, compute the shortest path, using Dijkstras algorithm, as an estimate of intrinsic distance on the data manifold. The weights of edges of the graphs are computed based on the Euclidean distance measure.

3. Classical Multi-Dimensional Scaling algorithm is computed using these pairwise distances to find a lower dimensional embedding y_i.

Bernstein et al. [22] have described the convergence properties of the estimation procedure for the intrinsic distances. For large and dense data sets, computation of pairwise distances is time consuming, and moreover the calculation of eigenvalues can be computationally intensive for large data sets. Such constraints have motivated researchers to find simpler variations of the Isomap algorithm. One such algorithm uses subsampled data called landmarks. Firstly, it calculates Isomap for random points called landmarks and between those landmarks a simple triangulation algorithm is applied.

Locally Linear Embedding (LLE) is an unsupervised learning method based on global and local optimization [11]. It is is similar to Isomap in the sense that it generates a graphical representation of the data set. However, it is different from Isomap as it only attempts to preserve local structures of the data. Because of the locality property used in LLE, the algorithm allows for successful embedding of nonconvex manifolds. An important point to be noted is that LLE creates the local properties of a manifold using the linear combinations of k nearest neighbors of the data x_i. LLE attempts to create a local regression like model and thereby tries to fit a hyperplane through the data point x_i. This appears to be reasonable for smooth manifolds where the nearest neighbors align themselves well in a linear space. For very non-smooth or noisy data sets, LLE does not perform well. It has been noted that LLE preserves the reconstruction weights in the space of lower dimensionality, as the reconstruction weights of a data point are invariant to linear transformational operations like translation, rotation, etc.

LLE is a popular algorithm for non-linear dimension reduction. A linear variant of this algorithm [20] has been proposed. Though there have been some successful applications [17, 21], certain experimental studies such as [23] show that this algorithm has its limitations. In another study [24] LLE failed to work on manifolds that had holes in them.

Zhang et. al. [16] proposed a method of finding Principal Manifolds using Local Tangent Space Alignment (LTSA). As the name suggests, this method uses local Tangent space information of high dimensional data. Again, there is an underlying assumption that manifolds are smooth and without kinks. LTSA is based on the observation that for smooth manifolds, it is possible to derive a linear mapping from a high dimensional data space to low dimensional local tangent space. A linear variant of LTSA is proposed in [26]. This algorithm has been used in applications like face recognition [18, 25].

Donoho and Grimes have proposed a method similar to LEM using Hessian Maps (HLLE) [5]. This algorithm is a variant of LLE. It uses a Hessian to compute the curvature of the manifold around each data point. Similar to LLE, the local Hessian in the low dimensional space is computed by using eigenvalue analysis. Also popular are Laplacian Eigenmaps that use spectral techniques to perform dimensionality reduction [2]. Finally, generalization of principal curves to principal surfaces have been proposed with several applications such as characterization of images of 3-dimensional objects with varying poses [19].

Bibliography

[1] Beardwood, J., Halton, H., and Hammersley, J. M. The shortest path through many points. *Proceedings of Cambridge Philosophical Society* 55:299–327. 1959.

[2] Belkin, M., and Niyogi, P. Laplacian eigenmaps for dimensionality reduction and data representation. *Neural Computation* 15:1373–1396. 2003.

[3] Chung, F. R. K. *Spectral Graph Theory.* Providence, RI: American Mathematical Society. 1997.

[4] Costa, J. A., and Hero, A. O. Geodesic entropic graphs for dimension and entropy estimation in manifold learning. *IEEE Trans. on Signal Processing* 52:2210–2221. 2004.

[5] Donoho, D. L., and Grimes, C. Hessian eigenmaps: Locally linear embedding techniques for high-dimensional data. *PNAS* 100(10):5591–5596. 2003.

[6] Fiedler, M. Algebraic connectivity of graphs. *Czech. Math. Journal* 23:298–305. 1973.

[7] Harary, F. Sum graphs and difference graphs. *Congress Numerantium* 72:101–108. 1990.

[8] Hastie, T., Tibshirani, R., and Friedman, J. *The Elements of Statistical Learning: Data Mining, Inference, and Prediction.* New York: Springer. 2001.

[9] ISOMAP. http://isomap.stanford.edu. 2009.

[10] Merris, R. Laplacian matrices of graphs: a survey. *Linear Algebra and its Applications* 197(1):143–176. 1994.

[11] Saul, L. K., Roweis, S. T., and Singer, Y. Think globally, fit locally: Unsupervised learning of low dimensional manifolds. *Journal of Machine Learning Research* 4:119–155. 2003.

[12] Seung, H., and Lee, D. The manifold ways of perception. *Science* 290:2268–2269. 2000.

[13] Steele, J. M. *Probability theory and combinatorial optimization*, volume 69 of *CBMF-NSF regional conferences in applied mathematics*. Society for Industrial and Applied Mathematics (SIAM). 1997.

[14] Tenenbaum, J. B., de Silva, V., and Langford, J. C. A global geometric framework for nonlinear dimensionality reduction. *Science* 290(5500):2319–2323. 2000.

[15] Yukich, J. E. *Probability theory of classical Euclidean optimization*, volume 1675 of *Lecture Notes in Mathematics*. Springer-Verlag, Berlin. 1998.

[16] Zhang, Z., and Zha, H. Principal manifolds and nonlinear dimension reduction via local tangent space alignment. *SIAM Journal of Scientific Computing* 26:313–338. 2002.

[17] Duraiswami, R., and Raykar, V.C. The manifolds of spatial hearing. In Proceedings of International Conference on Acoustics, Speech and Signal Processing, 3:285–288. 2005.

[18] Graf, A.B.A., and Wichmann, F.A. Gender classification of human faces. Biologically Motivated Computer Vision 2002, LNCS 2525:491–501. 2002.

[19] Chang, K., and Ghosh, J. A Unified Model for Probabilistic Principal Surfaces IEEE Trans. Pattern Anal. Mach. Intell., 23:22–41. 2001.

[20] He, X., Cai, D., Yan, S., and Zhang, H.-J. Neighborhood preserving embedding. In Proceedings of the 10th IEEE International Conference on Computer Vision, pages 1208–1213. 2005.

[21] Chang, H., Yeung, D.-Y., and Xiong, Y. Super-resolution through neighbor embedding. IEEE Computer Society Conference on Computer Vision and Pattern Recognition, volume 1, pages 275–282. 2004.

[22] Bernstein, M., de Silva, V., Langford, J., and Tenenbaum, J. Graph approximations to geodesics on embedded manifolds. Technical report, Department of Psychology, Stanford University. 2000.

[23] Mekuz, N. and Tsotsos, J.K. Parameterless Isomap with adaptive neighborhood selection. In Proceedings of the 28th AGM Symposium, pages 364–373, Berlin, Germany. 2006.

[24] Saul L.K., Weinberger, K.Q., Ham, J.H., Sha, F., and Lee, D.D. Spectral methods for dimensionality reduction. In Semisupervised Learning, The MIT Press, Cambridge, MA, USA. 2006.

[25] Zhang T., Yang, J., Zhao, D., and Ge, X. Linear local tangent space alignment and application to face recognition. Neurocomputing, 70: pages 1547–1533. 2007.

[26] Zhang Z., and Zha, H. Local linear smoothing for nonlinear manifold learning. Technical Report CSE-03-003, Department of Computer Science and Engineering, Pennsylvania State University, University Park, PA, USA. 2003.

Chapter 3

Density Preserving Maps

Arkadas Ozakin, Nikolaos Vasiloglou II, Alexander Gray

3.1 Introduction

Much of the recent work in manifold learning and nonlinear dimensionality reduction focuses on distance-based methods, i.e., methods that aim to preserve the local or global (geodesic) distances between data points on a submanifold of Euclidean space. While this is a promising approach when the data manifold is known to have no intrinsic curvature (which is the case for common examples such as the "Swiss roll"), classical results in Riemannian geometry show that it is impossible to map a d-dimensional data manifold with intrinsic curvature into $\mathbb{R}^d$ in a manner that preserves distances. Consequently, distance-based methods of dimensionality reduction distort intrinsically curved data spaces, and they often do so in unpredictable ways. In this chapter, we discuss an alternative paradigm of manifold learning. We show that it is possible to perform nonlinear dimensionality reduction by preserving the underlying *density* of the data, for a much larger class of data manifolds than intrinsically flat ones, and demonstrate a proof-of-concept algorithm demonstrating the promise of this approach.

Visual inspection of data after dimensional reduction to two or three dimensions is among the most common uses of manifold learning and nonlinear dimensionality reduction. Typically, what is sought by the user's eye in two or three-dimensional plots is clustering and other relationships in the data. Knowledge of the density, in principle, allows one to identify such basic structures as clusters and outliers, and even define nonparametric classifiers; the underlying density of a data set is arguably one of the most fundamental statistical objects that describe it. Thus, a method of dimensionality reduction that is guaranteed to preserve densities may well be preferable to methods that aim to preserve distances, but end up distorting them in uncontrolled ways.

Many of the manifold learning methods require the user to set a neighborhood radius h, or, for k-nearest neighbor approaches, a positive integer k, to be used in determining the neighborhood graph. Most of the time, there is no automatic way to pick the appropriate values of the tweak parameters h and k, and one resorts to trial and error, looking for values that result in reasonable-looking plots. Kernel density estimation, one of the most popular and useful methods of estimating the underlying density of a data set, comes with a natural way to choose h or k; it suggests to us to pick the value that maximizes a cross-validation score for the density estimate. While the usual kernel density estimation does not allow one to estimate the density of data on submanifolds of Euclidean space, a small modification allows one to do so. This modification and its ramifications are discussed below in the context of density-preserving maps.

The chapter is organized as follows. In Section 3.2, using a theorem of Moser, we prove the existence of density preserving maps into $\mathbb{R}^d$ for a large class of d-dimensional manifolds, and give an intuitive discussion on the nonuniqueness of such maps. In Section 3.3, we describe a method for *estimating* the underlying density of a data set on a Riemannian submanifold of Euclidean space. We state the main result on the consistency of this submanifold density estimator, and give a bound on its convergence rate, showing that the latter is determined by the intrinsic dimensionality of the data instead of the full dimensionality of the feature space. This, incidentally, shows that the curse of dimensionality in the widely-used method of kernel density estimation is not as severe as is generally believed, if the method is properly modified for data on submanifolds. In Section 3.4, using a modified version of the estimator defined in Section 3.3, we describe a proof-of-concept algorithm for density preserving maps based on semidefinite programming, and give experimental results. Finally, in Sections 7.7 and 3.6, we summarize the chapter and discuss relevant bibliography.

3.2 The Existence of Density Preserving Maps

In this section, we prove that it is always possible to find a density-preserving map from a closed d-dimensional submanifold M of $\mathbb{R}^D$ $(d < D)$ into a d-dimensional Riemannian manifold that is diffeomorphic to M, if the two manifolds have the same total volume. We also discuss the application of this theorem to maps into $\mathbb{R}^d$, which exist when M is "topologically simple," i.e., when it is diffeomorphic to a subset of $\mathbb{R}^d$. For the case of $D = 3$ and $d = 2$, this means that M is a surface that can be "stretched" onto the plane without tearing.[1] When a surface is intrinsically curved, stretching it onto the plane necessarily changes the distances between its points. The results of this section show that densities can, in fact, be preserved. The fundamental result we will use in the proof of this claim is a theorem by Moser [18].

3.2.1 Moser's Theorem and Its Corollary on Density Preserving Maps

Riemannian manifolds. We begin by restricting our attention to data subspaces which are Riemannian submanifolds of $\mathbb{R}^D$. Riemannian manifolds provide a generalization of the notion of a smooth surface in $\mathbb{R}^3$ to higher dimensions. As first clarified by Gauss in the two-dimensional case and by Riemann in the general case, it turns out that *intrinsic* features of the geometry of a surface, such as the lengths of its curves or intrinsic distances between its points, etc., can be given in terms of the so-called metric tensor[2] $\mathbf{g}$, without referring to the particular way the the surface is embedded in $\mathbb{R}^3$. A space whose geometry is defined in terms of a metric tensor is called a Riemannian manifold (for a rigorous definition, see, e.g., [12, 16, 2]).

The Gauss/Riemann result mentioned above states that if the intrinsic curvature of a Riemannian manifold $(M, \mathbf{g}_M)$ is not zero in an open set $U \in M$, it is not possible to find

[1]The assumption that M is simple in this sense is implicit in most works in the manifold learning literature. The class of 2-dimensional surfaces for which this holds includes intrinsically curved surfaces like a hemisphere, in addition to the intrinsically flat but extrinsically curved spaces like the Swiss roll, but excludes surfaces like the torus, which can't be stretched onto the plane without tearing or folding.

[2]The metric tensor with components g_{ij} can be thought of as giving the "infinitesimal distance" ds between two points whose coordinates differ by infinitesimal amounts $(dy^1, \ldots, dy^D)$, as $ds^2 = \sum_{ij} g_{ij} dy^i dy^j$. For the case of a unit hemisphere given in spherical coordinates as $\{(r, \theta, \phi) : \theta < \pi\}$, one can read off the metric tensor from the infinitesimal distance $ds^2 = d\theta^2 + \sin^2\theta d\phi^2$.

a map from M into $\mathbb{R}^d$ that preserves the distances between the points of U. Thus, there exists a *local obstruction*, namely, the curvature, to the existence of distance-preserving maps. It turns out that no such local obstruction exists for *volume*-preserving maps. The only invariant is a global one, namely, the total volume.[3] This is the content of Moser's theorem on volume-preserving maps, which we state next.

Theorem 3.2.1 *(Moser [18]) Let $(M, \mathbf{g}_M)$ and $(N, \mathbf{g}_N)$be two closed, connected, orientable, d-dimensional differentiable manifolds that are diffeomorphic to each other. Let τ_M and τ_N be volume forms, i.e., nowhere vanishing d-forms on these manifolds, satisfying $\int_M \tau_M = \int_N \tau_N$. Then, there exists a diffeomorphism $\phi : M \to N$ such that $\tau_M = \phi^* \tau_N$, i.e., the volume form on M is the same as the pull-back of the volume form on N by ϕ.*[4]

The meaning of this result is that, if two manifolds with the same "global shape" (i.e., two manifolds that are diffeomorphic) have the same *total* volume, one can find a map between them that preserves the volume *locally*. The surfaces of a mug and a torus are the classical examples used for describing global, topological equivalence. Although these objects have the same "global shape" (topology/smooth structure) their intrinsic, local geometries are different. Moser's theorem states that if their *total* surface areas are the same, one can find a map between them that preserves the areas locally, as well, i.e., a map that sends all small regions on one surface to regions in the other surface in a way that preserves the areas.

Using this theorem, we now show that it is possible to find density-preserving maps between Riemannian manifolds that have the same total volume. This is due to the fact that if local volumes are preserved under a map, the density of a distribution will also be preserved.

Corollary. Let $(M, \mathbf{g}_M)$ and $(N, \mathbf{g}_N)$ be two closed, connected, orientable, d-dimensional Riemannian manifolds that are diffeomorphic to each other, with the same *total* Riemannian volume. Let X be a random variable on M, i.e., a measurable map $X : \Omega \to M$ from a probability space $(\Omega, \mathcal{F}, P)$ to M. Assume that $X_*(P)$, the pushforward measure of P by X, is absolutely continuous with respect to the Riemannian volume measure μ_M on M, with a continuous density f on M. Then there exists a diffeomorphism $\phi : M \to N$ such that the pushforward measure $P_N := \phi_*(X_*(P))$ is absolutely continuous with respect to the Riemannian volume measure μ_N on N, and the density of P_N is given by $f \circ \phi^{-1}$.

Proof: Let the Riemannian volume forms on M and N be τ_M and τ_N, respectively. By Moser's theorem, there exists a diffeomorphism $\phi : M \to N$ that preserves the volume elements: $\tau_M = \phi^* \tau_N$. Thus, $\mu_N = \phi_* \mu_M$. Since $X_*(P) = (\phi^{-1})_* P_N$ is absolutely continuous with respect to $\mu_M = (\phi^{-1})_* \mu_N$, P_N is absolutely continuous with respect to μ_N. Let $B \in N$ be a measurable set in N, and let $A = \phi^{-1}(B)$. We have, $P_N[B] = \phi_*(X_*(P))[B] = X_*(P)[A] = \int_A f d\mu_M = \int_{\phi(A)} (f \circ \phi^{-1}) d(\phi_*(\mu_M)) = \int_B (f \circ \phi^{-1}) d\mu_N$. Thus, the density of $P_N = \phi_*(X_*(P))$ is given by $f \circ \phi^{-1}$.

The meaning of this corollary is that, if two diffeomorphic Riemannian manifolds M and N have the same total volume, we can find a map $\phi : M \to N$ that preserves the density f of the random variable X defined on M. Intuitively, this follows from the fact that density = number of points/volume, and preserving the volume covered by a set of points is sufficient for preserving the density.

---

[3]When considering maps into $\mathbb{R}^d$, the total volume is not an issue, since one can always find a subset of $\mathbb{R}^d$ with the appropriate volume, as long as there are no global, topological obstructions to embedding M in $\mathbb{R}^d$.

[4]As noted by Moser, the theorem can be generalized to d-forms "of odd kind" (which are also known as volume pseudo-forms, or twisted volume forms), hence allowing the theorem to be applied to the case of non-orientable manifolds.

3.2.2 Dimensional Reduction to $\mathbb{R}^d$

These results were formulated in terms of so-called *closed manifolds*, i.e., compact manifolds without boundary. The practical dimensionality reduction problem we would like to address, on the other hand, involves starting with a d-dimensional data submanifold M of $\mathbb{R}^D$ (where $d < D$), and dimensionally reducing to $\mathbb{R}^d$. In order to be able to do this diffeomorphically, M must be diffeomorphic to a subspace of $\mathbb{R}^d$, which is not generally the case for closed manifolds. For instance, although we can find a diffeomorphism from a hemisphere (a manifold with boundary, *not* a closed manifold) into the plane, we cannot find one from the unit sphere (a closed manifold) into the plane. This is a constraint on *all* dimensional reduction algorithms that preserve the global topology of the data space, not just density preserving maps. Any algorithm that aims to avoid "tearing" or "folding" the data subspace during the reduction will fail on problems like reducing a sphere to $\mathbb{R}^2$.[5]

Thus, in order to show that density preserving maps into $\mathbb{R}^d$ exist for a useful class of d-dimensional data manifolds, we have to make sure that the conclusion of Moser's theorem and our corollary work for certain manifolds with boundary, or for certain non-compact manifolds, as well. Fortunately, this is not so hard, at least for a simple class of manifolds that is enough to be useful. In proving his theorem for closed manifolds, Moser [18] first gives a proof for a single "coordinate patch" in such a manifold, which, basically, defines a compact manifold with boundary minus the boundary itself. Not all d-dimensional manifolds with boundary (minus their boundaries) can be given by atlases consisting of a single coordinate patch, but the ones that can be so given cover a wide range of curved Riemannian manifolds, including the hemisphere and the Swiss roll, possibly with punctures. In the following, we will assume that M consists of a single coordinate patch.

3.2.3 Intuition on Non-Uniqueness

Note that the results above claim the existence of volume (or density) preserving maps, but not uniqueness. In fact, the space of volume-preserving maps is very large. An intuitive way to see this is to consider the flow of an incompressible fluid in $\mathbb{R}^3$. The fluid may cover the same region in space at two given times, but the fluid particles may have gone through significant shuffling. The map from the original configuration of the fluid to the final one is a volume preserving diffeomorphism, assuming the flow is smooth. The infinity of ways a fluid can move shows the infinity of ways of preserving volume.

Distance-preserving maps may also have some non-uniqueness, but this is parametrized by a finite-dimensional group, namely, the isometry group of the Riemannian manifold under consideration.[6] The case of volume-preserving maps is much worse, the space of volume-preserving diffeomorphisms being infinite-dimensional. Since the aim of this chapter is to describe a manifold-learning method that preserves volumes/densities, we are faced with the following question: Given a data manifold with intrinsic dimension d that is diffeomorphic to a subset of $\mathbb{R}^d$, which map, in the infinite-dimensional space of volume-preserving maps from this manifold to $\mathbb{R}^d$, is the "best"? In Section 3.4, we will describe an approach to this problem by setting up a specific optimization procedure. But first, let us describe a method for *estimating* densities on submanifolds.

---

[5]If one is willing to do dimensional reduction to $\mathbb{R}^{d'}$ with $d' > d$, one can deal with more general d-dimensional data manifolds. For instance, if M is an ordinary sphere with intrinsic dimension 2 living in $\mathbb{R}^{10}$, one can do dimensional reduction to $\mathbb{R}^3$. Although interesting and possibly useful, this is a different problem from the one we are considering.

[6]E.g., if the data manifold under consideration is isometric to $\mathbb{R}^3$, the isometry group is generated by three rotations and three translations.

3.3 Density Estimation on Submanifolds

3.3.1 Introduction

Kernel density estimation (KDE) [21] is one of the most popular methods of estimating the underlying probability density function (PDF) of a data set. Roughly speaking, KDE consists of having the data points contribute to the estimate at a given point according to their distances from that point — closer the point, the bigger the contribution. More precisely, in the simplest multi-dimensional KDE [5], the estimate $\hat{f}_m(\mathbf{y}_0)$ of a PDF $f(\mathbf{y}_0)$ at a point $\mathbf{y}_0 \in \mathbb{R}^D$ is given in terms of a sample $\{\mathbf{y}_1, \ldots, \mathbf{y}_m\}$ as,

$$\hat{f}_m(\mathbf{y}_0) = \frac{1}{m} \sum_{i=1}^{m} \frac{1}{h_m^D} K\left(\frac{\|\mathbf{y}_i - \mathbf{y}_0\|}{h_m}\right), \tag{3.1}$$

where $h_m > 0$, the *bandwidth*, is chosen to approach to zero in a suitable manner as the number m of data points increases, and $K : [0, \infty) \to [0, \infty)$ is a *kernel function* that satisfies certain properties such as boundedness. Various theorems exist on the different types and rates of convergence of the estimator to the correct result. The earliest result on the pointwise convergence rate in the multivariable case seems to be given in [5], where it is stated that under certain conditions for f and K, assuming $h_m \to 0$ and $mh_m^D \to \infty$ as $m \to \infty$, the mean squared error in the estimate $\hat{f}(\mathbf{y}_0)$ of the density at a point goes to zero with the rate,

$$\text{MSE}[\hat{f}_m(\mathbf{y}_0)] = \text{E}\left[\left(\hat{f}_m(\mathbf{y}_0) - f(\mathbf{y}_0)\right)^2\right] = O\left(h_m^4 + \frac{1}{mh_m^D}\right) \tag{3.2}$$

as $m \to \infty$. If h_m is chosen to be proportional to $m^{-1/(D+4)}$, one gets,

$$\text{MSE}[\hat{f}_m(p)] = O\left(\frac{1}{m^{4/(D+4)}}\right), \tag{3.3}$$

as $m \to \infty$. The two conditions $h_m \to 0$ and $mh_m^D \to \infty$ ensure that, as the number of data points increases, the density estimate at a point is determined by the values of the density in a smaller and smaller region around that point, but the number of data points contributing to the estimate (which is roughly proportional to the volume of a region of size h_m) grows unboundedly, respectively.

3.3.2 Motivation for the Submanifold Estimator

We would like to estimate the values of a PDF that lives on an (unknown) d-dimensional Riemannian submanifold M of $\mathbb{R}^D$, where $d < D$. Usually, D-dimensional KDE does not work for such a distribution. This can be intuitively understood by considering a distribution on a line in the plane: 1-dimensional KDE performed on the line (with a bandwidth h_m satisfying the asymptotics given above) would converge to the correct density on the line, but 2-dimensional KDE, differing from the former only by a normalization factor that blows up as the bandwidth $h_m \to 0$ (compare (3.1) for the cases $D = 2$ and $D = 1$), diverges. This behavior is due to the fact that, similar to a "delta function" distribution on $\mathbb{R}$, the D-dimensional density of a distribution on a d-dimensional submanifold of $\mathbb{R}^D$ is, strictly speaking, undefined—the density is zero outside the submanifold, and in order to have proper normalization, it has to be infinite on the submanifold. More formally, the D-dimensional probability measure for a d-dimensional PDF supported on M is not absolutely continuous with respect to the Lebesgue measure on $\mathbb{R}^D$, and does not have a probability density function on $\mathbb{R}^D$. If one attempts to use D-dimensional KDE for data drawn from such a probability measure, the estimator will "attempt to converge" to a singular PDF; one that is infinite on M, zero outside.

For a distribution with support on a line in the plane, we can resort to 1-dimensional KDE to get the correct density on the line, but how could one estimate the density on an *unknown*, possibly curved submanifold of dimension $d < D$? Essentially the same approach works: *even for data that lives on an unknown, curved d-dimensional submanifold* of $\mathbb{R}^D$, it suffices to use the d-dimensional kernel density estimator with the *Euclidean distance* on $\mathbb{R}^D$ to get a consistent estimator of the submanifold density. Furthermore, the convergence rate of this estimator can be bounded as in (3.3), with D being replaced by d, the intrinsic dimension of the submanifold. [20]

The intuition behind this approach is based on three facts: 1) For small bandwidths, the main contribution to the density estimate at a point comes from data points that are nearby; 2) For small distances, a d-dimensional Riemannian manifold "looks like" $\mathbb{R}^d$, and densities in $\mathbb{R}^d$ should be estimated by a d-dimensional kernel, instead of a D-dimensional one; and 3) For points of M that are close to each other, the intrinsic distances as measured on M are close to Euclidean distances as measured in the surrounding $\mathbb{R}^D$. Thus, as the number of data points increases and the bandwidth is taken to be smaller and smaller, estimating the density by using a kernel normalized for d dimensions and distances as measured in $\mathbb{R}^D$ should give a result closer and closer to the correct value.

We will next give the formal definition of the estimator motivated by these considerations, and state the theorem on its asymptotics. As in the original work of Parzen [21], the pointwise consistence of the estimator can be proven by using a bias-variance decomposition. The asymptotic unbiasedness of the estimator follows from the fact that as the bandwidth converges to zero, the kernel function becomes a "delta function." Using this fact, it is possible to show that with an appropriate choice for the vanishing rate of the bandwidth, the variance also vanishes asymptotically, completing the proof of the pointwise consistency of the estimator.

3.3.3 Statement of the Theorem

Let $(M, \mathbf{g})$ be a d-dimensional, embedded, complete, compact Riemannian submanifold of $\mathbb{R}^D$ $(d < D)$ without boundary. Let the injectivity radius of M be $r_{inj} > 0$.[7] Let $d(p, q) = d_p(q)$ be the length of a length-minimizing geodesic in M between $p, q \in M$, and let $u(p, q) = u_p(q)$ be the geodesic distance between p and q as measured in $\mathbb{R}^D$ (thus, $u(p, q)$ is simply the Euclidean distance between p and q in $\mathbb{R}^D$). Note that $u(p, q) \leq d(p, q)$. We will denote the Riemannian volume measure on M by V, and the volume form by dV.[8]

Theorem 3.3.1 *Let $f : M \to [0, \infty)$ be a probability density function defined on M (so that the related probability measure is fV), and $K : [0, \infty) \to [0, \infty)$ be a continuous function that vanishes outside $[0, 1)$, is differentiable with a bounded derivative in $[0, 1)$, and satisfies the normalization condition, $\int_{\|\mathbf{z}\| \leq 1} K(\|\mathbf{z}\|) d^d\mathbf{z} = 1$. Assume f is differentiable to second order in a neighborhood of $p \in M$, and for a sample $q_1, \ldots, q_m$ of size m drawn from the*

[7] The injectivity radius r_{inj} of a Riemannian manifold is a distance such that all geodesic pieces (i.e., curves with zero intrinsic acceleration) of length less than r_{inj} minimize the length between their endpoints. On a complete Riemannian manifold, there exists a distance-minimizing geodesic between any given pair of points, however, an arbitrary geodesic need not be distance minimizing. For example, any two non-antipodal points on the sphere can be connected by two geodesics with different lengths, one of which is distance-minimizing, namely, the two pieces of the great circle passing throught the points. For a detailed discussion of these issues, see, e.g., [2].

[8] Note that we are making a slight abuse of notation here, denoting the corresponding points in M and $\mathbb{R}^D$ with the same symbols, p, q.

density f, define an estimator $\hat{f}_m(p)$ of $f(p)$ as,

$$\hat{f}_m(p) = \frac{1}{m}\sum_{j=1}^{m}\frac{1}{h_m^d}K\left(\frac{u_p(q_j)}{h_m}\right), \tag{3.4}$$

where $h_m > 0$. If h_m satisfies $\lim_{m\to\infty} h_m = 0$ and $\lim_{m\to\infty} mh_m^d = \infty$, then, there exist non-negative numbers m_*, C_b, and C_V such that for all $m > m_*$ the mean squared error of the estimator (3.4) satisfies,

$$\mathrm{MSE}\left[\hat{f}_m(p)\right] = \mathrm{E}\left[\left(\hat{f}_m(p) - f(p)\right)^2\right] < C_b h_m^4 + \frac{C_V}{mh_m^d}. \tag{3.5}$$

If h_m is chosen to be proportional to $m^{-1/(d+4)}$, this gives,

$$\mathrm{E}\left[\left(f_m(p) - f(p)\right)^2\right] = O\left(\frac{1}{m^{4/(d+4)}}\right), \tag{3.6}$$

as $m \to \infty$.

Thus, the bound on the convergence rate of the submanifold density estimator is as in (3.2), (3.3), with the dimensionality D replaced by the intrinsic dimension d of M. As mentioned above, the proof of this theorem follows from two lemmas on the convergence rates of the bias and the variance; the h_m^4 term in the bound corresponds to the bias, and the $1/mh_m^d$ term corresponds to the variance; see [20] for details. This approach to submanifold density estimation was previously mentioned in [11], and the thesis [10] contains the details, although in a more technical and general approach than the elementary one followed in [20].

3.3.4 Curse of Dimensionality in KDE

The convergence rate (3.3) is an example of the *curse of dimensionality*; the rate gets slower and slower as the dimensionality D of the data set increases. In Table 4.2 of [25], which we reproduce in Table 3.1, Silverman demonstrates how the sample size required for a given mean squared error in the estimate of a multivariate normal distribution increases with the dimensionality. The numbers look as discouraging as the formula (3.3).

Dimensionality	Required sample size
1	4
2	19
3	67
4	223
5	768
6	2790
7	10700
8	43700
9	187000
10	842000

Table 3.1: Sample size required to ensure that the relative mean squared error at zero is less than 0.1, when estimating a standard multivariate normal density using a normal kernel and the window width that minimizes the mean square error at zero.

One source of optimism towards various curses of dimensionality is the fact that real-life high-dimensional data sets usually lie on low-dimensional subspaces of the full space they are sampled from. If the performance of a method/algorithm under consideration depends only on the intrinsic dimensionality of the data, the curse of dimensionality can be avoided for data with low *intrinsic* dimensionality. Alternatively, even if the performance depends on the full dimensionality of the feature space, it may still be possible to avoid the curse for the case of low intrinsic dimension, by using dimensional reduction on the data in order to work with a low-dimensional, faithful representation.

One example of the former case is the results on nearest neighbor search [14, 3] indicating that the performance of certain nearest-neighbor search algortihms is determined not by the full dimensionality of the feature space, but by the intrinsic dimensionality of the data subspace. Here, we see that submanifold density estimation provides another result of this sort, showing that after a small modification, the pointwise convergence rate of kernel density estimation is bounded by the intrinsic dimension of the data subspace, not the dimension of the full feature space. As a result, we conclude that results such as those in Table 3.1 are overly pessimistic, leading one to expect convergence rates much slower than what one observes in reality, after the aforementioned modification.

3.4 Preserving the Estimated Density: The Optimization

3.4.1 Preliminaries

Now that we have a method to estimate the density on a submanifold of $\mathbb{R}^D$, we can proceed to define an algorithm for density preserving maps.[9] Suppose we are given a sample $X = \{x_1, x_2, \ldots, x_m\}$ of m data points $x_i \in \mathbb{R}^D$ that live on a d-dimensional submanifold M of $\mathbb{R}^D$. We first proceed to estimate the density at each one of the points, by using a slightly generalized version of the submanifold estimator that has variable bandwidths. Denoting the bandwidth for a given evaluation point x_j and a reference (data) point x_i by h_{ij}, the generalized, variable bandwidth estimator at x_j is,[10]

$$\hat{f}_j = \hat{f}(x_j) = \frac{1}{m} \sum_i \frac{1}{h_{ij}^d} K_d \left(\frac{\|x_j - x_i\|_D}{h_{ij}} \right). \tag{3.7}$$

Variable bandwidth methods allow the estimator to adapt to the inhomogeneities in the data. Various approaches exist for picking the bandwidths h_{ij} as functions of the query (evaluation) point x_j and/or the reference point x_i [25]. Here, we focus on the kth-nearest neighbor approach for evaluation points, i.e., we take h_{ij} to depend only on the evaluation point x_j, and we let $h_{ij} = h_j =$ the distance of the kth nearest data (reference) point to the evaluation point x_j. Here, k is a free parameter that needs to be picked by the user. However, instead of tuning it by hand, one can use a leave-one-out cross-validation score [25] such as the log-likelihood score for the density estimate to pick the best value. This is done by estimating the log-likelihood of each data point by using the leave-one-out version

[9]We do not claim that this is the only way to define algorithms for density preserving maps. DPMs that do not first estimate the submanifold density are also conceivable. For instance, for the case of intrinsically flat submanifolds, distance-preserving maps automatically preserve densities. For intrinsically curved manifolds, one can obtain density-preserving maps by aiming to preserve local volumes instead of dealing directly with densities. Certain area-preserving surface meshing algorithms can be thought of as two-dimensional examples of (approximate) density preserving maps. Generalizations of these meshing algorithms could provide another approach to DPMs.

[10]When evaluating the accuracy of the estimator via a leave-one-out log-likelihood cross-validation score [25], the sum in (3.7) is taken over all points except the evaluation point x_j, and the factor of $1/m$ in the front gets replaced with $1/(m-1)$.

of the density estimate (3.7) for a range of k values, and picking the k that gives the highest log-likelihood.

Now, given the estimates $\hat{f}_j = \hat{f}(x_j)$ of the submanifold density at the D-dimensional data points x_j, we want to find a d-dimensional representation $X' = \{x'_1, x'_2, \ldots, x'_m\}$, $x'_i \in \mathbb{R}^d$ such that the new estimates $\hat{f}'_i$ at the points $x'_i \in \mathbb{R}^d$ agree with the original density estimates, i.e.,

$$\hat{f}'_i = \hat{f}_i\,, \qquad i = 1, \ldots, m\,. \tag{3.8}$$

For this purpose, one can attempt, for example, to minimize the mean squared deviation of $\hat{f}'_i$ from $\hat{f}_i$ as a function of the x'_is, but such an approach would result in a non-convex optimization problem with many local minima. We formulate an alternative approach involving semidefinite programming, for the special case of the Epanechnikov kernel [25], which is known to be asymptotically optimal for density estimation, and is convenient for formulating a convex optimization problem for the matrix of inner products (the *Gram matrix*, or the *kernel matrix*) of the low dimensional data set X'.

3.4.2 The Optimization

The Epanechnikov kernel. The Epanechnikov kernel k_e in d dimensions is defined as,

$$k_e(\|x_i - x_j\|) = \begin{cases} N_e(1 - \|x_i - x_j\|^2), & 0 \leq \|x_i - x_j\| \leq 1 \\ 0, & 1 \leq \|x_i - x_j\| \end{cases} \tag{3.9}$$

where N_e is the normalization constant that ensures $\int_{\mathbb{R}^d} k_e(\|x - x'\|) d^d x' = 1$. We will assume that the kernel used in the estimates $\hat{f}_i$ and $\hat{f}'_i$ of the density via (3.7) is the Epanechnikov kernel. Owing to its quadratic form (3.9), this kernel facilitates the formulation of a convex optimization problem. Instead of seeking the dimensionally reduced version $X' = \{x'_1, \ldots, x'_n\}$ of the data set directly, we will first aim to obtain the kernel matrix $K_{ij} = x'_i \cdot x'_j$ for the low-dimensional data points. This is a common approach in the manifold learning literature, where one obtains the low-dimensional data points themselves from the K_{ij} via a singular value decomposition.

We next formulate the DPM optimization problem using the Epanechnikov kernel, and comment on the motivation behind it. As in the case of distance-based manifold learning methods, there will likely be various approaches to density-preserving dimensional reduction, some computationally more efficient than the one discussed here. We hope the discussions in this chapter will stimulate further research in this area.

Given the estimated densities $\hat{f}_i$, we seek a symmetric, positive semidefinite inner product matrix $K_{ij} = x'_i \cdot x'_j$ that results in d-dimensional density estimates that agree with $\hat{f}_i$. In order to deal with the non-uniqueness problem mentioned during our discussion of density-preserving maps between manifolds (which likely carries over to the discrete setting), we need to pick a suitable objective function to maximize. We choose the objective function to be the same as that of Maximum Variance Unfolding (MVU) [29], namely, trace(K). After getting rid of translations by constraining the center of mass of the dimensionally reduced data points to the origin, maximizing the objective function trace(K) becomes equivalent to maximizing the sum of the squared distances between the data points [29].

While the objective function for DPM is the same as that of MVU, the constraints of the former will be weaker. Instead of preserving the distances between k-nearest neighbors, the DPM optimization defined below preserves the total contribution of the original k-nearest neighbors to the density estimate at the data points. As opposed to MVU, this allows for local stretches of the data set, and results in optimal kernel matrices K that can be faithfully represented by a smaller number of dimensions than the intrinsic dimensionality suggested by MVU. For instance, while MVU is capable of unrolling data on the Swiss roll onto a flat plane, it is impossible to lay data from a spherical cap onto the plane while keeping the distances to the kth nearest neighbors fixed.[11] Thus, the constraints of the optimization in MVU are too stringent to give an inner product matrix K of rank 2, when the original data is on an intrinsically curved surface in $\mathbb{R}^3$. We will see below that the looser constraints of DPM allow it to do a better job in capturing the intrinsic dimensionality of a curved surface.

The precise statement of the DPM optimization problem is as follows. We use d_{ij} and ϵ_{ij} to denote the distance between x'_i and x'_j, and the (unnormalized) contribution $(1-d_{ij}^2/h_i^2)$ of x'_j to the density estimate at x'_i, respectively. These auxiliary variables are given in terms the kernel matrix K_{ij} directly.

$$\max_K \text{trace}(K) \tag{3.10}$$

$$\begin{aligned} \text{such that:} & \\ d_{ij}^2 &= K_{ii} + K_{jj} - K_{ij} - K_{ji} \\ \epsilon_{ij} &= (1 - d_{ij}^2/h_i^2) \ \ (j \in I_i) \\ \hat{f}_i &= \frac{N_e}{h_i^d} \sum_{j \in I_i} \epsilon_{ij} \\ K &\succeq 0 \\ \epsilon_{ij} &\geq 0 \ \ (j \in I_i) \\ \sum_{i,j=1}^{n} K_{ij} &= 0 \end{aligned}$$

Here, I_i is the index set for the k-nearest neighbors to the point x_i in the original data set X in $\mathbb{R}^D$. The last constraint ensures that the center of mass of the dimensionally reduced data set is at the origin, as in MVU. Since ϵ_{ij} and d_{ij} are fixed for a given K_{ij}, the unknown quantities in the optimization are the entries of the matrix K_{ij}. Once K_{ij} is found, we can get the eigenvalues/eigenvectors and obtain the dimensionally reduced data $\{x'_i\}$, as in MVU. Note that the optimization (3.10) is performed over the set of symmetric, positive semidefinite matrices.

With the condition $\epsilon_{ij} \geq 0$, we enforce the dimensionally reduced versions of the original k-nearest neighbors of x_i to stay close to x'_i, and at least marginally contribute to the new density estimate at that point. Thus, although we allow local stretches in the data set by allowing the distance values d_{ij} to be different from the original distances, we do not let the points in the neighbor set I_i move too far away from x'_i.[12,13]

At first sight, the optimization (3.10) seems to require the user to set a dimensionality d in the normalization factor $\frac{N_e}{h^d}$. However, the same factor occurs in the original submanifold

[11] In order to see this, think of a mesh on the hemisphere with rigid rods at the edges, joined to each other by junction points that allow rotation. It is impossible to open up such spherical mesh onto the plane without breaking the rods.

[12] Although the constraints $\epsilon_{ij} \geq 0$ enforce nearby points to stay nearby, it is possible in principle (as it is in MVU) for faraway points to get close upon dimensional reduction. In our case, this would result in the actual density estimate in the dimensionally reduced data set being different from the one estimated by using the original neighbor sets I_i. This can be avoided by including structure-preserving constraints. In [24], a structure-preserving version of MVU is presented, which gives a dimensional reduction that preserves the original neighborhood graph. Similarly, a structure-preserving version of the DPM presnted here can also be implemented, however, the large number of constraints will make it computationally less practical than the version given here.

[13] Note that h_i are the original bandwidth values fixed by data set X, and are not reevaluated in the dimensionally reduced version, X'. Thus, it is possible, in principle, to get slightly different estimates of the density if one reevaluates the bandwidth in the new data set. However, we expect the objective function to push the data points as far away from each other as possible, resulting in kth nearest neighbor distances being close to those of the original data set (since they can't go further out than h_{ij}, due to the constraints $\epsilon_{ij} \geq 0$).

density estimates $\hat{f}_i$, since the submanifold KDE algorithm [20] requires the kernel to be normalized according to the intrinsic dimensionality d, instead of D. Thus, the only place the dimensionality d comes up in the optimization problem, namely, the third line of (3.10), has the same d-dependent factor on both sides of the equation, and these factors can be canceled to perform the optimization without choosing a specific d.

Let us remark that, as is usual in methods that use a neighborhood graph to do manifold learning, DPM optimization does not use the directed graph of the original k-nearest neighbors, but uses a symmetrized, undirected version instead, basically by lifting the "arrows" in the original graph. In other words, we call two points neighbors if either one is a k-nearest neighbor of the other, and set the bandwidth h_i for a given evaluation point x_i to be the largest distance to the elements in its set of neighbors, which may have more than k elements.

3.4.3 Examples

We next compare the results of DPM to those of Maximum Variance Unfolding, Isomap, and Principal Components Analysis, all of which are methods that are based on kernel matrices. We use two data sets that live on intrinsically curved spaces, namely, a sphere cap and a surface with two peaks. For a given, fixed value of k, we obtain the kernel matrices from each one of the methods, and plot the eigenvalues of these matrices. The top d eigenvalues give a measure of the spread of the data one would encounter in each dimension, if one were to reduce to d dimensions. Thus, the number of eigenvalues that have appreciable magnitudes gives the dimensionality that the method "thinks" is required to represent the original data set in a manner consistent with the constraints imposed.

The results are given in the figures below. As can be seen from the eigenvalue plots in Figure 3.2, the kernel matrix learned by DPM captures the intrinsic dimensionality of the data sets under consideration more effectively than any of the other methods shown. In Figure 3.3, we demonstrate the capability of DPMs to pick an optimal neighborhood number k by using the maximum likelihood criterion and show the resulting dimensional reduction for data on a hemisphere. The DPM reduction can be compared with the Isomap reductions of the same data set for three different values of k, given in Figure 3.3. The results do depend on the value of k, and for a user of a method such as Isomap, there is no obvious way to pick a "best" k, whereas DPMs come with a natural quantitative way to evaluate and pick the optimal k.

For intrinsically flat cases such as the Swiss roll, there is a more or less canonical dimensionally reduced form, and we can judge the performance of various methods of nonlinear dimensional reduction according to how well they "unwind" the roll to its expected planar form. However, since there is no canonical dimensionally reduced form for an intrinsically curved surface like a sphere cap,[14] judging the quality of the dimensionally reduced forms is less straightforward. Two advantages of DPMs stand out. First, due to Moser's theorem and its corollary discussed in Section 3.2, density preserving maps are in principle capable of reducing data on intrinsically curved spaces to $\mathbb{R}^d$ with d being the intrinsic dimension of the data, whereas distance-preserving maps[15] require higher dimensionalities. This can be observed in the eigenvalue plot in Figure 3.2. Second, whereas methods that attempt to preserve distances of data on curved spaces end up distorting the distances in various ways, density preserving maps hold their promise of preserving density. Thus, when investigating a data set that was dimensionally reduced to its intrinsic dimensionality by DPM, we can be confident that the density we observe accurately represents the intrinsic density of the data manifold, whereas with distance-based methods, we do not know how the data is deformed.

[14] Think of the different ways of producing maps by using different projections of the spherical Earth.
[15] Even *locally* distance-preserving ones.

Perhaps the main disadvantage of the specific DPM discussed in this chapter is one of computational inefficiency; solving the semidefinite problem (3.10) is slow,[16] and the density estimation step is inefficient, as well. Both of these disadvantages can be partly remedied by using faster algorithms like the one presented in [4] for semidefinite programming, or an Epanechnikov version of the approach in [15] for KDE, but radically different approaches that possibly eliminate the density estimation step may turn out to be even more fruitful. We hope the discussion in this chapter will motivate the reader to consider alternative approaches to this problem.

In Figures 3.1 and 3.3 we show the two-peaks data set and the hemisphere data set, respectively, and their reduction to two dimensions by DPM.

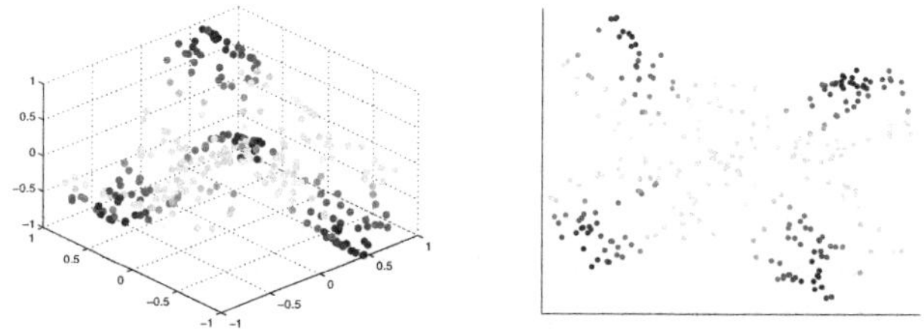

Figure 3.1: (See Color Insert.) The twin peaks data set, dimensionally reduced by density preserving maps.

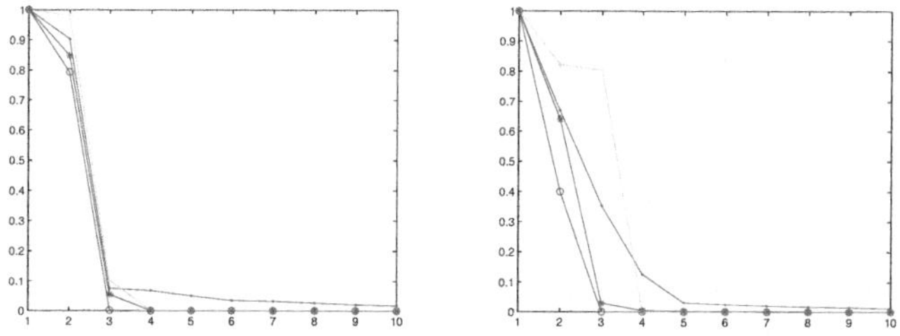

Figure 3.2: (See Color Insert.) The eigenvalue spectra of the inner product matrices learned by PCA. (green, '+'), Isomap (red, '.'), MVU (blue, '*'), and DPM (blue, 'o'). Left: A spherical cap. Right: The "twin peaks" data set. As can be seen, DPM suggests the lowest dimensional representation of the data for both cases.

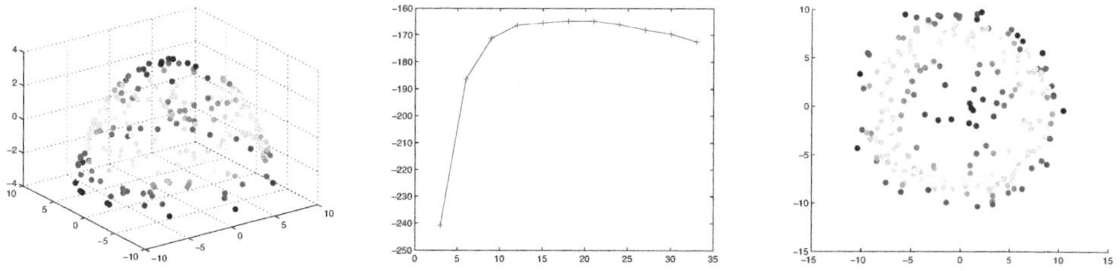

Figure 3.3: (See Color Insert.) The hemisphere data, log-likelihood of the submanifold KDE for this data as a function of k, and the resulting DPM reduction for the optimal k.

[16] We have implemented the optimization in SeDuMi [26].

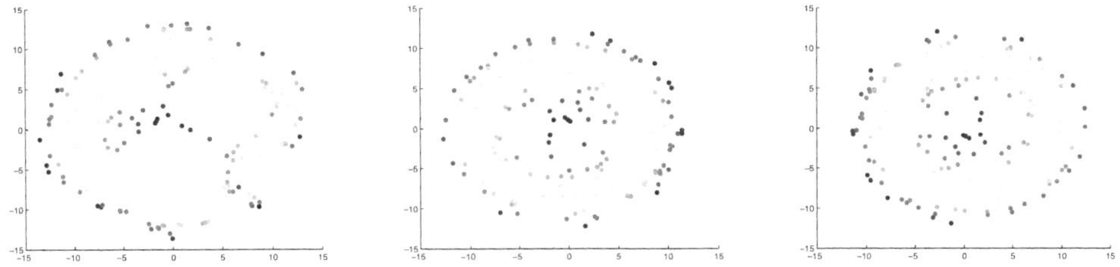

Figure 3.4: (See Color Insert.) Isomap on the hemisphere data, with $k = 5, 20, 30$.

3.5 Summary

In this chapter, we discussed density preserving maps, a density-based alternative to distance-based methods of manifold learning. This method aims to perform dimensionality reduction on large-dimensional data sets in a way that preservs their density. By using a classical result due to Moser, we proved that density preserving maps to $\mathbb{R}^d$ exist even for data on intrinsically curved d-dimensional submanifolds of $\mathbb{R}^D$ that are globally, or topologically "simple." Since the underlying probability density function is arguably one of the most fundamental statistical quantities pertaining to a data set, a method that preserves densities while performing dimensionality reduction is guaranteed to preserve much valuable structure in the data. While distance-preserving approaches distort data on intrinsically curved spaces in various ways, density preserving maps guarantee that certain fundamental statistical information is conserved.

We reviewed a method of estimating the density on a submanifold of Euclidean space. This method was a slightly modified version of the classical method of kernel density estimation, with the additional property that the convergence rate was determined by the intrinsic dimensionality of the data, instead of the full dimensionality of the Euclidean space the data was embedded in. We made a further modification on this estimator to allow for variable "bandwidths," and used it with a specific kernel function to set up a semidefinite optimization problem for a proof-of-concept approach to density preserving maps. The objective function used was identical to the one in Maximum Variance Unfolding [29], but the constraints were significantly weaker than the distance-preserving constraints in MVU. By testing the methods on two relatively small, synthetic data sets, we experimentally confirmed the theoretical expectations and showed that density preserving maps are better in detecting and reducing to the intrinsic dimensionality of the data than some of the commonly used distance-based approaches that also work by first estimating a kernel matrix.

While the initial formulation presented in this chapter is not yet scalable to large data sets, we hope our discussion will motivate our readers to pursue the idea of density preserving maps further, and explore alternative, superior formulations. One possible approach to speeding up the computation is to use fast semidefinite programming techniques [4].

3.6 Bibliographical and Historical Remarks

The aim of manifold learning methods is to find representations of large-dimensional data sets in low-dimensional Euclidean spaces while preserving some aspects of the original data as accurately as possible. Some of the more successful manifold learning methods are as follows.

Isomap [27] begins by forming a neighborhood graph of the data set, either by picking the k-nearest neighbors, or using a neighborhood distance cutoff. It then estimates the geodesic distances between data points on the data manifold by finding paths of minimal length on the neighborhood graph. The estimated geodesic distances are then used to calculate a kernel matrix that gives Euclidean distances that are equal to these geodesic distances. Singular value decomposition then allows one to reproduce the data set from this kernel matrix, by picking the most significant eigenvalues. When used to reduce the data to its intrinsic dimensionality, Isomap unavoidably distorts the distances between points that lie on a curved manifold.

Locally Linear Embedding (LLE) [23] also begins by forming a neighborhood graph for the data set. It then computes a set of weights for each point so that the point is given as an approximate linear combination of its neighbors. This is done by minimizing a cost function which quantifies the reconstruction error. Once the weights are obtained, one seeks a low-dimensional representation of the data set that satistifes the same approximate linear relations between the points as in the original data. Once again, a cost function that measures the reconstruction error is used. The minimization of the cost function is not done explicitly, but is done by solving a sparse eigenvalue problem. A modified version of LLE called Hessian LLE [7] produces results of higher quality, but has a higher computational complexity.

As in LLE and Isomap, the method of Laplacian EigenMaps [1] begins by obtaining a neighborhood graph. This time, the graph is used to define a graph Laplacian, whose truncated eigenvectors are used to construct a dimensionally reduced form of the data set.

Among the existing manifold learning algorithms, Maximum Variance Unfolding (MVU) [29] is the most similar to the specific approach to density preserving maps described in this chapter. After the neighborhood graph is obtained, MVU maximizes the mean squared distances between the data points, while keeping the distances to the nearest neighbors fixed, by using a semidefinite programming approach. As mentioned in Section 3.4, this method results in a more strongly constrained optimization problem than that of DPM, and ends up suggesting higher intrinsic dimensionalities for data sets on intrinsically curved spaces.

Other prominent approaches to manifold learning include Local Tangent Space Alignment [30], Diffusion Maps [6], and Manifold Sculpting [8].

The problem of preserving the density of a data set that lives on a submanifold of Euclidean space has led us to the more basic problem of estimating the density of such a data set. Pelletier [22] defined and proved the consistency of a version of kernel density estimation for Riemannian manifolds; however, his approach cannot be used directly in the submanifold problem, since one needs to know in advance the manifold the data lives on, and be able to calculate various intricate geometric quantities pertaining to it. In [28], the authors provide a method for estimating submanifold densities in $\mathbb{R}^d$, but do not give a proof of consistency for the method proposed. For other related work, see [19, 13].

The method used in this chapter is based on the work in [20], where a submanifold kernel density estimator was defined, and a theorem on its consistency was proven. The convergence rate was bounded in terms of the intrinsic dimension of the data submanifold, showing that the usual assumptions on the behavior of KDE in large dimensions is overly pessimistic. In this chapter, we have modified the estimator in [20] slightly by allowing a variable bandwidth. The submanifold KDE approach was previously described in [11], and the thesis [10] contains the details of the proof of consistency.

The existence of density preserving maps in the continuous case was proved by using a result due to Moser [18] on the existence of volume-preserving maps. Moser's result was generalized to non-compact manifolds in [9]. We mentioned the abundance of volume-preserving maps and the the need to fix a criterion for picking the "best one." The group of volume-preserving maps of $\mathbb{R}$ was investigated in [17].

Bibliography

[1] M. Belkin and P. Niyogi. Laplacian Eigenmaps for Dimensionality Reduction and Data Representation, 2003.

[2] M. Berger. *A panoramic view of Riemannian geometry.* New York: Springer Verlag, 2003.

[3] A. Beygelzimer, S. Kakade, and J. Langford. Cover trees for nearest neighbor. In *Proceedings of the 23rd International Conference on Machine Learning*, 97–104. ACM New York, 2006.

[4] S. Burer and R. Monteiro. A nonlinear programming algorithm for solving semidefinite programs via low-rank factorization. *Mathematical Programming*, 95(2):329–357, 2003.

[5] T. Cacoullos. Estimation of a multivariate density. *Annals of the Institute of Statistical Mathematics*, 18(1):179–189, 1966.

[6] R. Coifman and S. Lafon. Diffusion maps. *Applied and Computational Harmonic Analysis*, 21(1):5–30, 2006.

[7] D. Donoho and C. Grimes. Hessian eigenmaps: Locally linear embedding techniques for high-dimensional data. *Proceedings of the National Academy of Sciences*, 100(10):5591–5596, 2003.

[8] M. Gashler, D. Ventura, and T. Martinez. Iterative non-linear dimensionality reduction with manifold sculpting. *Advances in Neural Information Processing Systems*, 19, 2007.

[9] R. Greene and K. Shiohama. Diffeomorphisms and volume-preserving embeddings of noncompact manifolds. *American Mathematical Society*, 255, 1979.

[10] M. Hein. Geometrical aspects of statistical learning theory. 2006.

[11] M. Hein. Uniform convergence of adaptive graph-based regularization. In *Proceedings of the 19th Annual Conference on Learning Theory (COLT), 50–64*. New York: Springer, 2006.

[12] J. Jost. *Riemannian geometry and geometric analysis.* Springer, 2008.

[13] V. Koltchinskii. Empirical geometry of multivariate data: A deconvolution approach. *Annals of statistics*, 28(2):591–629, 2000.

[14] F. Korn, B. Pagel, and C. Faloutsos. On dimensionality and self-similarity. *IEEE Transactions on Knowledge and Data Engineering*, 13(1):96–111, 2001.

[15] D. Lee, A. Gray, and A. Moore. Dual-tree fast gauss transforms. *Arxiv preprint arXiv:1102.2878*, 2011.

[16] J. Lee. *Riemannian manifolds: an introduction to curvature.* New York: Springer Verlag, 1997.

[17] D. McDuff. On the group of volume-preserving diffeomorphisms of R. *American Mathematical Society*, 261(1), 1980.

[18] J. Moser. On the volume elements on a manifold. *Transactions of the American Mathematical Society*, 1965.

[19] K. Nijmegen. Nonparametric estimation of a probability density on a Riemannian manifold using Fourier expansions. *The Annals of Statistics*, 18(2):832–849, 1990.

[20] A. Ozakin and A. Gray. Submanifold density estimation. *Advances in neural information processing systems*, 2009.

[21] E. Parzen. On the estimation of a probability density function and mode. *Annals of Mathematical Statistics*, 33:1065–1076, 1962.

[22] B. Pelletier. Kernel density estimation on Riemannian manifolds. *Statistics and Probability Letters*, 73(3):297–304, 2005.

[23] S. Roweis and L. Saul. Nonlinear Dimensionality Reduction by Locally Linear Embedding. www.sciencemag.org, 290: 2323–2326, 2000.

[24] B. Shaw and T. Jebara. Structure preserving embedding. Proceedings of the 26th Annual International Conference on Machine Learning, 937–944, ACM, Montréal, 2009.

[25] B. Silverman. *Density Estimation for Statistics and Data Analysis.* Chapman & Hall/CRC, 1986.

[26] J. Sturm. Using SeDuMi 1.02, a MATLAB toolbox for optimization over symmetric cones. *Optimization methods and software*, 11(1):625–653, 1999.

[27] J. B. Tenenbaum, V. Silva, and J. C. Langford. A Global Geometric Framework for Nonlinear Dimensionality Reduction. *Science*, 290(5500):2319–2323, 2000.

[28] P. Vincent and Y. Bengio. Manifold Parzen Windows. *Advances in neural information processing systems*, 849–856, 2003.

[29] K. Weinberger, F. Sha, and L. Saul. Learning a kernel matrix for nonlinear dimensionality reduction. In *Proceedings of the Twenty-first International Conference on Machine Learning.* ACM, New York, 2004.

[30] Z. Zhang and H. Zha. Principal Manifolds and Nonlinear Dimension Reduction via Local Tangent Space Alignment. *Arxiv preprint cs.LG/0212008*, 2002.

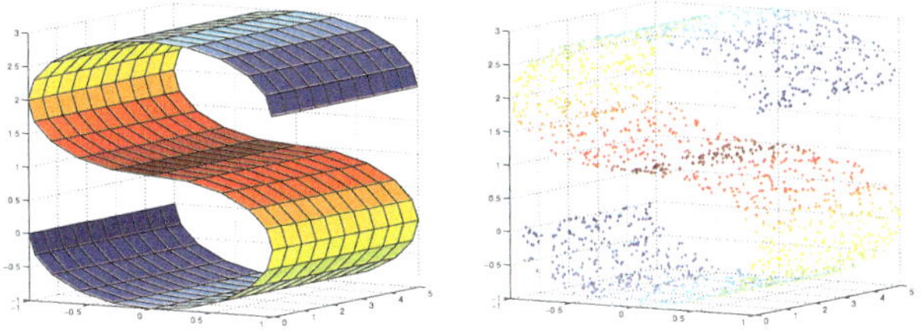

Figure 1.1: Left panel: The S-curve, a two-dimensional S-shaped manifold embedded in three-dimensional space. Right panel: 2,000 data points randomly generated to lie on the surface of the S-shaped manifold. Reproduced from Izenman (2008, Figure 16.6) with kind permission from Springer Science+Business Media.

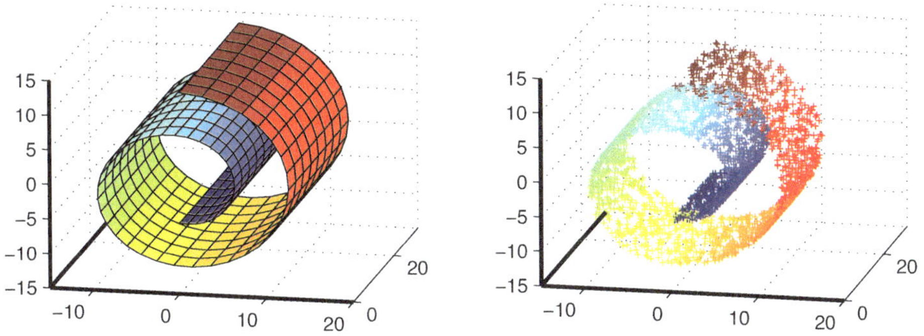

Figure 1.2: Left panel: The Swiss Roll: a two-dimensional manifold embedded in three-dimensional space. Right panel: 20,000 data points lying on the surface of the Swiss-roll manifold. Reproduced from Izenman (2008, Figure 16.7) with kind permission from Springer Science+Business Media.

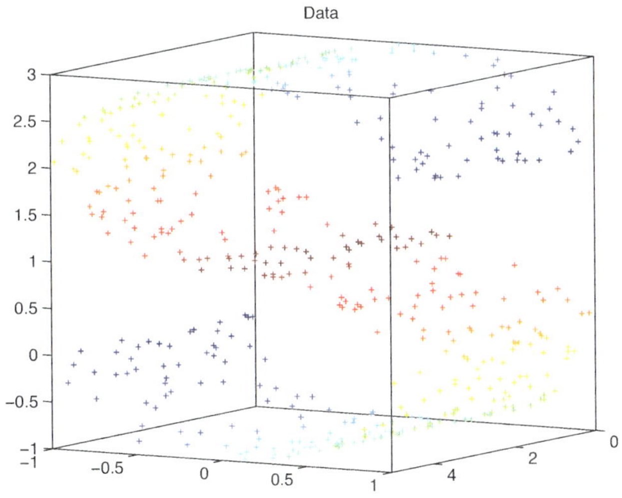

Figure 2.2: S-Curve manifold data. The graph is easier to understand in color.

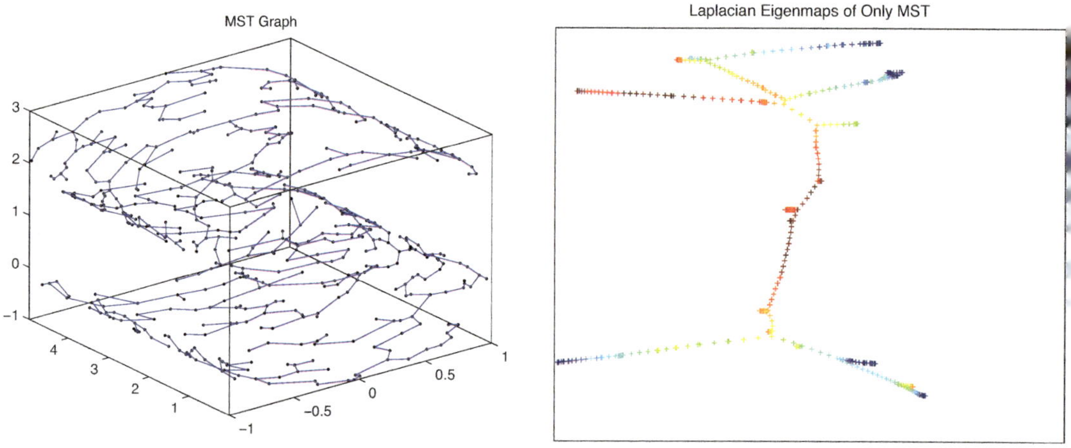

Figure 2.3: The MST graph and the embedded representation.

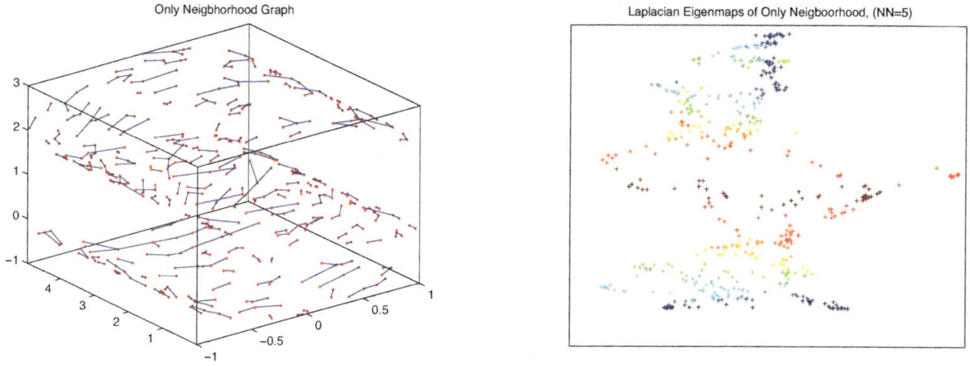

Figure 2.5: The graph with $k = 5$ and its embedding using LEM. Increasing the neighborhood information to 5 neighbors better represents the continuity of the original manifold.

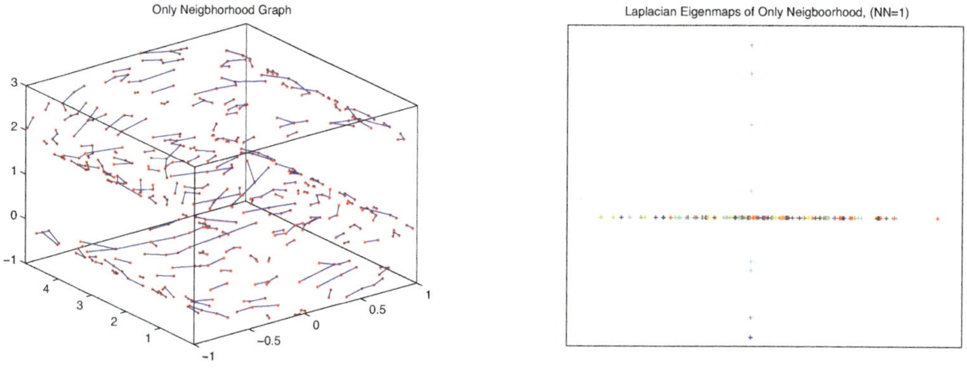

Figure 2.7: The graph with $k = 1$ and its embedding using LEM. Because of very limited neighborhood information the embedded representation cannot capture the continuity of the original manifold.

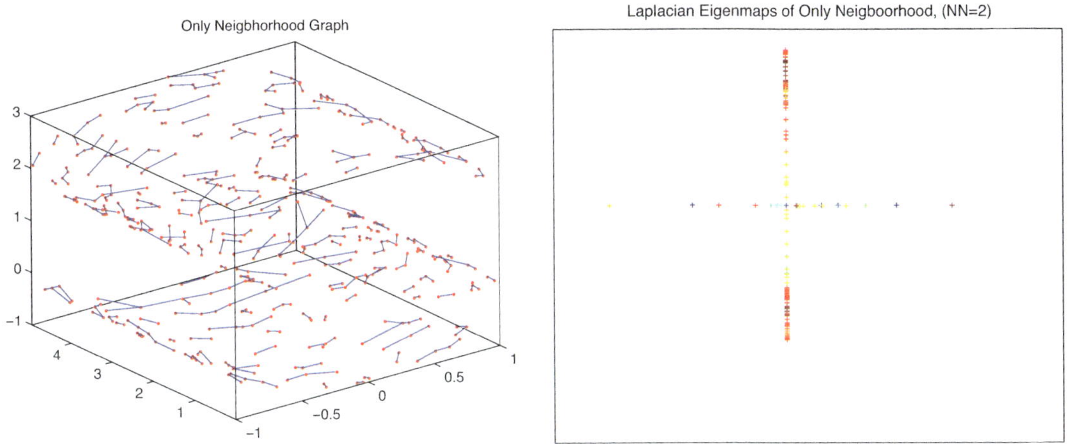

Figure 2.8: The graph with $k = 2$ and its embedding using LEM. Increasing the neighborhood information to 2 neighbors is still not able to represent the continuity of the original manifold.

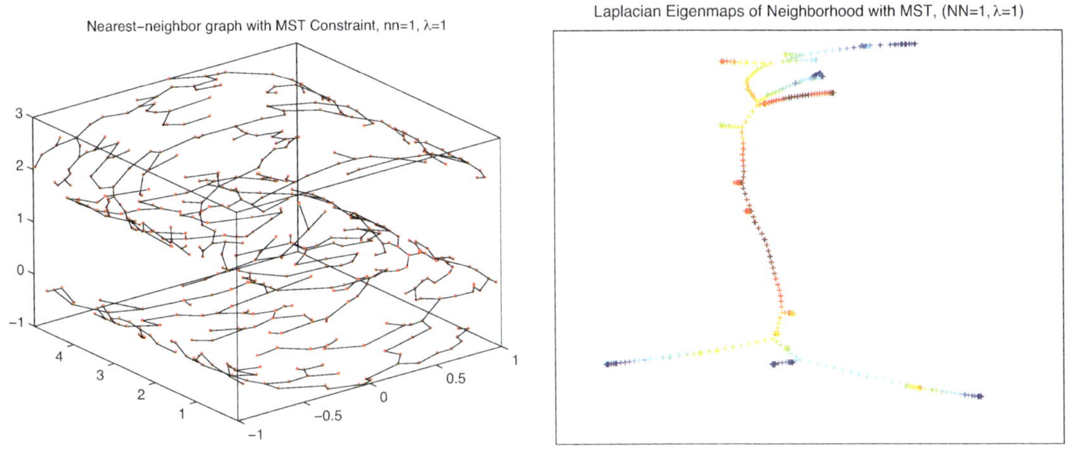

Figure 2.9: The graph sum of a graph with neighborhood of $k = 1$ and MST; and its embedding. In spite of very limited neighborhood information the GLEM is able to preserve the continuity of the original manifold and is primarily due to MST's contribution.

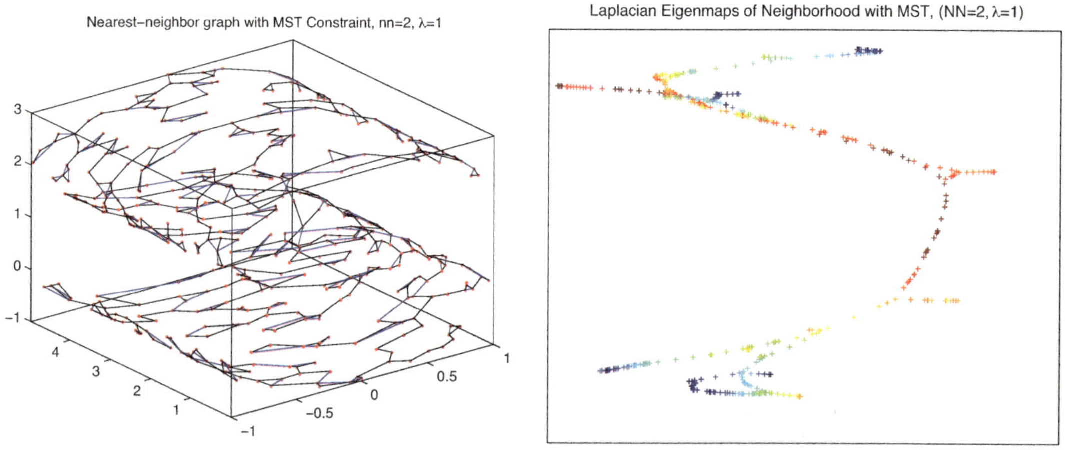

Figure 2.10: GLEM results for $k = 2$ and MST; and its embedding GLEM. In this case also, embedding's continuity is dominated by the MST.

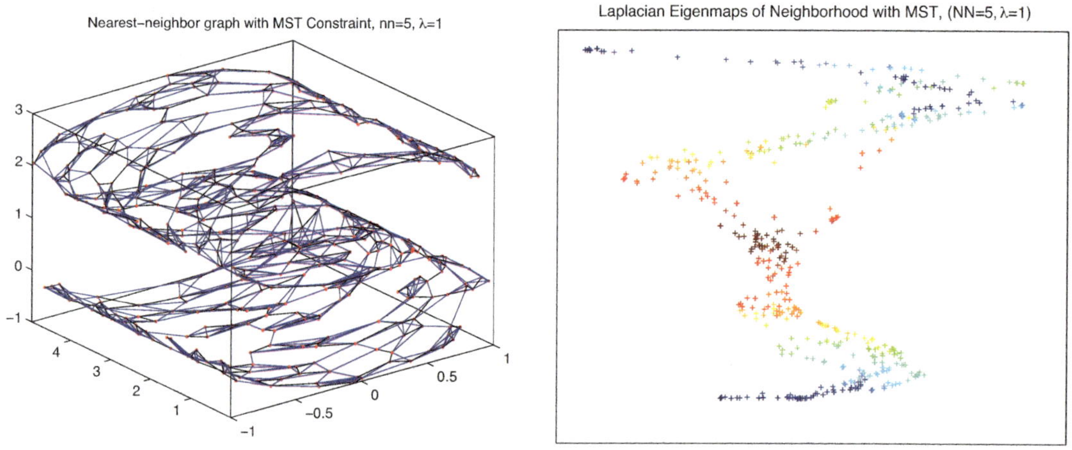

Figure 2.11: Increase the neighbors to $k = 5$ and the neighborhood graph starts dominating and the embedded representation is similar to Figure.2.5

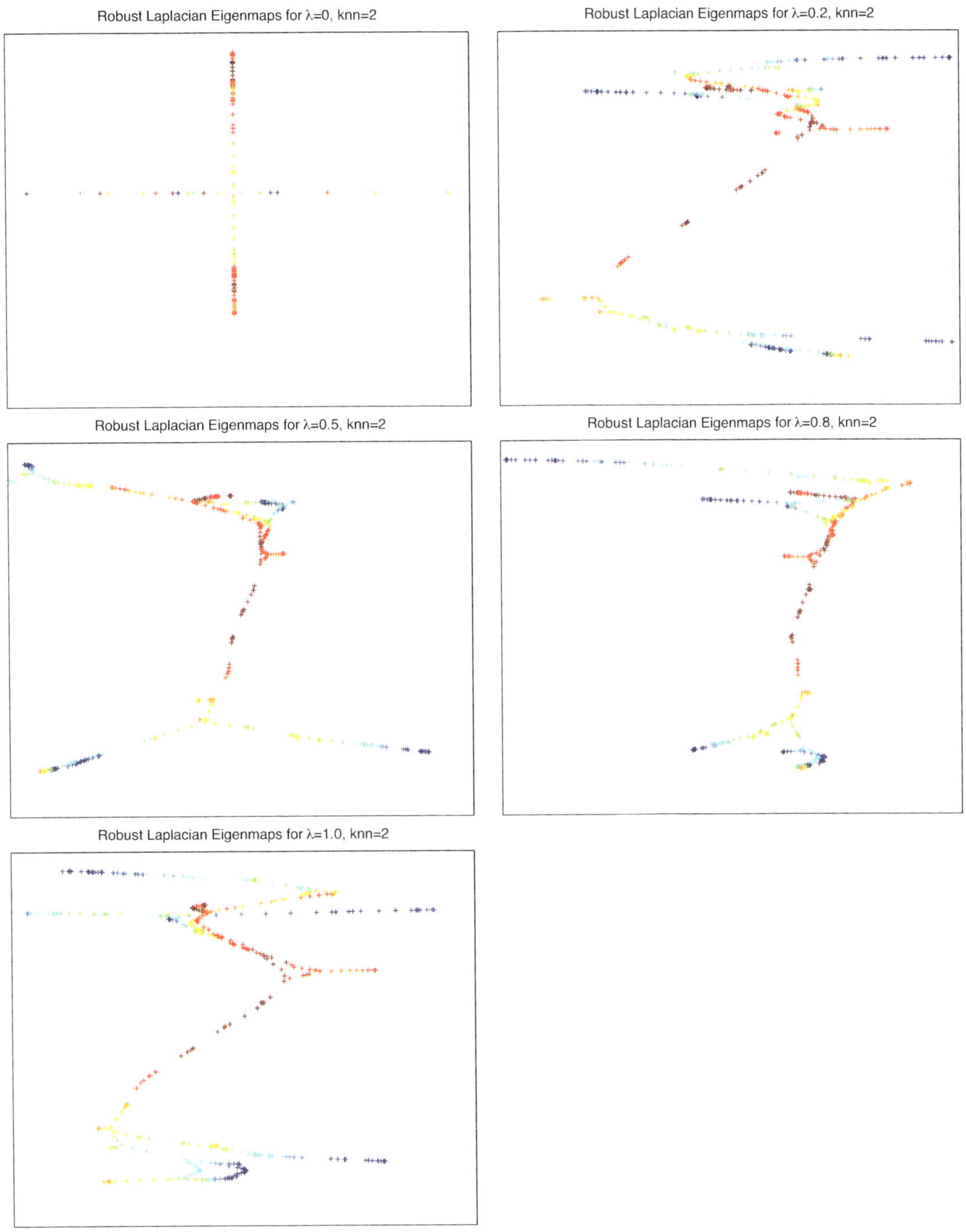

Figure 2.12: Change in regularization parameter $\lambda \in \{0, 0.2, 0.5, 0.8, 1.0\}$ for $k = 2$. In fact the results here show that the embedded representation is controlled by the MST.

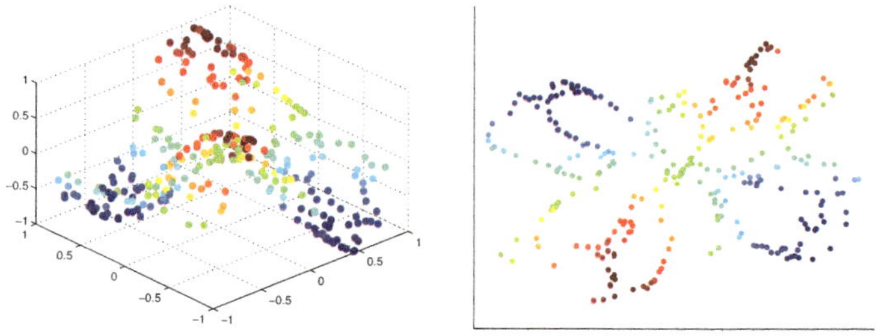

Figure 3.1: The twin peaks data set, dimensionally reduced by density preserving maps.

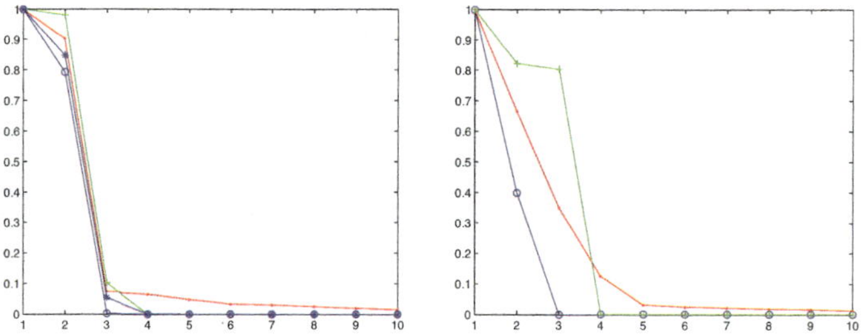

Figure 3.2: The eigenvalue spectra of the inner product matrices learned by PCA (green, '+'), Isomap (red, '.'), MVU (blue, '*'), and DPM (blue, 'o'). Left: A spherical cap. Right: The "twin peaks" data set. As can be seen, DPM suggests the lowest dimensional representation of the data for both cases.

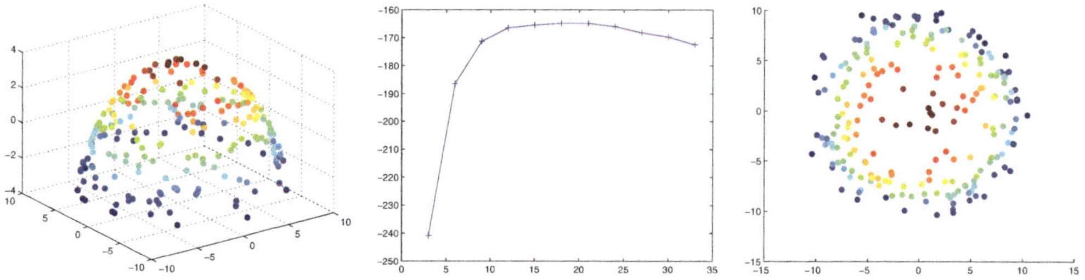

Figure 3.3: The hemisphere data, log-likelihood of the submanifold KDE for this data as a function of k, and the resulting DPM reduction for the optimal k.

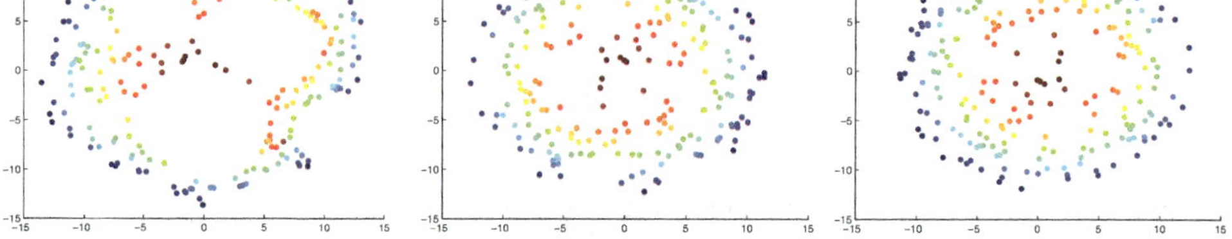

Figure 3.4: Isomap on the hemisphere data, with $k = 5$, 20, 30.

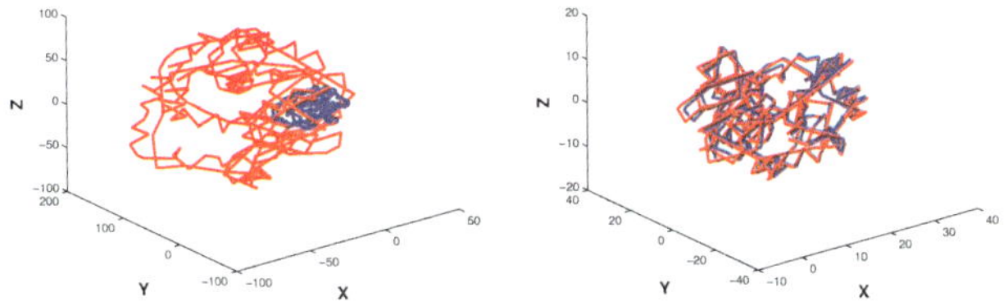

Figure 5.1: A simple example of alignment involving finding correspondences across protein tertiary structures. Here two related structures are aligned. The smaller blue structure is a scaling and rotation of the larger red structure in the original space shown on the left, but the structures are equated in the new coordinate frame shown on the right.

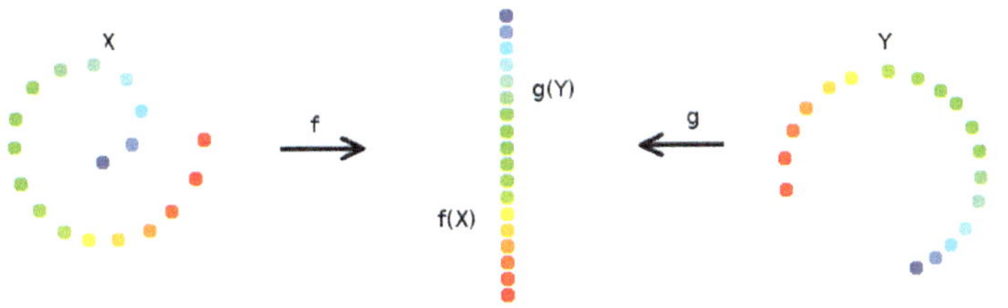

Figure 5.3: An illustration of the problem of manifold alignment. The two datasets X and Y are embedded into a single space where the corresponding instances are equal and local similarities within each dataset are preserved.

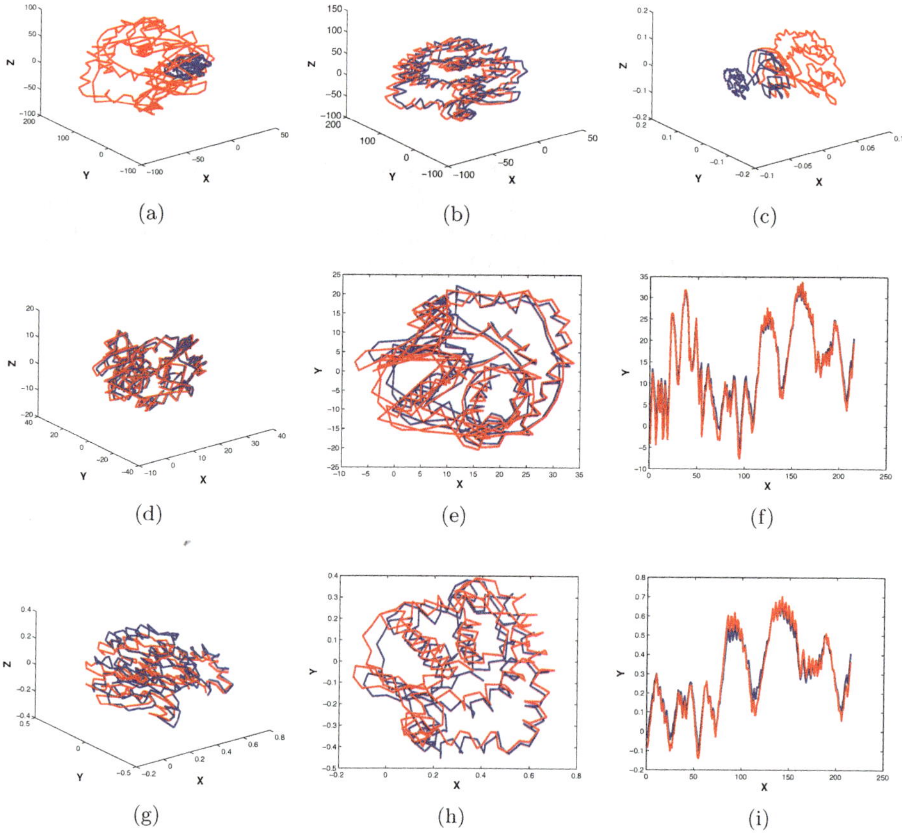

Figure 5.5: (a): Comparison of proteins $X^{(1)}$ (red) and $X^{(2)}$ (blue) before alignment; (b): Procrustes manifold alignment; (c): Semi-supervised manifold alignment; (d): 3D alignment using manifold projections; (e): 2D alignment using manifold projections; (f): 1D alignment using manifold projections; (g): 3D alignment using manifold projections without correspondence; (h): 2D alignment using manifold projections without correspondence; (i): 1D alignment using manifold projections without correspondence.

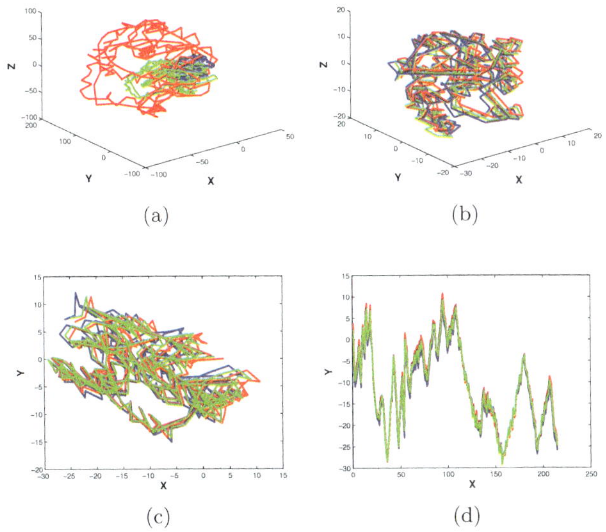

Figure 5.6: (a): Comparison of the proteins $X^{(1)}$ (red), $X^{(2)}$ (blue) and $X^{(3)}$ (green) before alignment; (b): 3D alignment using multiple manifold alignment; (c): 2D alignment using multiple manifold alignment; (d): 1D alignment using multiple manifold alignment.

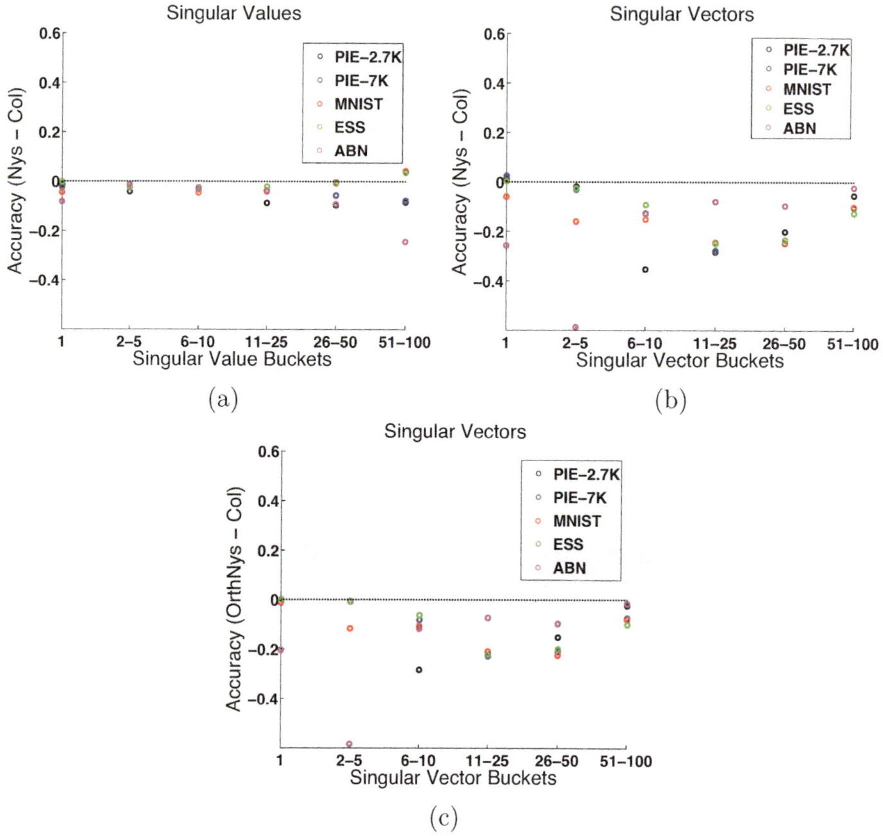

Figure 6.1: Differences in accuracy between Nyström and column sampling. Values above zero indicate better performance of Nyström and vice-versa. (a) Top 100 singular values with $l = n/10$. (b) Top 100 singular vectors with $l = n/10$. (c) Comparison using orthogonalized Nyström singular vectors.

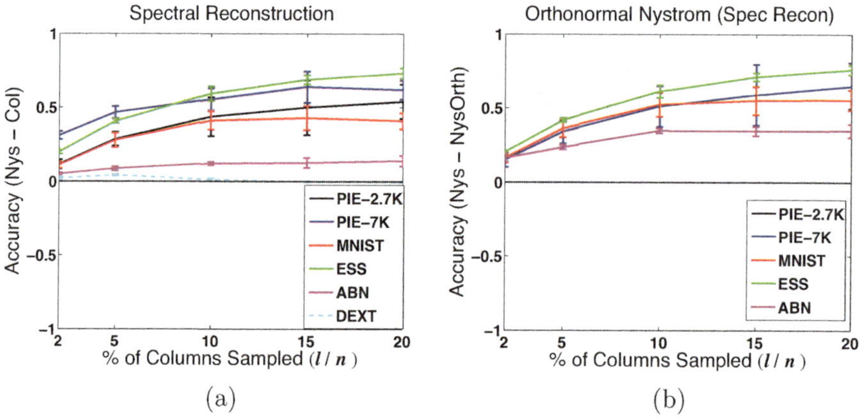

Figure 6.2: Performance accuracy of spectral reconstruction approximations for different methods with $k = 100$. Values above zero indicate better performance of the Nyström method. (a) Nyström versus Column sampling. (b) Nyström versus Orthonormal Nyström.

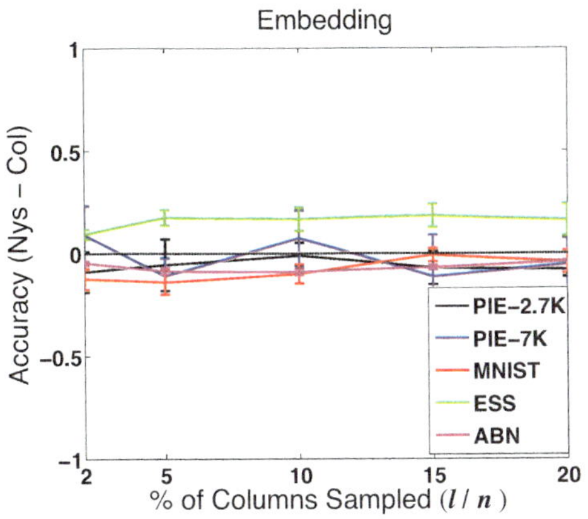

Figure 6.3: Embedding accuracy of Nyström and column sampling. Values above zero indicate better performance of Nyström and vice-versa.

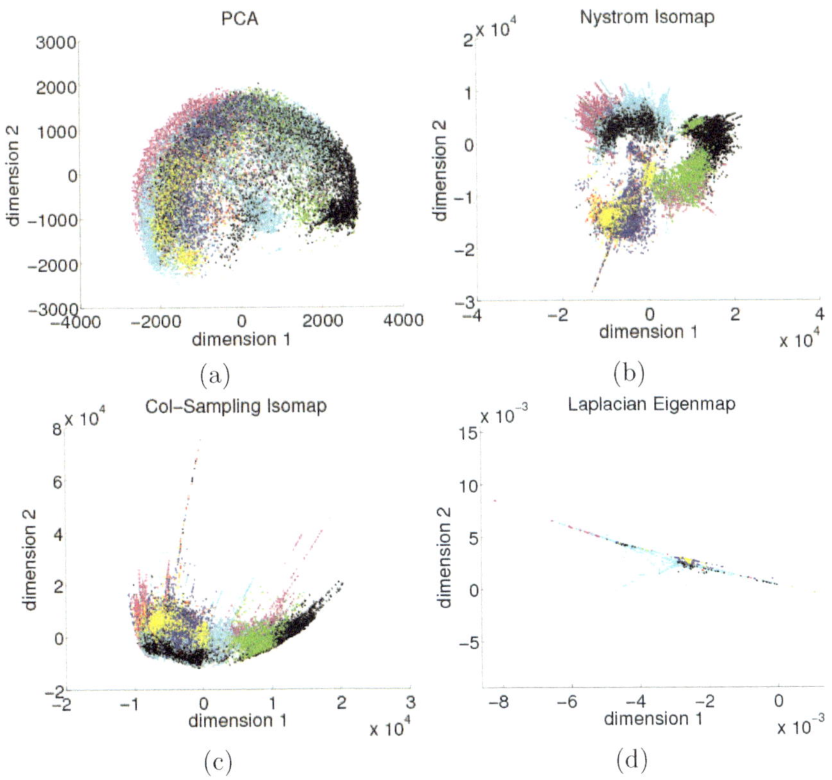

Figure 6.8: Optimal 2D projections of PIE-35K where each point is color coded according to its pose label. (a) PCA projections tend to spread the data to capture maximum variance. (b) Isomap projections with Nyström approximation tend to separate the clusters of different poses while keeping the cluster of each pose compact. (c) Isomap projections with column sampling approximation have more overlap than with Nyström approximation. (d) Laplacian Eigenmaps projects the data into a very compact range.

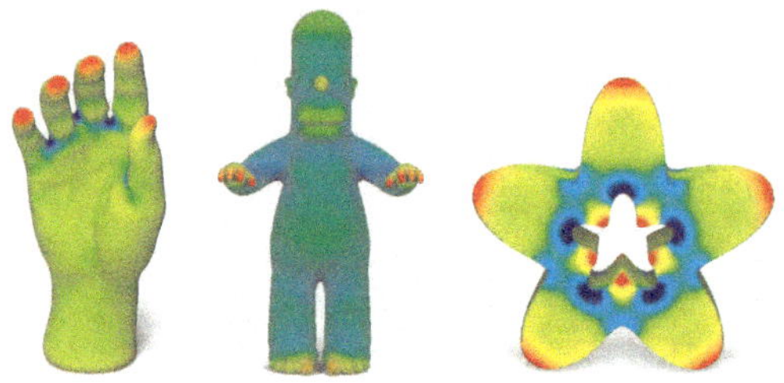

Figure 7.3: Heat kernel function $k_t(x, x)$ for a small fixed t on the hand, Homer, and trim-star models. The function values increase as the color goes from blue to green and to red, with the mapping consistent across the shapes. Note that high and low values of $k_t(x, x)$ correspond to areas with positive and negative Gaussian curvatures, respectively.

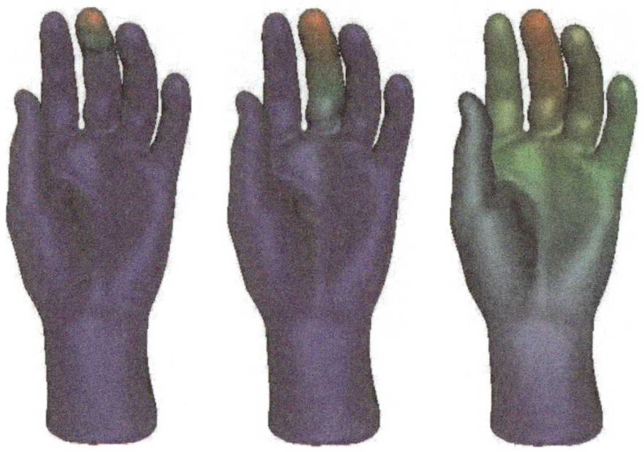

Figure 7.5: From left to right, the function $k_t(x, \cdot)$ with $t = 0.1, 1, 10$ where x is at the tip of the middle figure.

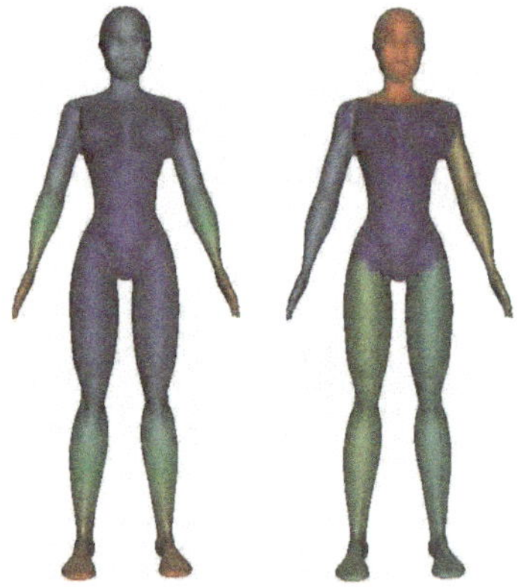

(a) (b)

Figure 7.7: (a) the function of $k_t(x, x)$ for a fixed scale t over a human (b) the segmentation of the human based on the stable manifold of extreme points of the function shown in (a).

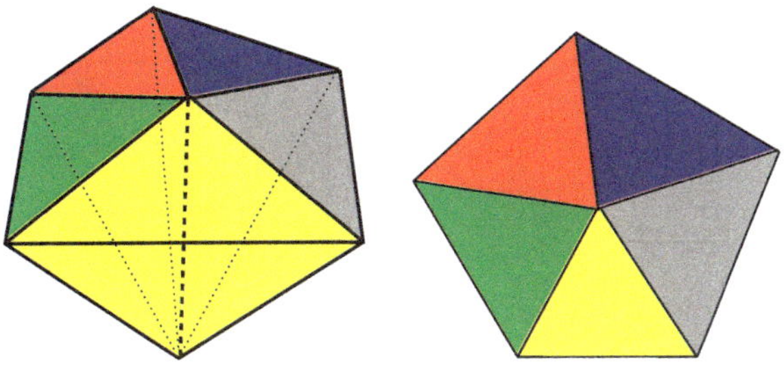

Figure 8.18: Edge collapse in tetrahedron mesh.

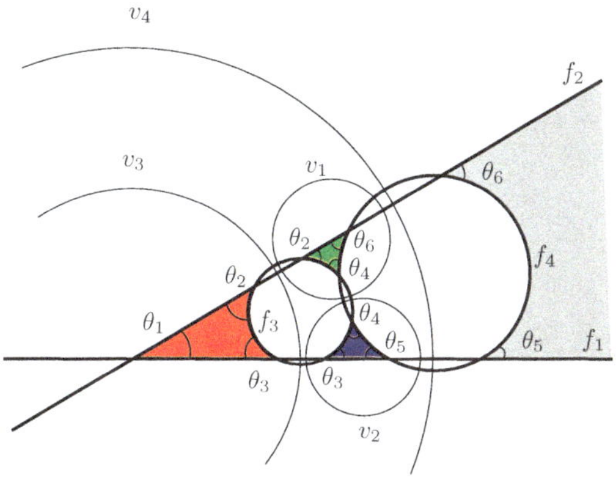

Figure 8.19: Circle packing for the truncated tetrahedron.

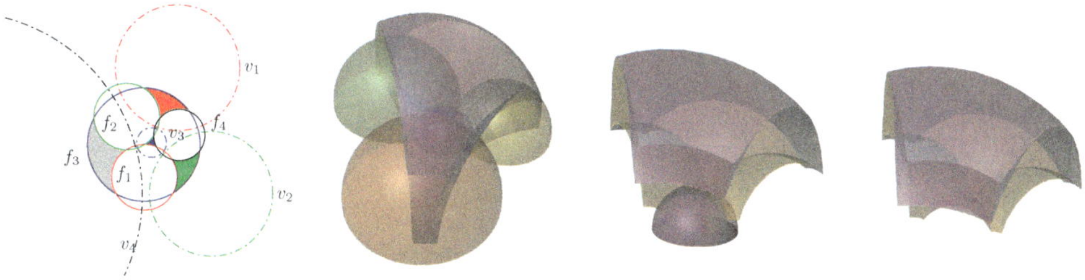

Figure 8.20: Constructing an ideal hyperbolic tetrahedron from circle packing using CSG operators.

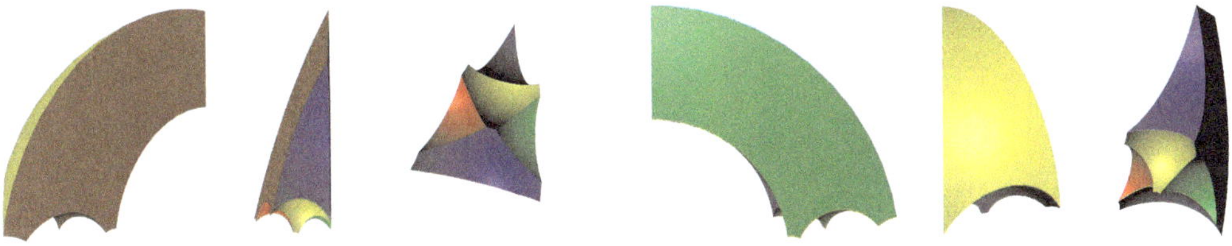

Figure 8.21: Realization of a truncated hyperbolic tetrahedron in the upper half space model of $\mathbb{H}^3$, based on the circle packing in Figure 8.19.

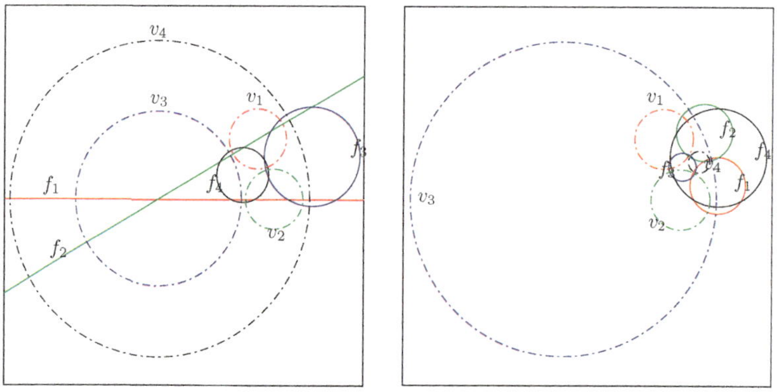

Figure 8.23: Glue two tetrahedra by using a Möbius transformation to glue their circle packings, such that $f_3 \to f_4$, $v_1 \to v_1, v_2 \to v_2, v_4 \to v_3$.

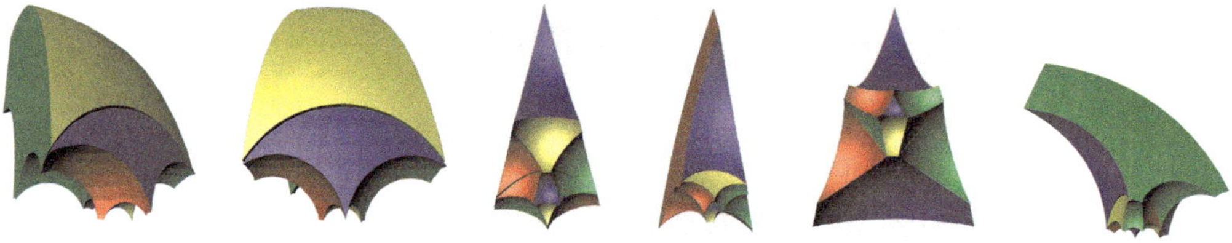

Figure 8.24: Glue T_1 and T_2. Frames (a)(b)(c) show different views of the gluing $f_3 \to f_4$, $\{v_1, v_2, v_4\} \to \{v_1, v_2, v_3\}$. Frames (d) (e) (f) show different views of the gluing $f_4 \to f_3$,$\{v_1, v_2, v_3\} \to \{v_2, v_1, v_4\}$.

Figure 8.25: Embed the 3-manifold periodically in the hyperbolic space $\mathbb{H}^3$.

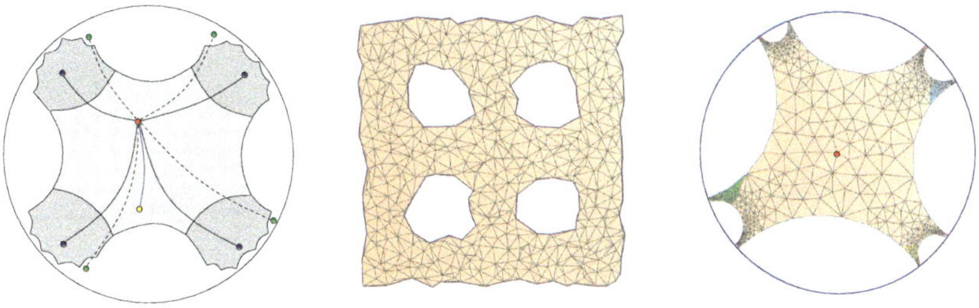

(a) Hyperbolic universal covering space

(b) Euclidean covering space

Figure 8.36: Ricci flow for greedy routing and load balancing in wireless sensor network.

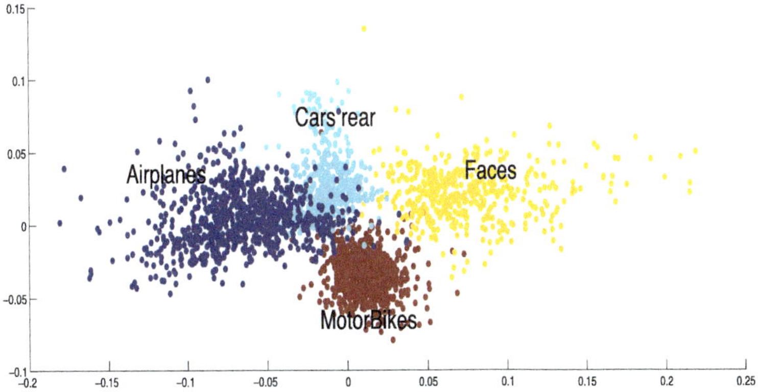

Figure 10.5: Manifold Embedding for all images in Caltech-4-II. Only first two dimensions are shown.

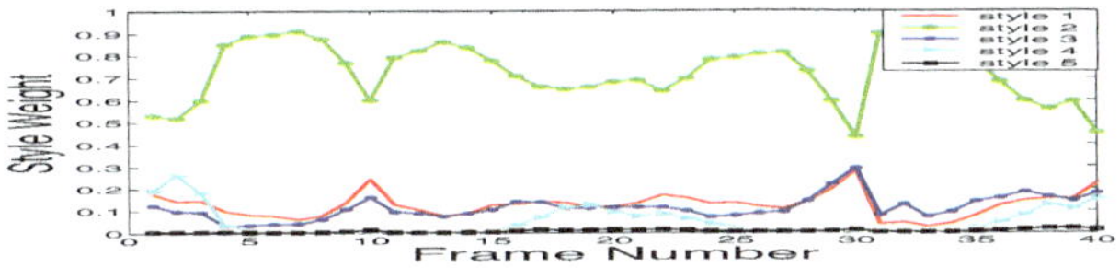

c

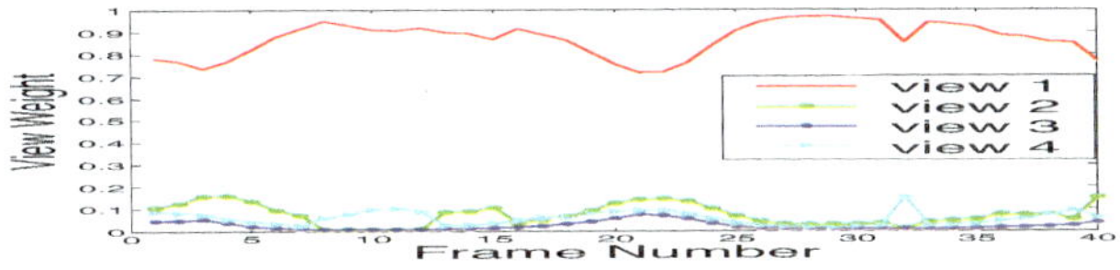

d

Figure 11.11: c) Style weights. d) View weights.

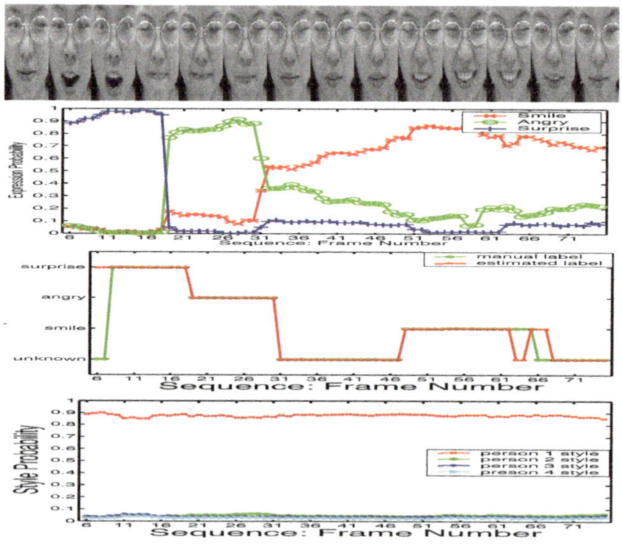

Figure 11.14: From top to bottom: Samples of the input sequences; Expression probabilities; Expression classification; Style probabilities

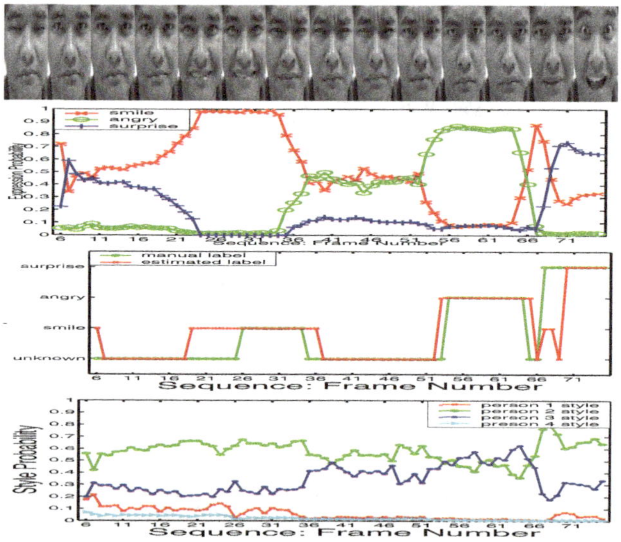

Figure 11.15: Generalization to new people: expression recognition for a new person. From top to bottom: Samples of the input sequences; expression probabilities; expression classification; style probabilities

Chapter 4

Sample Complexity in Manifold Learning

Hariharan Narayanan

4.1 Introduction

Manifold Learning may be defined as a collection of methods and associated analysis motivated by the hypothesis that high dimensional data lie in the vicinity of a low dimensional manifold. A rationale often provided to justify this hypothesis (which we term the "manifold hypothesis") is that high dimensional data, in many cases of interest, are generated by a process that possesses few essential degrees of freedom. The manifold hypothesis is a way of circumventing the "curse of dimensionality," i. e. the exponential dependence of critical quantities such as computational complexity (the amount of computation needed) and sample complexity (the number of samples needed), as a function of the dimensionality of data. Some other hypotheses which can allow one to avoid the curse of dimensionality are sparsity (i. e. the assumption that that the number of non-zero coordinates in a typical data point is small) and the assumption that data is generated from a Markov random field in which the number of hyper-edges is small.

As an illustration of how the curse of dimensionality affects the task of data analysis in high dimensions, consider the following situation. Suppose data $x_1, x_2, \ldots, x_s$ are i.i.d draws from the uniform probability distribution in a unit ball in $\mathbb{R}^m$ and the value of a $1-$Lipschitz function f is revealed at these points. If we wish to learn with probability bounded away from 0, the value f takes at a fixed point x within an error of ϵ from the values taken at the random samples, the number of samples needed would have to be at least of the order of ϵ^{-m}, since if the number of samples were less than this, the probability that there is a sample point x_i within ϵ of x would not be bounded below by a constant as ϵ tends to zero.

In this chapter, we first describe some quantitative results about the sense in which the manifold hypothesis allows us to avoid the curse of dimensionality for the task of classification [14]. We then consider the basic question of whether the manifold hypothesis can be tested in high dimensions using a small amount of data. In an appropriate setting, we show that the number of samples needed for this task is independent of the ambient dimension [13].

4.2 Sample Complexity of Classification on a Manifold

4.2.1 Preliminaries

For us, C will denote a universal constant.

Definition 1 (reach) *Let $\mathcal{M}$ be a smooth d-dimensional submanifold of $\mathbb{R}^m$. We define* $\text{reach}(\mathcal{M})$ *to be τ, where τ is the largest number to have the property that any point at a distance $r < \tau$ from $\mathcal{M}$ has a unique nearest point (with respect to the Euclidean norm) in $\mathcal{M}$.*

Suppose that $\mathcal{P}$ is a probability measure supported on a d-dimensional Riemannian submanifold $\mathcal{M}$ of $\mathbb{R}^m$ having reach $\geq \kappa$. Suppose that data samples $\{x_i\}_{i\geq 1}$ are randomly drawn from $\mathcal{P}$ in an i.i.d fashion. Let each data point x be associated with a label $f(x) \in \{0, 1\}$.

We first introduce the notion of annealed entropy due to Vapnik [17].

Definition 2 (Annealed Entropy) *Let $\mathcal{P}$ be a probability measure supported on a manifold $\mathcal{M}$. Given a class of indicator functions Λ and a set of points $Z = \{z_1, \ldots, z_\ell\} \subset \mathcal{M}$, let $N(\Lambda, Z)$ be the number of ways of partitioning $z_1, \ldots, z_\ell$ into two sets using indicators belonging to Λ. We define $G(\Lambda, \mathcal{P}, \ell)$ to be the expected value of $N(\Lambda, Z)$. Thus*

$$G(\Lambda, \mathcal{P}, \ell) := \mathbb{E}_{Z \vdash \mathcal{P}^{\times \ell}} N(\Lambda, Z),$$

where expectation is with respect to Z and $\vdash$ signifies that Z is drawn from the Cartesian product of ℓ copies of $\mathcal{P}$. The annealed entropy of Λ with respect to ℓ samples from $\mathcal{P}$ is defined to be

$$H_{ann}(\Lambda, \mathcal{P}, \ell) := \ln G(\Lambda, \mathcal{P}, \ell).$$

Definition 3 *The risk $R(\alpha)$ of a classifier α is defined as the probability that α misclassifies a random data point x drawn from $\mathcal{P}$. Formally, $R(\alpha) := \mathbb{E}_{\mathcal{P}}[\alpha(x) \neq f(x)]$. Given a set of ℓ labeled data points $(x_1, f(x_1)), \ldots, (x_\ell, f(x_\ell))$, the empirical risk is defined to be $R_{emp}(\alpha, \ell) := \frac{\sum_{i=1}^{\ell} \mathcal{I}[\alpha(x_i) \neq f(x_i)]}{\ell}$, where $\mathcal{I}[\cdot]$ denotes the indicator of the respective event and $f(x)$ is the label of point x.*

Theorem 1 (Vapnik [17], Thm 4.2) *For any ℓ the inequality*

$$\mathbb{P}\left[\sup_{\alpha\in\Lambda} \frac{R(\alpha) - R_{emp}(\alpha, \ell)}{\sqrt{R(\alpha)}} > \epsilon\right] \quad < \quad 4e^{\left(\frac{H_{ann}(\Lambda, \mathcal{P}, 2\ell)}{\ell} - \frac{\epsilon^2}{4}\right)\ell}$$

holds true, where random samples are drawn from the distribution $\mathcal{P}$.

4.2.2 Remarks

Our setting is the natural generalization of half-space learning applied to data on a d-dimensional sphere. In fact, when the sphere has radius τ, $\mathcal{C}_\tau$ corresponds to half-spaces, and the VC dimension is $d + 2$. However, when $\tau < \kappa$, as we show in Lemma 2, on a d-dimensional sphere of radius κ, the VC dimension of $\mathcal{C}_\tau$ is infinite.

4.3 Learning Smooth Class Boundaries

Definition 4 *Let*

$$\mathcal{S}_\tau := \left\{S \middle| S = \overline{S} \subseteq \mathcal{M} \text{ and } \text{reach}(S \cap \overline{\mathcal{M} \setminus S}) \geq \tau\right\},$$

where $\overline{S}$ is the closure of S. Let

$$\mathcal{C}_\tau := \{f | f : \mathcal{M} \to \{0,1\} \text{ and } f^{-1}(1) \in \mathcal{S}_\tau\}.$$

Thus, the concept class $\mathcal{C}_\tau$ is the collection of indicators of all closed sets in $\mathcal{M}$ whose boundaries are $d-1$ dimensional submanifolds of $\mathbb{R}^m$ whose reach is at least τ.

Following Definition 4, let $\mathcal{C}_\tau$ be the collection of indicators of all open sets in $\mathcal{M}$ whose boundaries are submanifolds of $\mathbb{R}^m$ of dimension $d-1$, whose reach is at least τ.
Our main theorem is the following.

Definition 5 (Packing number) *Let $N_p(\epsilon_r)$ be the largest number N such that $\mathcal{M}$ contains N disjoint balls $B_{\mathcal{M}}(x_i, \epsilon_r)$, where $B_{\mathcal{M}}(x, \epsilon_r)$ is a geodesic ball in $\mathcal{M}$ around x of radius ϵ_r.*

Notation 1 *Without loss of generality, let ρ_{max} be greater or equal to 1. Let*

$$\epsilon_r = \min(\frac{\tau}{4}, \frac{\kappa}{4}, 1)\epsilon/(2\rho_{max}).$$

Let

$$\ell := C\left(\frac{\ln\frac{1}{\delta} + N_p(\epsilon_r/2) d \ln(d\rho_{max}/\epsilon)}{\epsilon^2}\right).$$

Theorem 2 *Let $\mathcal{M}$ be a d-dimensional submanifold of $\mathbb{R}^m$ whose reach is $\geq \kappa$. Let $\mathcal{P}$ be a probability measure on $\mathcal{M}$, whose density relative to the uniform probability measure on $\mathcal{M}$ is bounded above by ρ_{max}. Then the number of random samples needed before the empirical risk and the true risk are uniformly close over $\mathcal{C}_\tau$ can be bounded above as follows. Let ℓ be defined as in Notation 1. Then*

$$\mathbb{P}\left[\sup_{\alpha \in \mathcal{C}_\tau} \frac{R(\alpha) - R_{emp}(\alpha, \ell)}{\sqrt{R(\alpha)}} > \sqrt{\epsilon}\right] < \delta$$

Lemma 1 provides a lower bound on the sample complexity that shows that some dependence on the packing number cannot be avoided in Theorem 2. Further, Lemma 2 shows that it is impossible to learn an element of $\mathcal{C}_\tau$ in a distribution-free setting in general.

Lemma 1 *Let $\mathcal{M}$ be a d-dimensional sphere in $\mathbb{R}^m$. Let the $\mathcal{P}$ have a uniform density over the disjoint union of $N_p(2\tau)$ identical spherical caps*

$$S = \{B_{\mathcal{M}}(x_i, \tau)\}_{1 \leq i \leq N_p(2\tau)}$$

of radius τ, whose mutual distances are all $\geq 2\tau$. Then, if $s < (1-\epsilon)N_p(2\tau)$,

$$\mathbb{P}\left[\sup_{\alpha \in \mathcal{C}_\tau} \frac{R(\alpha) - R_{emp}(\alpha, s)}{\sqrt{R(\alpha)}} > \sqrt{\epsilon}\right] = 1.$$

Proof 1 *Suppose that the labels are given by $f : \mathcal{M} \to \{0,1\}$, such that $f^{-1}(1)$ is the union of some of the caps in S as depicted in Figure 4.1. Suppose that s random samples $z_1, \dots, z_s$ are chosen from $\mathcal{P}$. Then at least $\epsilon N_p(2\tau)$ of the caps in S do not contain any of the z_i. Let X be the union of these caps. Let $\alpha : \mathcal{M} \to \{0,1\}$ satisfy $\alpha(x) = 1 - f(x)$ if $x \in X$ and $\alpha(x) = f(x)$ if $x \in \mathcal{M} \setminus X$. Note that $\alpha \in \mathcal{C}_\tau$. However, $R_{emp}(\alpha, s) = 0$ and $R(\alpha) \geq \epsilon$. Therefore $\frac{R(\alpha) - R_{emp}(\alpha,s)}{\sqrt{R(\alpha)}} > \sqrt{\epsilon}$, which completes the proof.*

Figure 4.1: This illustrates the distribution from Lemma 1. The intersections of $f^{-1}(1)$ and $f^{-1}(0)$ with the support of $\mathcal{P}$ are, respectively, black and grey.

Lemma 2 *For any $m > d \geq 2$, and $\tau > 0$, there exist compact d-dimensional manifolds on which the VC dimension of $\mathcal{C}_\tau$ is infinite. In particular, this is true for the standard d-dimensional Euclidean sphere of radius κ embedded in $\mathbb{R}^m$, where $m > d \geq 2$ and $\kappa > \tau$.*

1. Partition the manifold into small pieces $\mathcal{M}_i$ that are almost Euclidean, such that the restrictions of any cut hypersurface is almost linear.

2. Let the probability measure $\frac{\mathcal{P}|_{\mathcal{M}_i}}{\mathcal{P}(\mathcal{M}_i)}$ be denoted $\mathcal{P}_i$ for each i. Lemma 8 allows us to show, roughly, that

$$\frac{H_{ann}(\mathcal{C}_\tau, \mathcal{P}, n)}{n} \lesssim \sup_i \frac{H_{ann}(\mathcal{C}_\tau, \mathcal{P}_i, \lfloor n\mathcal{P}(\mathcal{M}_i)\rfloor)}{\lfloor n\mathcal{P}(\mathcal{M}_i)\rfloor},$$

   thereby allowing us to focus on a single piece $\mathcal{M}_i$.

3. We use a projection π_i, to map $\mathcal{M}_i$ orthogonally onto the tangent space to $\mathcal{M}_i$ at a point $x_i \in \mathcal{M}_i$ and then reduce the question to a sphere inscribed in a cube $\Box$ of Euclidean space.

4. We cover $\mathcal{C}_\tau|_\Box$ by the union of classes of functions, each class having the property that there is a thin slab such that any two functions in the class are identical in the complement of the slab (See Figure 4.2).

5. Finally, we bound the annealed entropy of each of these classes using Lemma 9.

4.3.1 Volumes of Balls in a Manifold

The following lower bound on the volume of a manifold with bounded reach is from (Lemma 5.3, [15]).

Lemma 3 *Let $\mathcal{M}$ be a d-dimensional submanifold of $\mathbb{R}^n$, whose reach is $\geq \tau$. For $p \in \mathcal{M}$, let $B(p, \epsilon)$ be the ball in $\mathbb{R}^n$ of radius ϵ centered at p, and let ω_d be the volume of the d-dimensional unit ball. Suppose $\theta := \arcsin(\frac{\epsilon}{2\tau})$ for $\epsilon \leq 2\tau$, then, the volume of $\mathcal{M} \cap B(p, \epsilon)$ is greater or equal to $\epsilon^d(\cos^d \theta)\omega_d$.*

As a consequence of Lemma 3, we obtain an upper bound of

$$\frac{V}{\epsilon^d(\cos^d(\arcsin \frac{\epsilon}{2\tau}))\omega_d}$$

on the number of disjoint sets of the form $\mathcal{M} \cap B(p, \epsilon)$ that can be packed in $\mathcal{M}$. If $\{\mathcal{M} \cap B(p_1, \epsilon), \dots, \mathcal{M} \cap B(p_k, \epsilon)\}$ is a maximal family of disjoint sets of the form $\mathcal{M} \cap B(p, \epsilon)$, then there is no point $p \in \mathcal{M}$ such that $\min_i \|p - p_i\| > 2\epsilon$. Therefore, $\mathcal{M}$ is contained in the union of balls,

$$\bigcup_i B(p_i, 2\epsilon).$$

The geodesic ball $B_{\mathcal{M}}(x_i, \epsilon_r)$ is contained inside $B(x_i, \epsilon) \cap \mathcal{M}$. This allows us to get an explicit upper bound on the packing number $N_p(\epsilon_r/2)$, namely

$$N_p(\epsilon_r/2) \leq \frac{2^d \text{vol}\mathcal{M}}{\epsilon_r^d(1 - (\frac{\epsilon_r}{4\tau})^2)^{d/2}\omega_d}.$$

4.3.2 Partitioning the Manifold

The next step is to partition the manifold $\mathcal{M}$ into disjoint pieces $\{\mathcal{M}_i\}$ such that each piece $\mathcal{M}_i$ is contained in the geodesic ball $B_{\mathcal{M}}(x_i, \epsilon_r)$. Such a partition can be constructed by the following natural greedy procedure.

- Choose $N_p(\epsilon_r/2)$ disjoint balls $B_{\mathcal{M}}(x_i, \epsilon_r/2)$, $1 \leq i \leq N_p(\epsilon_r/2)$ where $N_p(\epsilon_r/2)$ is the packing number as in Definition 5.
- Let $\mathcal{M}_1 := B_{\mathcal{M}}(x_1, \epsilon_r)$.
- Iteratively, for each $i \geq 2$, let $\mathcal{M}_i := B_{\mathcal{M}}(x_i, \epsilon_r) \setminus \{\cup_{k=1}^{i-1} \mathcal{M}_k\}$.

4.3.3 Constructing Charts by Projecting onto Euclidean Balls

In this section, we show how the question can be reduced to Euclidean space using a family of charts. The strategy is the following. Let ϵ_r be as defined in Notation 1. Choose a set of points $X = \{x_1, \dots, x_N\}$ belonging to $\mathcal{M}$ such that the union of geodesic balls in $\mathcal{M}$ (measured in the intrinsic Riemannian metric) of radius ϵ_r centered at these points in $\mathcal{M}$ covers all of $\mathcal{M}$.

$$\bigcup_{i \in [N]} B_{\mathcal{M}}(x_i, \epsilon_r) = \mathcal{M}.$$

Definition 6 *For each $i \in [N_p(\epsilon_r/2)]$, let the d-dimensional affine subspace of $\mathbb{R}^m$ tangent to $\mathcal{M}$ at x_i be denoted $\mathbb{A}_i$, and let the d-dimensional ball of radius ϵ_r contained in $\mathbb{A}_i$, centered at x_i be $B_{\mathbb{A}_i}(x_i, \epsilon_r)$. Let the orthogonal projection from $\mathbb{R}^m$ onto $\mathbb{A}_i$ be denoted π_i.*

Lemma 4 *The image of $B_{\mathcal{M}}(x_i, \epsilon_r)$ under the projection π_i is contained in the corresponding ball $B_{\mathcal{M}}(x_i, \epsilon_r)$ in $\mathbb{A}_i$.*

$$\pi_i(B_{\mathcal{M}}(x_i, \epsilon_r)) \subseteq B_{\mathbb{A}_i}(x_i, \epsilon_r).$$

Proof 2 *This follows from the fact that the length of a geodesic segment on $B_{\mathcal{M}}(x_i, \epsilon_r)$ is greater or equal to the length of its image under a projection.*

Let P be a smooth boundary (i. e. reach$(P) \geq \tau$) separating $\mathcal{M}$ into two parts and reach$(\mathcal{M}) \geq \kappa$.

Lemma 5 *Let $\epsilon_r \leq \min(1, \tau/4, \kappa/4)$. Let $\pi_i(B_{\mathcal{M}}(x_i, \epsilon_r) \cap P)$ be the image of P restricted to $B_{\mathcal{M}}(x_i, \epsilon_r)$ under the projection π_i. Then, the reach of $\pi_i(B_{\mathcal{M}}(x_i, \epsilon_r) \cap P)$ is bounded below by $\frac{2}{\tau}$.*

Proof 3 *Let $T_{\pi_i(x)}$ and $T_{\pi_i(y)}$ be the spaces tangent to*

$$\pi_i(B_{\mathcal{M}}(x_i, \epsilon_r) \cap P)$$

at $\pi_i(x)$ and $\pi_i(y)$ respectively. Then, for any $x, y \in B_{\mathcal{M}}(x_i, \epsilon_r) \cap P$, because the kernel of π_i is nearly orthogonal to $T_{\pi_i(x)}$ and $T_{\pi_i(y)}$, if $A_{\pi(x)}$ is the orthogonal projection onto $T_{\pi_i(x)}$ in the image of π_i,

$$\frac{\|A_{\pi(x)}(\pi_i(x) - \pi_i(y))\|}{\|\pi_i(x) - \pi_i(y)\|^2} \leq \frac{\sqrt{2}\|A_x(x-y)\|}{\|x-y\|^2}. \tag{4.1}$$

$B_{\mathcal{M}}(x_i, \epsilon_r) \cap P$ is contained in a neighborhood of the affine space tangent to $B_{\mathcal{M}}(x_i, \epsilon_r) \cap P$ at x_i, which is orthogonal to the kernel of π_i. This can be used to show that for all $x, y \in B_{\mathcal{M}}(x_i, \epsilon_r) \cap P$,

$$\frac{1}{\sqrt{2}} \leq \frac{\|\pi_i(x) - \pi_i(y)\|}{\|x-y\|} \leq 1. \tag{4.2}$$

The reach of a manifold is determined by local curvature and the nearness to self-intersection.

Both of these issues are taken care of by Equations (4.1) and (4.2) respectively, thus completing the proof.

4.3.4 Proof of Theorem 2

We shall organize this proof into several Lemmas, which will be proved immediately after their respective statements. The following Lemma allows us to work with a random rather than deterministic number of samples. The purpose of allowing the number of samples to be a Poisson random variable is that we are able make the set of numbers of samples $\{\nu_i\}$ from different $\mathcal{M}_i$, a collection of independent random variables.

Lemma 6 (Poissonization) *Let ν be a Poisson random variable with mean λ, where $\lambda > 0$. Then, for any $\epsilon > 0$ the expected value of the annealed entropy of a class of indicators with respect to ν random samples from a distribution $\mathcal{P}$ is asymptotically greater or equal to the annealed entropy of $\lfloor(1-\epsilon)\lambda\rfloor$ random samples from the distribution $\mathcal{P}$. More precisely, for any $\epsilon > 0$, $\ln \mathbb{E}_\nu G(\Lambda, \mathcal{P}, \nu) \geq \ln G(\Lambda, \mathcal{P}, \lfloor\lambda(1-\epsilon)\rfloor) - \exp\left(-\epsilon^2\lambda + \frac{\ln(2\pi\lambda)}{2}\right)$.*

Proof 4

$$\begin{aligned}
\ln \mathbb{E}_\nu G(\Lambda, \mathcal{P}, \nu) &= \ln \sum_{n \in \mathbb{N}} \mathbb{P}[\nu = n] H_{ann}(\Lambda, \mathcal{P}, n) \\
&\geq \ln \sum_{n \geq \lfloor\lambda(1-\epsilon)\rfloor} \mathbb{P}[\nu = n] G(\Lambda, \mathcal{P}, n).
\end{aligned}$$

$G(\Lambda, \mathcal{P}, n)$ is monotonically increasing as a function of n. Therefore the above expression can be lower bounded by $\ln \mathbb{P}[\nu \geq \lfloor\lambda(1-\epsilon)\rfloor] G(\Lambda, \mathcal{P}, \nu) \geq H_{ann}(\Lambda, \mathcal{P}, \lfloor\lambda(1-\epsilon)\rfloor)$ $- \exp\left(-\epsilon^2\lambda + \frac{\ln(2\pi\lambda)}{2}\right)$.

Definition 7 *For each $i \in [N_p(\epsilon_r/2)]$, let $\mathcal{P}_i$ be the restriction of $\mathcal{P}$ to $\mathcal{M}_i$. Let $|\mathcal{P}_i|$ denote the total measure of $\mathcal{P}_i$. Let λ_i denote $\lambda|\mathcal{P}_i|$. Let $\{\nu_i\}$ be a collection of independent Poisson random variables such that for each $i \in [N_p(\epsilon_r/2)]$, the mean of ν_i is λ_i.*

The following Lemma allows us to focus our attention to small pieces $\mathcal{M}_i$ which are almost Euclidean.

Lemma 7 (Factorization) *The quantity $\ln \mathbb{E}_\nu G(\mathcal{C}_\tau, \mathcal{P}, \nu)$ is less than or equal to the sum over i of the corresponding quantities $\mathcal{C}_\tau$ with respect to ν_i random samples from $\mathcal{P}_i$. i. e.*

$$\ln \mathbb{E}_\nu G(\mathcal{C}_\tau, \mathcal{P}, \nu) \leq \sum_{i \in N_p(\epsilon_r/2)} \ln \mathbb{E}_{\nu_i} G(\mathcal{C}_\tau, \mathcal{P}_i, \nu_i).$$

Proof 5

$$G(\mathcal{C}_\tau, \mathcal{P}, \ell) := \ln \mathbb{E}_{X \vdash \mathcal{P}^{\times \ell}} N(\mathcal{C}_\tau, X),$$

where expectation is with respect to X and $\vdash$ signifies that X is drawn from the Cartesian product of ℓ copies of $\mathcal{P}$. The number of ways of splitting $X = \{x_1, \dots, x_k, \dots, x_\ell\}$ using elements of $\mathcal{C}_\tau$, $N(\mathcal{C}_\tau, X)$ satisfies a sub-multiplicative property, namely

$$N(\mathcal{C}_\tau, \{x_1, \dots, x_\ell\}) \leq$$

$$N(\mathcal{C}_\tau, \{x_1, \dots, x_k\}) N(\mathcal{C}_\tau, \{x_{k+1}, \dots, x_\ell\}).$$

This can be iterated to generate inequalities where the right side involves a partition with any integer number of parts. Note that $\mathcal{P}$ is a mixture of the $\mathcal{P}_i$, and can be expressed as

$$\mathcal{P} = \sum_i \frac{\lambda_i}{\lambda} \mathcal{P}_i.$$

A draw from $\mathcal{P}$ of a Poisson number of samples can be decomposed as the union of independently chosen sets of samples. The ith set is a draw of size ν_i from $\mathcal{P}_i$, ν_i being a Poisson random variable having mean λ_i. These facts imply that

$\ln \mathbb{E}_\nu G(\mathcal{C}_\tau, \mathcal{P}, \nu) \leq \sum_{i \in N_p(\epsilon_r/2)} \ln \mathbb{E}_{\nu_i} G(\mathcal{C}_\tau, \mathcal{P}_i, \nu_i)$.

Lemma 7 can be used together with an upper bound on annealed entropy based on the number of samples to obtain

Lemma 8 (Localization) *For any $\epsilon' > 0$*

$$\frac{\ln \mathbb{E}_\nu G(\mathcal{C}_\tau, \mathcal{P}, \nu)}{\lambda} \leq \sup_{i \ s.t \ |\mathcal{P}_i| \geq \frac{\epsilon'}{N_p(\epsilon_r/2)}} \frac{\ln \mathbb{E}_{\nu_i} G(\mathcal{C}_\tau, \mathcal{P}_i, \nu_i)}{\lambda_i} + \epsilon'.$$

Proof 6 *Lemma 8 allows us to reduce the question to a single $\mathcal{M}_i$ in the following way.*

$$\frac{\ln \mathbb{E}_\nu G(\mathcal{C}_\tau, \mathcal{P}, \nu)}{\lambda} \leq \sum_{i \in N_p(\epsilon_r/2)} \frac{\lambda_i}{\lambda} \frac{\ln \mathbb{E}_{\nu_i} G(\mathcal{C}_\tau, \mathcal{P}_i, \nu_i)}{\lambda_i}$$

Allowing all summations to be over i s.t $|\mathcal{P}_i| \geq \frac{\epsilon'}{N_p(\epsilon_r/2)}$, the right side can be split into

$$\sum_i \frac{\lambda_i}{\lambda} \frac{\ln \mathbb{E}_{\nu_i} G(\mathcal{C}_\tau, \mathcal{P}_i, \nu_i)}{\lambda_i} + \sum_i \ln \mathbb{E}_{\nu_i} G(\mathcal{C}_\tau, \mathcal{P}_i, \nu_i).$$

$G(\mathcal{C}_\tau, \mathcal{P}_i, \nu_i)$ must be less or equal to the expression obtained in the case of complete shattering, which is 2^{ν_i}. Therefore the second term in the above expression can be bounded above as follows,

$$\begin{aligned}\sum_i \ln \mathbb{E}_{\nu_i} G(\mathcal{C}_\tau, \mathcal{P}_i, \nu_i) &\leq \sum_i \ln \mathbb{E}_{\nu_i} 2^{\nu_i} \\ &= \sum_i \lambda_i \\ &\leq \epsilon'.\end{aligned}$$

Therefore,

$$\begin{aligned}\frac{\ln \mathbb{E}_\nu G(\mathcal{C}_\tau, \mathcal{P}, \nu)}{\lambda} &\leq \sum_i \frac{\lambda_i}{\lambda} \frac{\ln \mathbb{E}_{\nu_i} G(\mathcal{C}_\tau, \mathcal{P}_i, \nu_i)}{\lambda_i} + \epsilon' \\ &\leq \sup_i \frac{\ln \mathbb{E}_{\nu_i} G(\mathcal{C}_\tau, \mathcal{P}_i, \nu_i)}{\lambda_i} + \epsilon'.\end{aligned}$$

As mentioned earlier, Lemma 8 allows us to reduce the proof to a question concerning a single piece $\mathcal{M}_i$. This is more convenient because $\mathcal{M}_i$ can be projected onto a single Euclidean ball in the way described in Section 4.3.3 without incurring significant distortion. By Lemmas 4 and 5, the question can be transferred to one about the annealed entropy of the induced function class $\mathcal{C}_\tau \circ \pi_i^{-1}$ on chart $B_{\mathbb{A}_i}(x_i, \epsilon_r)$ with respect to ν_i random samples from the projected probability distribution $\pi_i(\nu_i)$. $\mathcal{C}_\tau \circ \pi_i^{-1}$ is contained in $\mathcal{C}_{\tau/2}(\mathbb{A}_i)$ which is the analogue of $\mathcal{C}_{\tau/2}$ on $\mathbb{A}_i$. For simplicity, henceforth we shall abbreviate $\mathcal{C}_{\tau/2}(\mathbb{A}_i)$ as $\mathcal{C}_{\tau/2}$. Then,

$$\begin{aligned}\frac{\ln \mathbb{E}_{\nu_i} G(\mathcal{C}_\tau, \mathcal{P}_i, \nu_i)}{\lambda_i} &= \frac{\ln \mathbb{E}_{\nu_i} G(\mathcal{C}_\tau \circ \pi_i^{-1}, \pi_i(\mathcal{P}_i), \nu_i)}{\lambda_i} \\ &\leq \frac{\ln \mathbb{E}_{\nu_i} G(\mathcal{C}_{\tau/2}, \pi_i(\mathcal{P}_i), \nu_i)}{\lambda_i}.\end{aligned}$$

We inscribe $B_{\mathbb{A}_i}(x_i, \epsilon_r)$ in a cube of side $2\epsilon_r$ for convenience, and proceed to find the desired upper bound on $G(\mathcal{C}_{\tau/2}, \pi_i(\mathcal{P}_i), \nu_i)$. We shall indicate how to achieve this using covers. For convenience, let this cube be dilated until we have the cube of side 2. The measure $\pi_i(\mathcal{P}_i)$ assigns to it must be scaled to a probability measure that we call $\mathcal{P}_\circ$, which is actually supported on the inscribed ball. We shall normalize all quantities appropriately when the calculations are over. The $\tau_\square$ that we shall work with below is a rescaled version of the original, $\tau_\square = \tau/\epsilon_r$. Let B_∞^d be the cube of side 2 centered at the origin and ι_∞^d be its indicator. Let B_2^d be the unit ball inscribed in B_∞^d.

Definition 8 *Let $\tilde{\mathcal{C}}_{\tau_\square}$ be defined to be the set of all indicators of the form $\iota_\infty^d \cdot \iota$, where ι is the indicator of some set in $\mathcal{C}_{\tau_\square}$.*

In other words, $\tilde{\mathcal{C}}_{\tau_\square}$ is the collection of all functions that are indicators of sets that can be expressed as the intersection of the unit cube and an element of $\mathcal{C}_{\tau_\square}$.

$$\tilde{\mathcal{C}}_{\tau_\square} = \{f \mid \exists c \in \mathcal{C}_{\tau_\square}, \text{ for which } f = \mathcal{I}_c \cdot \iota_\infty^d\}, \tag{4.3}$$

where $\mathcal{I}_c$ is the indicator of c.

Definition 9 *For every $v \in \mathbb{R}^d$ where $\|v\| = 1$, $t \in \mathbb{R}$ and $\epsilon > 0$ and $\epsilon_s = \epsilon^2/\rho_{max}$. Let $\tilde{\mathcal{C}}_{\epsilon_s}^{(v,t)}$ be a class of indicator functions consisting of all those measurable indicators ι that satisfy the following.*

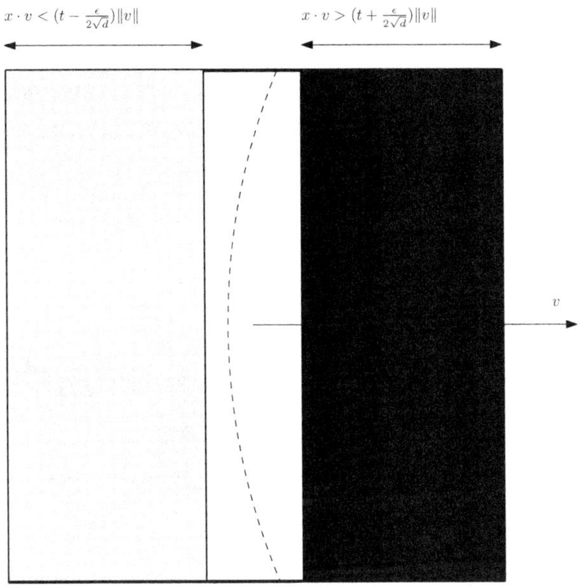

Figure 4.2: Each class of the form $\tilde{\mathcal{C}}_{\epsilon_s}^{(v,t)}$ contains a subset of the set of indicators of the form $\mathcal{I}_c \cdot \iota_\infty^d$

.

1. $x \cdot v < (t - \frac{\epsilon_s}{2\sqrt{d}})\|v\|$ *or* $x \notin B_\infty^d \Rightarrow \iota(x) = 0$ *and*

2. $x \cdot v > (t + \frac{\epsilon_s}{2\sqrt{d}})\|v\|$ *and* $x \in B_\infty^d \Rightarrow \iota(x) = 1$.

The VC dimension of the above class is clearly infinite since any samples lying within the slab of thickness $\epsilon_s/\sqrt{d}$ get shattered. However, if a distribution is sufficiently uniform, most samples would lie outside the slab and so the annealed entropy can be bounded from above. We shall construct a finite set W of tuples (v, t) such that the union of the corresponding classes $\tilde{\mathcal{C}}_{\epsilon_s}^{(v,t)}$ contains $\tilde{\mathcal{C}}_{\tau_\square}$. Let tv take values in an $\frac{\tau_\square}{2}$-grid contained in B_∞^d, i. e. $tv \in \frac{\epsilon_s}{2\sqrt{d}}\mathbb{Z}^d \cap B_\infty^d$. It is then the case (see Figure 4.2) that any indicator in $\tilde{\mathcal{C}}_{\tau_\square}$ agrees over B_2^d with a member in some class $\tilde{\mathcal{C}}_{\epsilon_s}^{(v,t)}$, if $\epsilon_s \geq \frac{2}{\tau_\square}$, i. e.

$$\tilde{\mathcal{C}}_{\tau_\square} \subseteq \bigcup_{tv \in \frac{\epsilon_s}{2\sqrt{d}}\mathbb{Z}^d \cap B_\infty^d} \tilde{\mathcal{C}}_{\epsilon_s}^{(v,t)}.$$

A bound on the volume of the band where $(t - \frac{\epsilon_s}{2\sqrt{d}})\|v\| < x \cdot v < (t + \frac{\epsilon_s}{2\sqrt{d}})\|v\|$ in B_2^d follows from the fact that the maximum volume hyperplane section is a bisecting hyperplane, whose volume is $< 2\sqrt{d}\,\mathrm{vol}(B_2^d)$.

This allows us to bound the annealed entropy of a single class $\tilde{\mathcal{C}}_{\epsilon_s}^{(v,t)}$ in the following lemma, where ρ_{max} is the same maximum density with respect to the uniform density on B_2^d. (Rescaling was unnecessary because that was with respect to the Lebesgue measure normalized to be a probability measure).

Lemma 9 *The logarithm of the expected growth function of a class* $\tilde{\mathcal{C}}_{\epsilon_s}^{(v,t)}$ *with respect to* $\nu_\circ$ *random samples from* $\mathcal{P}_\circ$ *is* $< 2\epsilon_s\rho_{max}\lambda_\circ$, *where* $\nu_\circ$ *is a Poisson random variable of mean* $\lambda_\circ$*; i. e.*

$$\ln \mathbb{E}_{\nu_\circ} G(\mathcal{C}_{\tau_\square}, \mathcal{P}_\circ, \nu_\circ) < 2\epsilon_s\rho_{max}\lambda_\circ.$$

Proof 7 *A bound on the volume of the band where* $(t-\frac{\epsilon_s}{2\sqrt{d}})\|v\| < x\cdot v < (t+\frac{\epsilon_s}{2\sqrt{d}})\|v\|$ *in* B_2^d *follows from the fact that the maximum volume hyperplane section is a bisecting hyperplane, whose* $d-1$*-dimensional volume is* $< 2\sqrt{d}\, vol(B_2^d)$*. Therefore, the number of samples that fall in this band is a Poisson random variable whose mean is less than* $2\epsilon_s\rho_{max}\lambda_\circ$*. This implies the Lemma.*

Therefore the expected annealed entropy of

$$\bigcup_{tv\in\frac{\epsilon_s}{2\sqrt{d}}\mathbb{Z}^d\cap B_\infty^d} \tilde{C}_{\epsilon_s}^{(v,t)}$$

with respect to $\nu_\circ$ random samples from $\mathcal{P}_\circ$ is bounded above by $2\epsilon_s\rho_{max}\lambda_\circ + \ln|\frac{\epsilon}{2\sqrt{d}}\mathbb{Z}^d \cap B_\infty^d|$. Putting these observations together,

$$\begin{aligned}\ln \mathbb{E}_\nu G\left(\mathcal{C}_\mathcal{T}, \mathcal{P}, \nu\right)/\lambda &\leq \frac{\ln \mathbb{E}_{\nu_\circ} G(\mathcal{C}_{\mathcal{T}_\square}, \mathcal{P}_\circ, \nu_\circ)}{\lambda_\circ} + \epsilon\\ &\leq 2\epsilon_s\rho_{max} + \frac{d\ln(2\sqrt{d}/\epsilon_s)}{\lambda_\circ} + \epsilon.\end{aligned}$$

We know that $\lambda_\circ N_p(\epsilon_r/2) \geq \epsilon\lambda$. Then,

$$2\epsilon_s\rho_{max} + \frac{d\ln(2\sqrt{d}/\epsilon_s)}{\lambda_\circ} + \epsilon \leq$$

$$2\epsilon + N_p(\epsilon_r/2)\frac{d\ln(2\sqrt{d}\rho_{max}/\epsilon_s)}{\epsilon\lambda} + \epsilon,$$

which is

$$\leq 2\epsilon + N_p(\epsilon_r/2)\frac{d\ln(2\sqrt{d}\rho_{max}^2/\epsilon)}{\epsilon\lambda} + \epsilon.$$

Therefore, if $\lambda \geq N_p(\epsilon_r/2)\frac{d\ln(2\sqrt{d}\rho_{max}^2/\epsilon)}{\epsilon^2}$, then,

$$\ln \mathbb{E}_\nu G\left(\mathcal{C}_\mathcal{T}, \mathcal{P}, \nu\right)/\lambda \leq 4\epsilon.$$

Together with Lemma 6, this shows that for any $\epsilon_1 > 0$,
if

$$\lambda \geq N_p(\epsilon_r/2)\frac{d\ln(2\sqrt{d}\rho_{max}^2/\epsilon)}{\epsilon^2},$$

then

$$\begin{aligned}H_{ann}(\Lambda, \mathcal{P}, \lfloor\lambda(1-\epsilon_1)\rfloor) &\leq \ln \mathbb{E}_\nu G(\Lambda, \mathcal{P}, \nu)\\ &+ \exp\left(-\epsilon_1^2\lambda + \frac{\ln(2\pi\lambda)}{2}\right)\\ &\leq 4\epsilon\lambda + \exp\left(-\epsilon_1^2\lambda + \frac{\ln(2\pi\lambda)}{2}\right).\end{aligned}$$

Setting ϵ_1 to $\sqrt{\frac{\ln(2\pi\lambda)}{\lambda}}$, $\exp\left(-\epsilon_1^2\lambda + \frac{\ln(2\pi\lambda)}{2}\right)$ is less than 1. Therefore,

$$H_{ann}(\Lambda, \mathcal{P}, \lfloor\lambda - \sqrt{\lambda\ln(2\pi\lambda)}\rfloor) \leq 4\epsilon\lambda + 1.$$

This completes the proof of Theorem 2.

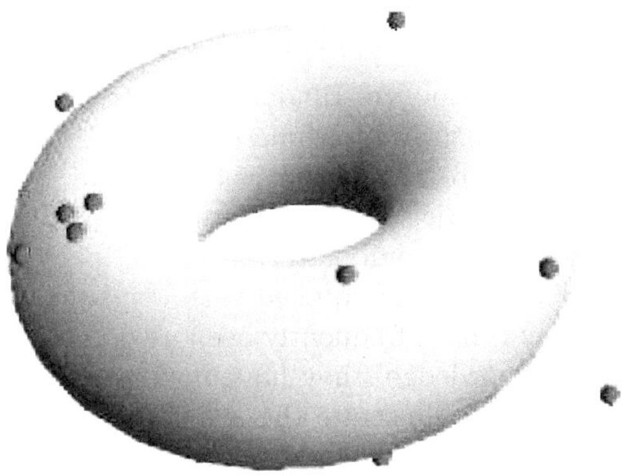

Figure 4.3: Fitting a torus to data.

4.4 Sample Complexity of Testing the Manifold Hypothesis

As we just saw, the sample complexity of classification is independent of the ambient dimension [14] in a natural setting assuming the manifold hypothesis, thus allowing us to avoid the curse of dimensionality. A recent empirical study [5], of a large number of 3×3 images, represented as points in $\mathbb{R}^9$, revealed that they approximately lie on a two dimensional manifold known as the Klein bottle. On the other hand, knowledge that the manifold hypothesis is false with regard to certain data would give us reason to exercise caution in applying algorithms from manifold learning and would provide an incentive for further study.

It is thus of considerable interest to know whether given data lie in the vicinity of a low dimensional manifold. Our primary technical results are the following.

1. We obtain uniform bounds relating the empirical squared loss and the true squared loss over a class $\mathcal{F}$ consisting of manifolds whose dimensions, volumes, and curvatures are bounded in Theorems 3 and 4. These bounds imply upper bounds on the sample complexity of Empirical Risk Minimization (ERM) that are *independent* of the ambient dimension, exponential in the intrinsic dimension, polynomial in the curvature, and almost linear in the volume.

2. We obtain a minimax lower bound on the sample complexity of *any* rule for learning a manifold from $\mathcal{F}$ in Theorem 8 showing that for a fixed error, the dependence of the sample complexity on intrinsic dimension, curvature, and volume must be at least exponential, polynomial, and linear, respectively.

3. We improve the best currently known upper bound [12] on the sample complexity of Empirical Risk minimization on k-means applied to data in a unit ball of arbitrary dimension from $O(\frac{k^2}{\epsilon^2} + \frac{\log \frac{1}{\delta}}{\epsilon^2})$ to $O\left(\frac{k}{\epsilon^2}\left(\min\left(k, \frac{\log^4 \frac{k}{\epsilon}}{\epsilon^2}\right)\right) + \frac{\log \frac{1}{\delta}}{\epsilon^2}\right)$. Whether the known

lower bound of $O(\frac{k}{\epsilon^2} + \frac{\log \frac{1}{\delta}}{\epsilon^2})$ is tight has been an open question since 1997 [3]. Here ϵ is the desired bound on the error and δ is a bound on the probability of failure.

We will use dimensionality reduction via random projections in the proof of Theorem 7 to bound the Fat-Shattering dimension of a function class, elements of which roughly correspond to the squared distance to a low dimensional manifold. The application of the probabilistic method involves a projection onto a low dimensional random subspace. This is then followed by arguments of a combinatorial nature involving the VC dimension of halfspaces, and the Sauer-Shelah Lemma applied with respect to the low dimensional subspace. While random projections have frequently been used in machine learning algorithms, for example in [2, 6], to our knowledge, they have not been used as a tool to bound the complexity of a function class. We illustrate the algorithmic utility of our uniform bound by devising an algorithm for k-means and a convex programming algorithm for fitting a piecewise linear curve of bounded length. For a fixed error threshold and length, the dependence on the ambient dimension is linear, which is optimal since this is the complexity of reading the input.

4.5 Connections and Related Work

In the context of curves, [8] proposed "Principal Curves," where it was suggested that a natural curve that may be fit to a probability distribution is one where every point on the curve is the center of mass of all those points to which it is the nearest point. A different definition of a principal curve was proposed by [10], where they attempted to find piecewise linear curves of bounded length which minimize the expected squared distance to a random point from a distribution. This paper studies the decay of the error rate as the number of samples tends to infinity, but does not analyze the dependence of the error rate on the ambient dimension and the bound on the length. We address this in a more general setup in Theorem 6, and obtain sample complexity bounds that are independent of the ambient dimension, and depend linearly on the bound on the length. There is a significant amount of recent research aimed at understanding topological aspects of data, such as its homology [18, 15]. It has been an open question since 1997 [3], whether the known lower bound of $O(\frac{k}{\epsilon^2} + \frac{\log \frac{1}{\delta}}{\epsilon^2})$ for the sample complexity of Empirical Risk minimization on k-means applied to data in a unit ball of arbitrary dimension is tight. Here ϵ is the desired bound on the error and δ is a bound on the probability of failure. The best currently known upper bound is $O(\frac{k^2}{\epsilon^2} + \frac{\log \frac{1}{\delta}}{\epsilon^2})$ and is based on Rademacher complexities. We improve this bound to $O\left(\frac{k}{\epsilon^2}\left(\min\left(k, \frac{\log^4 \frac{k}{\epsilon}}{\epsilon^2}\right)\right) + \frac{\log \frac{1}{\delta}}{\epsilon^2}\right)$, using an argument that bounds the Fat-Shattering dimension of the appropriate function class using random projections and the Sauer–Shelah Lemma. Generalizations of principal curves to parameterized principal manifolds in certain regularized settings have been studied in [16]. There, the sample complexity was related to the decay of eigenvalues of a Mercer kernel associated with the regularizer. When the manifold to be fit is a set of k points (k-means), we obtain a bound on the sample complexity s that is independent of m and depends at most linearly on k, which also leads to an approximation algorithm with additive error, based on sub-sampling. If one allows a multiplicative error of 4 in addition to an additive error of ϵ, a statement of this nature has been proven by Ben-David (Theorem 7, [4]).

4.6 Sample Complexity of Empirical Risk Minimization

For any submanifold $\mathcal{M}$ contained in, and probability distribution $\mathcal{P}$ supported on the unit ball B in $\mathbb{R}^m$, let $\mathcal{L}(\mathcal{M}, \mathcal{P}) := \int \mathbf{d}(\mathcal{M}, x)^2 d\mathcal{P}(x)$. Given a set of i.i.d points $x = \{x_1, \dots, x_s\}$ from $\mathcal{P}$, a tolerance ϵ and a class of manifolds $\mathcal{F}$, Empirical Risk Minimization (ERM) outputs a manifold in $\mathcal{M}_{erm}(x) \in \mathcal{F}$ such that $\sum_{i=1}^{s} \mathbf{d}(x_i, \mathcal{M}_{erm})^2 \leq \epsilon/2 + \inf_{\mathcal{N} \in \mathcal{F}} \mathbf{d}(x_i, \mathcal{N})^2$.

Definition 10 *Given error parameters ϵ, δ, and a rule $\mathbb{A}$ that outputs a manifold in $\mathcal{F}$ when provided with a set of samples, we define the sample complexity $s = s(\epsilon, \delta, \mathbb{A})$ to be the least number such that for any probability distribution $\mathcal{P}$ in the unit ball, if the result of $\mathbb{A}$ applied to a set of at least s i.i.d random samples from $\mathcal{P}$ is $\mathcal{N}$, then*

$$\mathbb{P}\left[\mathcal{L}(\mathcal{N}, \mathcal{P}) < \inf_{\mathcal{M} \in \mathcal{F}} \mathcal{L}(\mathcal{M}, \mathcal{P}) + \epsilon\right] > 1 - \delta.$$

4.6.1 Bounded Intrinsic Curvature

Let $\mathcal{M}$ be a Riemannian manifold and let $p \in \mathcal{M}$. Let ζ be a geodesic starting at p.

Definition 11 *The first point on ζ where ζ ceases to minimize distance is called the cut point of p along $\mathcal{M}$. The cut locus of p is the set of cut points of $\mathcal{M}$. The injectivity radius is the minimum taken over all points of the distance between the point and its cut locus. $\mathcal{M}$ is complete if it is complete as a metric space.*

Let $\mathcal{G}_i = \mathcal{G}_i(d, V, \lambda, \iota)$ be the family of all isometrically embedded, complete Riemannian submanifolds of B having dimension less or equal to d, induced d-dimensional volume less than or equal to V, sectional curvature less than or equal to λ, and injectivity radius greater or equal to ι. Let $U_{int}(\frac{1}{\epsilon}, d, V, \lambda, \iota) := V\left(C\left(\frac{d}{\min(\epsilon, \iota, \lambda^{-1/2})}\right)\right)^d$, which for brevity, we denote U_{int}.

Theorem 3 *Let ϵ and δ be error parameters. If*

$$s \geq C\left(\min\left(\left(\frac{1}{\epsilon^2}\right)\log^4\left(\frac{U_{int}}{\epsilon}\right), U_{int}\right)\frac{U_{int}}{\epsilon^2} + \frac{1}{\epsilon^2}\log\frac{1}{\delta}\right),$$

and $x = \{x_1, \dots, x_s\}$ is a set of i.i.d points from $\mathcal{P}$, then,

$$\mathbb{P}\left[\mathcal{L}(\mathcal{M}_{erm}(x), \mathcal{P}) - \inf_{\mathcal{M} \in \mathcal{G}_i} \mathcal{L}(\mathcal{M}, \mathcal{P}) < \epsilon\right] > 1 - \delta.$$

The proof of this theorem is deferred to Section 4.7.

4.6.2 Bounded Extrinsic Curvature

We will consider submanifolds of B that have the property that around each of them, for any radius $r < \tau$, the boundary of the set of all points within a distance r is smooth. This class of submanifolds has appeared in the context of manifold learning [15, 14].

Let $\mathcal{G}_e = \mathcal{G}_e(d, V, \tau)$ be the family of Riemannian submanifolds of B having dimension $\leq d$, volume $\leq V$ and condition number $\leq \frac{1}{\tau}$. Let ϵ and δ be error parameters. Let $U_{ext}(\frac{1}{\epsilon}, d, \tau) := V\left(C\left(\frac{d}{\min(\epsilon, \tau)}\right)\right)^d$, which for brevity, we denote by U_{ext}.

Theorem 4 *If*

$$s \geq C\left(\min\left(\left(\frac{1}{\epsilon^2}\right)\log^4\left(\frac{U_{ext}}{\epsilon}\right), U_{ext}\right)\frac{U_{ext}}{\epsilon^2} + \frac{1}{\epsilon^2}\log\frac{1}{\delta}\right),$$

and $x = \{x_1, \ldots, x_s\}$ is a set of i.i.d points from $\mathcal{P}$, then,

$$\mathbb{P}\left[\mathcal{L}(\mathcal{M}_{erm}(x), \mathcal{P}) - \inf_{\mathcal{M}} \mathcal{L}(\mathcal{M}, \mathcal{P}) < \epsilon\right] > 1 - \delta. \tag{4.4}$$

4.7 Relating Bounded Curvature to Covering Number

In this subsection, we note that that bounds on the dimension, volume, sectional curvature and injectivity radius suffice to ensure that they can be covered by relatively few $\epsilon-$balls. Let $V_p^{\mathcal{M}}$ be the volume of a ball of radius $\mathcal{M}$ centered around a point p. See ([7], page 51) for a proof of the following theorem.

Theorem 5 (Bishop-Günther Inequality) *Let $\mathcal{M}$ be a complete Riemannian manifold and assume that r is not greater than the injectivity radius ι. Let $K^{\mathcal{M}}$ denote the sectional curvature of $\mathcal{M}$ and let $\lambda > 0$ be a constant. Then, $K^{\mathcal{M}} \leq \lambda$ implies $V_p^{\mathcal{M}}(r) \geq \frac{2\pi^{\frac{n}{2}}}{\Gamma(\frac{n}{2})}\int_0^r \left(\frac{\sin(t\sqrt{\lambda})}{\sqrt{\lambda}}\right)^{n-1} dt.$*

Thus, if $\epsilon < \min(\iota, \frac{\pi\lambda^{-\frac{1}{2}}}{2})$, then, $V_p^{\mathcal{M}}(\epsilon) > \left(\frac{\epsilon}{Cd}\right)^d$.

Proof 8 (Proof of Theorem 3) *As a consequence of Theorem 5, we obtain an upper bound of $V\left(\frac{Cd}{\epsilon}\right)^d$ on the number of disjoint sets of the form $\mathcal{M} \cap B_{\epsilon/32}(p)$ that can be packed in $\mathcal{M}$. If $\{\mathcal{M} \cap B_{\epsilon/32}(p_1), \ldots, \mathcal{M} \cap B_{\epsilon/32}(p_k)\}$ is a maximal family of disjoint sets of the form $\mathcal{M} \cap B_{\epsilon/32}(p)$, then there is no point $p \in \mathcal{M}$ such that $\min_i \|p - p_i\| > \epsilon/16$. Therefore, $\mathcal{M}$ is contained in the union of balls, $\bigcup_i B_{\epsilon/16}(p_i)$. Therefore, we may apply Theorem 6 with $U\left(\frac{1}{\epsilon}\right) \leq V\left(\frac{Cd}{\min(\epsilon, \lambda^{-\frac{1}{2}}, \iota)}\right)^d$.*

The proof of Theorem 4 is along the lines of that of Theorem 3, so it has been deferred to the journal version.

4.8 Class of Manifolds with a Bounded Covering Number

In this section, we show that uniform bounds relating the empirical squares loss and the expected squared loss can be obtained for a class of manifolds whose covering number at a different scale ϵ has a specified upper bound. Let $U : \mathbb{R}^+ \to \mathbb{Z}^+$ be any integer valued function. Let $\mathcal{G}$ be any family of subsets of B such that for all $r > 0$ every element $\mathcal{M} \in \mathcal{G}$ can be covered using open Euclidean balls of radius r centered around $U(\frac{1}{r})$ points; let this set be $\Lambda_{\mathcal{M}}(r)$. Note that if the subsets consist of k-tuples of points, $U(1/r)$ can be taken to be the constant function equal to k and we recover the k-means question. A priori, it is unclear if

$$\sup_{\mathcal{M} \in \mathcal{G}} \left|\frac{\sum_{i=1}^s \mathbf{d}(x_i, \mathcal{M})^2}{s} - \mathbb{E}_{\mathcal{P}}\mathbf{d}(x, \mathcal{M})^2\right|, \tag{4.5}$$

is a random variable, since the supremum of a set of random variables is not always a random variable (although if the set is countable this is true). However (4.5) is equal to

$$\lim_{n\to\infty}\sup_{\mathcal{M}\in\mathcal{G}}\left|\frac{\sum_{i=1}^{s}\mathbf{d}(x_i,\Lambda_{\mathcal{M}}(1/n))^2}{s}-\mathbb{E}_{\mathcal{P}}\mathbf{d}(x,\Lambda_{\mathcal{M}}(1/n))^2\right|, \tag{4.6}$$

and for each n, the supremum in the limits is over a set parameterized by $U(n)$ points, which without loss of generality we may take to be countable (due to the density and countability of rational points). Thus, for a fixed n, the quantity in the limits is a random variable. Since the limit as $n\to\infty$ of a sequence of bounded random variables is a random variable as well, (4.5) is a random variable too.

Theorem 6 *Let ϵ and δ be error parameters. If*

$$s\geq C\left(\frac{U(16/\epsilon)}{\epsilon^2}\min\left(U(16/\epsilon),\left(\frac{1}{\epsilon^2}\right)\log^4\left(\frac{U(16/\epsilon)}{\epsilon}\right)\right)+\frac{1}{\epsilon^2}\log\frac{1}{\delta}\right),$$

Then,

$$\mathbb{P}\left[\sup_{\mathcal{M}\in\mathcal{G}}\left|\frac{\sum_{i=1}^{s}\mathbf{d}(x_i,\mathcal{M})^2}{s}-\mathbb{E}_{\mathcal{P}}\mathbf{d}(x,\mathcal{M})^2\right|<\frac{\epsilon}{2}\right]>1-\delta. \tag{4.7}$$

Proof 9 *For every $g\in\mathcal{G}$, let $\mathbf{c}(g,\epsilon)=\{c_1,\dots,c_k\}$ be a set of $k:=U(16/\epsilon)$ points in $g\subseteq B$, such that g is covered by the union of balls of radius $\epsilon/16$ centered at these points. Thus, for any point $x\in B$,*

$$\mathbf{d}^2(x,g) \leq \left(\frac{\epsilon}{16}+\mathbf{d}(x,\mathbf{c}(g,\epsilon))\right)^2 \tag{4.8}$$

$$\leq \frac{\epsilon^2}{256}+\frac{\epsilon\min_i\|x-c_i\|}{8}+\mathbf{d}(x,\mathbf{c}(g,\epsilon))^2. \tag{4.9}$$

Since $\min_i\|x-c_i\|$ is less than or equal to 2, the last expression is less than $\frac{\epsilon}{2}+\mathbf{d}(x,\mathbf{c}(g,\epsilon))^2$. Our proof uses the "kernel trick" in conjunction with Theorem 7. Let $\Phi:(x_1,\dots,x_m)^T\mapsto 2^{-1/2}(x_1,\dots,x_m,1)^T$ map a point $x\in\mathbb{R}^m$ to one in $\mathbb{R}^{m+1}$. For each i, let $c_i:=(c_{i1},\dots,c_{im})^T$, and $\tilde{c}_i:=2^{-1/2}(-c_{i1},\dots,-c_{im},\frac{\|c_i\|^2}{2})^T$. The factor of $2^{-1/2}$ is necessitated by the fact that we wish the image of a point in the unit ball to also belong to the unit ball. Given a collection of points $\mathbf{c}:=\{c_1,\dots,c_k\}$ and a point $x\in B$, let $f_{\mathbf{c}}(x):=\mathbf{d}(x,\mathbf{c}(g,\epsilon))^2$. Then,

$$f_{\mathbf{c}}(x)=\|x\|^2+4\min(\Phi(x)\cdot\tilde{c}_1,\dots,\Phi(x)\cdot\tilde{c}_k).$$

For any set of s samples $x_1,\dots,x_s$,

$$\sup_{f_{\mathbf{c}}\in\mathcal{G}}\left|\frac{\sum_{i=1}^{s}f_{\mathbf{c}}(x_i)}{s}-\mathbb{E}_{\mathcal{P}}f_{\mathbf{c}}(x)\right| \leq \left|\frac{\sum_{i=1}^{s}\|x_i\|^2}{s}-\mathbb{E}_{\mathcal{P}}\|x\|^2\right| \tag{4.10}$$

$$+\ 4\sup_{f_{\mathbf{c}}\in\mathcal{G}}\left|\frac{\sum_{i=1}^{s}\min_i\Phi(x_i)\cdot\tilde{c}_i}{s}-\mathbb{E}_{\mathcal{P}}\min_i\Phi(x)\cdot\tilde{c}_i\right|. \tag{4.11}$$

By Hoeffding's inequality,

$$\mathbb{P}\left[\left|\frac{\sum_{i=1}^{s}\|x_i\|^2}{s}-\mathbb{E}_{\mathcal{P}}\|x\|^2\right|>\frac{\epsilon}{4}\right]<2e^{-\left(\frac{1}{8}\right)s\epsilon^2}, \tag{4.12}$$

which is less than $\frac{\delta}{2}$.

By Theorem 7, $\mathbb{P}\left[\sup_{f_{\mathbf{c}}\in\mathcal{G}}\left|\frac{\sum_{i=1}^{s}\min_i\Phi(x_i)\cdot\tilde{c}_i}{s}-\mathbb{E}_{\mathcal{P}}\min_i\Phi(x)\cdot\tilde{c}_i\right|>\frac{\epsilon}{16}\right]<\frac{\delta}{2}$.

Therefore, $\mathbb{P}\left[\sup_{f_{\mathbf{c}}\in\mathcal{G}}\left|\frac{\sum_{i=1}^{s}f_{\mathbf{c}}(x_i)}{s}-\mathbb{E}_{\mathcal{P}}f_{\mathbf{c}}(x)\right|\leq\frac{\epsilon}{2}\right]\geq 1-\delta$.

4.9 Fat-Shattering Dimension and Random Projections

The core of the uniform bounds in Theorems 3 and 4 is the following uniform bound on the minimum of k linear functions on a ball in $\mathbb{R}^m$.

Theorem 7 *Let $\mathcal{F}$ be the set of all functions f from $B := \{x \in \mathbb{R}^m : \|x\| \leq 1\}$ to $\mathbb{R}$, such that for some k vectors $v_1, \dots, v_k \in B$,*

$$f(x) := \min_i (v_i \cdot x).$$

Independent of m, if

$$s \geq C\left(\frac{k}{\epsilon^2}\min\left(\frac{1}{\epsilon^2}\log^4\left(\frac{k}{\epsilon}\right), k\right) + \frac{1}{\epsilon^2}\log\frac{1}{\delta}\right),$$

then

$$\mathbb{P}\left[\sup_{F \in \mathcal{F}}\left|\frac{\sum_{i=1}^s F(x_i)}{s} - \mathbb{E}_{\mathcal{P}}F(x)\right| < \epsilon\right] > 1 - \delta. \tag{4.13}$$

It has been open since 1997 [3], whether the known lower bound of $C\left(\frac{k}{\epsilon^2} + \frac{1}{\epsilon^2}\log\frac{1}{\delta}\right)$ on the sample complexity s is tight. Theorem 5 in [12], uses Rademacher complexities to obtain an upper bound of

$$C\left(\frac{k^2}{\epsilon^2} + \frac{1}{\epsilon^2}\log\frac{1}{\delta}\right). \tag{4.14}$$

(The scenarios in [3, 12] are that of $k-$means, but the argument in Theorem 6 reduces k-means to our setting.) Theorem 7 improves this to

$$C\left(\frac{k}{\epsilon^2}\min\left(\frac{1}{\epsilon^2}\log^4\left(\frac{k}{\epsilon}\right), k\right) + \frac{1}{\epsilon^2}\log\frac{1}{\delta}\right) \tag{4.15}$$

by putting together (4.14) with a bound of

$$C\left(\frac{k}{\epsilon^4}\log^4\left(\frac{k}{\epsilon}\right) + \frac{1}{\epsilon^2}\log\frac{1}{\delta}\right) \tag{4.16}$$

obtained using the Fat-Shattering dimension. Due to constraints on space, the details of the proof of Theorem 7 will appear in the journal version, but the essential ideas are summarized here.

Let $u := \mathrm{fat}_{\mathcal{F}}(\frac{\epsilon}{24})$ and $x_1, \dots, x_u$ be a set of vectors that is γ-shattered by $\mathcal{F}$. We would like to use VC theory to bound u, but doing so directly leads to a linear dependence on the ambient dimension m. In order to circumvent this difficulty, for $g := C\frac{\log(u+k)}{\epsilon^2}$, we consider a g-dimensional random linear subspace and the image (Figure 4.4) under an appropriately scaled orthogonal projection R of the points $x_1, \dots, x_u$ onto it. We show that the expected value of the $\frac{\gamma}{2}$-shatter coefficient of $\{Rx_1, \dots, Rx_u\}$ is at least 2^{u-1} using the Johnson–Lindenstrauss Lemma [9] and the fact that $\{x_1, \dots, x_u\}$ is γ-shattered. Using Vapnik–Chervonenkis theory and the Sauer–Shelah Lemma, we then show that $\frac{\gamma}{2}$-shatter coefficient cannot be more than $u^{k(g+2)}$. This implies that $2^{u-1} \leq u^{k(g+2)}$, allowing us to conclude that $\mathrm{fat}_{\mathcal{F}}(\frac{\epsilon}{24}) \leq \frac{Ck}{\epsilon^2}\log^2\left(\frac{k}{\epsilon}\right)$. By a well-known theorem of [1], a bound of $\frac{Ck}{\epsilon^2}\log^2\left(\frac{k}{\epsilon}\right)$ on $\mathrm{fat}_{\mathcal{F}}(\frac{\epsilon}{24})$ implies the bound in (4.16) on the sample complexity, which implies Theorem 7.

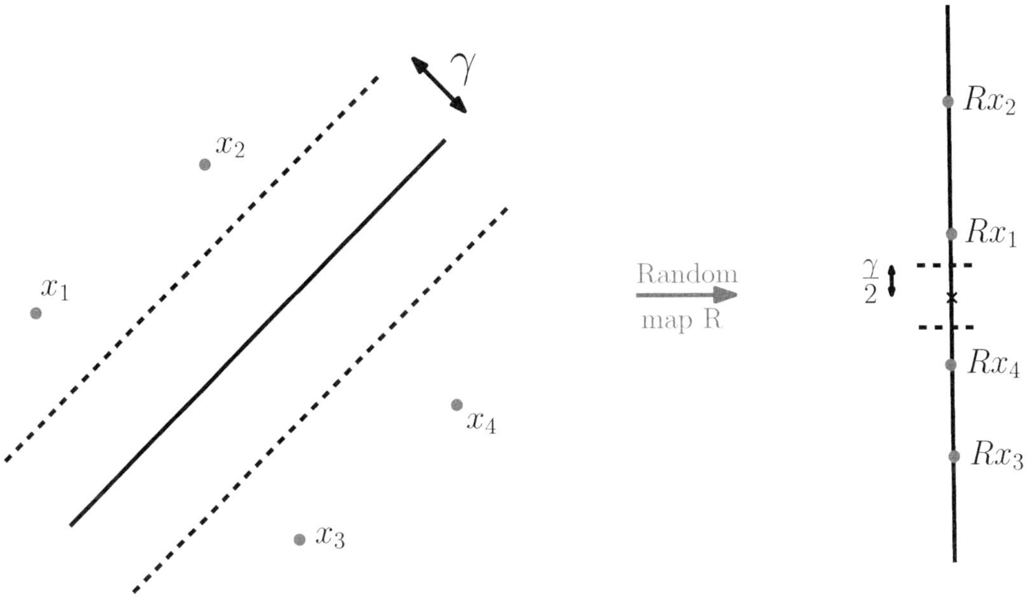

Figure 4.4: Random projections are likely to preserve linear separations.

4.10 Minimax Lower Bounds on the Sample Complexity

Let K be a universal constant whose value will be fixed throughout this section. In this section, we will state lower bounds on the number of samples needed for the minimax decision rule for learning from high dimensional data, with high probability, a manifold with a squared loss that is within ϵ of the optimal. We will construct a carefully chosen prior on the space of probability distributions and use an argument that can either be viewed as an application of the probabilistic method or of the fact that the Minimax risk is at least the risk of a Bayes optimal manifold computed with respect to this prior. Let U be a $K^{2d}k$-dimensional vector space containing the origin, spanned by the basis $\{e_1, \dots, e_{K^{2d}k}\}$ and S be the surface of the ball of radius 1 in $\mathbb{R}^m$. We assume that m be greater or equal to $K^{2d}k + d$. Let W be the d-dimensional vector space spanned by $\{e_{K^{2d}k+1}, \dots, e_{K^{2d}k+d}\}$. Let $S_1, \dots, S_{K^{2d}k}$ denote spheres, such that for each i, $S_i := S \cap (\sqrt{1-\tau^2}e_i + W)$, where $x + W$ is the translation of W by x. Note that each S_i has radius τ. Let $\ell = \binom{K^{2d}k}{K^d k}$ and $\{\mathcal{M}_1, \dots, \mathcal{M}_\ell\}$ consist of all $K^d k$-element subsets of $\{S_1, \dots, S_{K^{2d}k}\}$. Let ω_d be the volume of the unit ball in $\mathbb{R}^d$. The following theorem shows that no algorithm can produce a nearly optimal manifold with high probability unless it uses a number of samples that depends linearly on volume, exponentially on intrinsic dimension, and polynomially on the curvature.

Theorem 8 *Let $\mathcal{F}$ be equal to either $\mathcal{G}_e(d, V, \tau)$ or $\mathcal{G}_i(d, V, \frac{1}{\tau^2}, \pi\tau)$. Let $k = \lfloor \frac{V}{d\omega_d(K^{\frac{5}{4}}\tau)^d} \rfloor$. Let $\mathbb{A}$ be an arbitrary algorithm that takes as input a set of data points $x = \{x_1, \dots, x_k\}$*

and outputs a manifold $\mathcal{M}_{\mathbb{A}}(x)$ in $\mathcal{F}$. If $\epsilon + 2\delta < \frac{1}{3}\left(\frac{1}{2\sqrt{2}} - \tau\right)^2$ then,

$$\inf_{\mathcal{P}} \mathbb{P}\left[\mathcal{L}(\mathcal{M}_{\mathbb{A}}(x), \mathcal{P}) - \inf_{\mathcal{M}\in\mathcal{F}} \mathcal{L}(\mathcal{M}, \mathcal{P}) < \epsilon\right] < 1 - \delta,$$

where $\mathcal{P}$ ranges over all distributions supported on B and $x_1, \dots, x_k$ are i.i.d draws from $\mathcal{P}$.

Proof 10 *Observe from Lemma 3 and Theorem 5 that $\mathcal{F}$ is a class of a manifolds such that each manifold in $\mathcal{F}$ is contained in the union of $K^{\frac{3d}{2}}k$ m-dimensional balls of radius τ, and $\{\mathcal{M}_1, \dots, \mathcal{M}_\ell\} \subseteq \mathcal{F}$. (The reason why we have $K^{\frac{3d}{2}}$ rather than $K^{\frac{5d}{4}}$ as in the statement of the theorem is that the parameters of $\mathcal{G}_i(d, V, \tau)$ are intrinsic, and to transfer to the extrinsic setting of the last sentence, one needs some leeway.) Let $\mathcal{P}_1, \dots, \mathcal{P}_\ell$ be probability distributions that are uniform on $\{\mathcal{M}_1, \dots, \mathcal{M}_\ell\}$ with respect to the induced Riemannian measure. Suppose A is an algorithm that takes as input a set of data points $x = \{x_1, \dots, x_t\}$ and outputs a manifold $\mathcal{M}_{\mathbb{A}}(x)$. Let r be chosen uniformly at random from $\{1, \dots, \ell\}$. Then,*

$$\begin{aligned}
&\inf_{\mathcal{P}} \mathbb{P}\left[\left|\mathcal{L}(\mathcal{M}_{\mathbb{A}}(x), \mathcal{P}) - \inf_{\mathcal{M}\in\mathcal{F}} \mathcal{L}(\mathcal{M}, \mathcal{P})\right| < \epsilon\right] \\
&\leq \mathbb{E}_{\mathcal{P}_r}\mathbb{P}_x\left[\left|\mathcal{L}(\mathcal{M}_{\mathbb{A}}(x), \mathcal{P}_r) - \inf_{\mathcal{M}\in\mathcal{F}} \mathcal{L}(\mathcal{M}, \mathcal{P}_r)\right| < \epsilon\right] \\
&= \mathbb{E}_x\mathbb{P}_{\mathcal{P}_r}\left[\left|\mathcal{L}(\mathcal{M}_{\mathbb{A}}(x), \mathcal{P}_r) - \inf_{\mathcal{M}\in\mathcal{F}} \mathcal{L}(\mathcal{M}, \mathcal{P}_r)\right| < \epsilon \middle| x\right] \\
&= \mathbb{E}_x\mathbb{P}_{\mathcal{P}_r}\left[\mathcal{L}(\mathcal{M}_{\mathbb{A}}(x), \mathcal{P}_r) < \epsilon \middle| x\right].
\end{aligned}$$

We first prove a lower bound on $\inf_x \mathbb{E}_r\left[\mathcal{L}(\mathcal{M}_{\mathbb{A}}(x), \mathcal{P}_r)|x\right]$.
We see that

$$\mathbb{E}_r\left[\mathcal{L}(\mathcal{M}_{\mathbb{A}}(x), \mathcal{P}_r)\middle|x\right] = \mathbb{E}_{r,x_{k+1}}\left[\mathbf{d}(\mathcal{M}_{\mathbb{A}}(x), x_{k+1})^2\middle|x\right]. \tag{4.17}$$

Conditioned on x, the probability of the event (say E_{dif}) that x_{k+1} does not belong to the same sphere as one of the $x_1, \dots, x_k$ is at least $\frac{1}{2}$.

Conditioned on E_{dif} and $x_1, \dots, x_k$, the probability that x_{k+1} lies on a given sphere S_j is equal to 0 if one of $x_1, \dots, x_k$ lies on S_j and $\frac{1}{K^2k - k'}$ otherwise, where $k' \leq k$ is the number of spheres in $\{S_i\}$ which contain at least one point among $x_1, \dots, x_k$.

By construction, $A(x_1, \dots, x_k)$ can be covered by $K^{\frac{3d}{2}}k$ balls of radius τ; let their centers be $y_1, \dots, y_{K^{\frac{3d}{2}}k}$.

However, it is easy to check that for any dimension m, the cardinality of the set $\mathcal{S}_y$ of all S_i that have a nonempty intersection with the balls of radius $\frac{1}{2\sqrt{2}}$ centered around $y_1, \dots, y_{K^{\frac{3d}{2}}k}$, is at most $K^{\frac{3d}{2}}k$. Therefore, $\mathbb{P}\left[\mathbf{d}(\mathcal{M}_{\mathbb{A}}(x), x_{k+1}) \geq \frac{1}{2\sqrt{2}} - \tau \middle| x\right]$ is at least

$$\begin{aligned}
\mathbb{P}\left[\mathbf{d}(\{y_1, \dots, y_{K^{\frac{3d}{2}}k}\}, x_{k+1}) \geq \frac{1}{2\sqrt{2}}\middle|x\right] &\geq \mathbb{P}\left[E_{dif}\right]\mathbb{P}\left[x_{k+1} \notin \mathcal{S}_y | E_{dif}\right] \\
&\geq \frac{1}{2}\frac{K^{2d}k - k' - K^{\frac{3d}{2}}k}{K^{2d}k - k'} \\
&\geq \frac{1}{3}.
\end{aligned}$$

Therefore, $\mathbb{E}_{r,x_{k+1}}\left[\mathbf{d}(\mathcal{M}_{\mathbb{A}}(x), x_{k+1})^2\middle|x\right] \geq \frac{1}{3}\left(\frac{1}{2\sqrt{2}} - \tau\right)^2$. Finally, we observe that it is not possible for $\mathbb{E}_x\mathbb{P}_{\mathcal{P}_r}\left[\mathcal{L}(\mathcal{M}_{\mathbb{A}}(x), \mathcal{P}_r) < \epsilon\middle|x\right]$ to be more than $1-\delta$ if $\inf_x \mathbb{P}_{\mathcal{P}_r}\left[\mathcal{L}(\mathcal{M}_{\mathbb{A}}(x), \mathcal{P}_r)\middle|x\right] > \epsilon + 2\delta$, because $\mathcal{L}(\mathcal{M}_{\mathbb{A}}(x), \mathcal{P}_r)$ is bounded above by 2.

4.11 Algorithmic Implications

4.11.1 k-Means

Applying Theorem 6 to the case when $\mathcal{P}$ is a distribution supported equally on n specific points (that are part of an input) in a unit ball of $\mathbb{R}^m$, we see that in order to obtain an additive ϵ approximation for the k-means problem with probability $1-\delta$, it suffices to sample

$$s \geq C\left(\frac{k}{\epsilon^2}\left(\frac{\log^4\left(\frac{k}{\epsilon}\right)}{\epsilon^2}, k\right) + \frac{1}{\epsilon^2}\log\frac{1}{\delta}\right)$$

points uniformly at random (which would have a cost of $O(s \log n)$ if the cost of one random bit is $O(1)$) and exhaustively solve k-means on the resulting subset. Supposing that a dot product between two vectors x_i, x_j can be computed using $\tilde{m}$ operations, the total cost of sampling and then exhaustively solving k-means on the sample is $O(\tilde{m} s k^s \log n)$. In contrast, if one asks for a multiplicative $(1+\epsilon)$ approximation, the best running time known depends linearly on n [11]. If $\mathcal{P}$ is an unknown probability distribution, the above algorithm improves upon the best results in a natural statistical framework for clustering [4].

4.11.2 Fitting Piecewise Linear Curves

In this subsection, we illustrate the algorithmic utility of the uniform bound in Theorem 6 by obtaining an algorithm for fitting a curve of length no more than L, to data drawn from an unknown probability distribution $\mathcal{P}$ supported in B, whose sample complexity is independent of the ambient dimension. This curve, with probability $1-\delta$, achieves a mean squared error of less than ϵ more than the optimum. The proof of its correctness and analysis of its run-time have been deferred to the journal version. The algorithm is as follows:

---

1. Let $k := \lceil \frac{L}{\epsilon} \rceil$ and $s \geq C\left(\frac{k}{\epsilon^2}\left(\frac{\log^4\left(\frac{k}{\epsilon}\right)}{\epsilon^2}, k\right) + \frac{1}{\epsilon^2}\log\frac{1}{\delta}\right)$. Sample points $x_1, \dots, x_s$ i.i.d from $\mathcal{P}$ for $s =$, and set $J := \text{span}(\{x_i\}_{i=1}^s)$.
2. For every permutation σ of $[s]$, minimize the convex objective function $\sum_{i=1}^n d(x_{\sigma(i)}, y_i)^2$ over the convex set of all s-tuples of points $(y_1, \dots, y_s)$ in J, such that $\sum_{i=1}^{s-1} \|y_{i+1} - y_i\| \leq L$.
3. If the minimum over all $(y_1, \dots, y_s)$ (and σ) is achieved for $(z_1, \dots, z_s)$, output the curve obtained by joining z_i to z_{i+1} for each i by a straight line segment.

4.12 Summary

In this chapter, we discussed the sample complexity of classification, when data is drawn i.i.d from a probability distribution supported on a low dimensional submanifold of Euclidean space, and showed that this is independent of the ambient dimension based on work with P. Niyogi [14]. We also discussed the problem of fitting a manifold to data, when the manifold has prescribed bounds on its reach, its volume, and its dimension based on work with S.

Mitter [13]. We showed that the number of samples needed has no dependence on the ambient dimension, if the data were to be drawn i.i.d from a distribution supported in a unit ball.

Bibliography

[1] Noga Alon, Shai Ben-David, Nicolò Cesa-Bianchi, and David Haussler. Scale-sensitive dimensions, uniform convergence, and learnability. *J. ACM*, 44(4):615–631, 1997.

[2] Rosa I. Arriaga and Santosh Vempala. An algorithmic theory of learning: Robust concepts and random projection. In *FOCS*, pages 616–623, 1999.

[3] Peter Bartlett, Tamás Linder, and Gabor Lugosi. The minimax distortion redundancy in empirical quantizer design. *IEEE Transactions on Information Theory*, 44:1802–1813, 1997.

[4] Shai Ben-David. A framework for statistical clustering with constant time approximation algorithms for k-median and k-means clustering. *Mach. Learn.*, 66(2-3):243–257, 2007.

[5] Gunnar Carlsson. Topology and data. *Bulletin of the American Mathematical Society*, 46:255–308, January 2009.

[6] Sanjoy Dasgupta. Learning mixtures of Gaussians. In *FOCS*, pages 634–644, 1999.

[7] Alfred Gray. *Tubes*. Birkhauser Verlag, 2004.

[8] Trevor J. Hastie and Werner Stuetzle. Principal curves. *Journal of the American Statistical Association*, 84:502–516, 1989.

[9] William Johnson and Joram Lindenstrauss. Extensions of lipschitz mappings into a Hilbert space. *Contemporary Mathematics*, 26:419–441, 1984.

[10] Balázs Kégl, Adam Krzyzak, Tamás Linder, and Kenneth Zeger. Learning and design of principal curves. *IEEE Transactions on Pattern Analysis and Machine Intelligence*, 22:281–297, 2000.

[11] Amit Kumar, Yogish Sabharwal, and Sandeep Sen. A simple linear time $(1+\epsilon)-$approximation algorithm for k-means clustering in any dimensions. In *FOCS*, pages 454–462, 2004.

[12] Andreas Maurer and Massimiliano Pontil. Generalization bounds for k-dimensional coding schemes in hilbert spaces. In *ALT*, pages 79–91, 2008.

[13] Hariharan Narayanan and Sanjoy Mitter. On the sample complexity of testing the manifold hypothesis. In *NIPS*, 2010.

[14] Hariharan Narayanan and Partha Niyogi. On the sample complexity of learning smooth cuts on a manifold. In *Proc. of the 22nd Annual Conference on Learning Theory (COLT)*, June 2009.

[15] Partha Niyogi, Stephen Smale, and Shmuel Weinberger. Finding the homology of submanifolds with high confidence from random samples. *Discrete & Computational Geometry*, 39(1-3):419–441, 2008.

[16] Alexander J. Smola, Sebastian Mika, Bernhard Schölkopf, and Robert C. Williamson. Regularized principal manifolds. *J. Mach. Learn. Res.*, 1:179–209, 2001.

[17] Vladimir Vapnik. *Statistical Learning Theory.* Wiley. 1998.

[18] Afra Zomorodian and Gunnar Carlsson. Computing persistent homology. *Discrete & Computational Geometry*, 33(2):249–274, 2005.

Chapter 5

Manifold Alignment

Chang Wang, Peter Krafft, and Sridhar Mahadevan

Manifold alignment, the topic of this chapter, is simultaneously a solution to the problem of alignment and a framework for discovering a unifying representation of multiple datasets. The fundamental ideas of manifold alignment are to utilize the relationships of instances within each dataset to strengthen knowledge of the relationships between the datasets and ultimately to map initially disparate datasets to a joint latent space. At the algorithmic level, the approaches described in this chapter assume that the disparate datasets being aligned have the same underlying manifold structure. The underlying low-dimensional representation is extracted by modeling the local geometry using a graph Laplacian associated with each dataset. After constructing each of these Laplacians, standard manifold learning algorithms are then invoked on a joint Laplacian matrix constructed by concatenating the various Laplacians to obtain a joint latent representation of the original datasets. Manifold alignment can therefore be viewed as a form of constrained joint dimensionality reduction where the goal is to find a low-dimensional embedding of multiple datasets that preserves any known correspondences between them.

5.1 Introduction

This chapter addresses the fundamental problem of *aligning* multiple datasets to extract shared latent semantic structure. Specifically, the goal of the methods described here is to create a more meaningful representation by aligning multiple datasets. Domains of applicability range across the field of engineering, humanities, and science. Examples include automatic machine translation, bioinformatics, cross-lingual information retrieval, perceptual learning, robotic control, and sensor-based activity modeling.

What makes the data alignment problem challenging is that the multiple data streams that need to be coordinated are represented using disjoint features. For example, in cross-lingual information retrieval, it is often desirable to search for documents in a target language (e.g., Italian or Arabic) by typing in queries in English. In activity modeling, the motions of humans engaged in everyday indoor or outdoor activities, such as cooking or walking, is recorded using diverse sensors including audio, video, and wearable devices. Furthermore, as real-world datasets often lie in a high-dimensional space, the challenge is to construct a common semantic representation across heterogeneous datasets by automatically discovering a shared latent space. This chapter describes a *geometric* framework for data alignment, building on recent advances in manifold learning and nonlinear dimensionality reduction using spectral graph-theoretic methods.

The problem of alignment can be formalized as dimensionality reduction with constraints induced by the correspondences between datasets. In many application domains of interest, data appears high-dimensional, but often lies on low-dimensional structures, such as a *manifold*, which can be discretely approximated by a graph [1]. Nonlinear dimensionality reduction methods have recently emerged that empirically model and recover the underlying manifold, including diffusion maps [2], Isomap [3], Laplacian eigenmaps [4], LLE [5], Locality Preserving Projections (LPP) [6], and semi-definite embedding (SDE) [7]. When data lies on a manifold, these nonlinear techniques are much more effective than traditional linear methods, such as principal components analysis (PCA) [8]. This chapter describes a geometric framework for transfer using *manifold alignment* [9, 10]. Rather than constructing mappings on surface features, which may be difficult due to the high dimensionality of the data, manifold alignment constructs lower-dimensional mappings between two or more disparate datasets by aligning their underlying *learned* manifolds.

Many practical problems, ranging from bioinformatics to information retrieval and robotics, involve modeling multiple datasets that contain significant shared underlying structure. For example, in protein alignment, a set of proteins from a shared family are clustered together by finding correspondences between their three-dimensional structures. In information retrieval, it is often desirable to search documents in a target language, say Italian, given queries in a source language such as English. In robotics, activities can be modeled using parallel data streams such as visual input, audio input, and body posture. Even individual datasets can often be represented using multiple points of view. One example familiar to any calculus student is whether to represent a point in the plane with Cartesian coordinates or polar coordinates. This choice can be the difference between being able to solve a problem and not being able to solve that problem.

Generally, these examples of alignment problems fall into two categories: the stronger case of when different datasets have a single underlying meaning and the weaker case of when those datasets have related underlying structure. In the first case, finding the "original dataset," the underlying meaning that all of the observed datasets share, may be challenging. Manifold alignment solves this problem by finding a common set of features for those disparate datasets. These features provide coordinates in a single space for the instances in all the related datasets. In other words, manifold alignment discovers a unifying representation of all the initially separate datasets that preserves the qualities of each individual dataset and highlights the similarities between the datasets. Though this new representation may not be the actual "original dataset," if such an entity exists, it should reflect the structure that the original dataset would have.

For example, the Europarl corpus [11] contains a set of documents translated into eleven languages. A researcher may be interested in finding a language-invariant representation of these parallel corpora, for instance, as preprocessing for information retrieval, where different translations of corresponding documents should be close to one another in the joint representation. Using this joint representation, the researcher could easily identify identical or similar documents across languages. Section 5.4.2 describes how manifold alignment solves this problem using a small subset of the languages.

In the less extreme case, the multiple datasets do not have exactly the same underlying meaning but have related underlying structures. For example, two proteins may have related but slightly different tertiary structures, whether from measurement error or because the proteins are actually different but are evolutionarily related (see Figure 5.1). The initial representations of these two datasets, the locations of some points along each protein's structure, may be different, but comparing the local similarities within each dataset reveals that they lie on the same underlying manifold, that is, the relationships between the instances in each dataset are the same. In this case, if the proteins are actually different, there may be no "original dataset" that both observed datasets represent, there may only

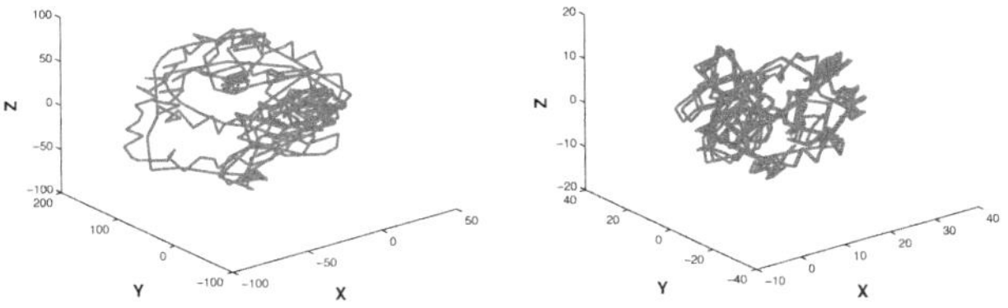

Figure 5.1: (See Color Insert.) A simple example of alignment involving finding correspondences across protein tertiary structures. Here two related structures are aligned. The smaller blue structure is a scaling and rotation of the larger red structure in the original space shown on the left, but the structures are equated in the new coordinate frame shown on the right.

be a structural similarity between the two datasets which allows them to be represented in similar locations in a new coordinate frame.

Manifold alignment is useful in both of these cases. Manifold alignment preserves similarities within each dataset being aligned and correspondences between the datasets being aligned by giving each dataset a new coordinate frame that reflects that dataset's underlying manifold structure. As such, the main assumption of manifold alignment is that any datasets being aligned must lie on the same low dimensional manifold. Furthermore, the algorithm requires a similarity function that returns the similarity of any two instances within the same dataset with respect to the geodesic distance along that manifold. If these assumptions are met, the new coordinate frames for the aligned manifolds will be consistent with each other and will give a unifying representation.

In some situations, such as the Europarl example, the required similarity function may reflect semantic similarity. In this case, the unifying representation discovered by manifold alignment represents the semantic space of the input datasets. Instances that are close with respect to Euclidean distance in the latent space will be semantically similar, regardless of their original dataset. In other situations, such as the protein example, the underlying manifold is simply a common structure to the datasets, such as related covariance matrices or related local similarity graphs. In this case, the latent space simply represents a new coordinate system for all the instances that is consistent with geodesic similarity along the manifold.

From an algorithmic perspective, manifold alignment is closely related to other manifold learning techniques for dimensionality reduction such as Isomap, LLE, and Laplacian eigenmaps. Given a dataset, these algorithms attempt to identify the low-dimensional manifold structure of that dataset and preserve that structure in a low dimensional embedding of the dataset. Manifold alignment follows the same paradigm but embeds multiple datasets simultaneously. Without any correspondence information (given or inferred), manifold alignment finds independent embeddings of each given dataset, but with some given or inferred correspondence information, manifold alignment includes additional constraints on these embeddings that encourage corresponding instances across datasets to have similar locations in the embedding. Figure 5.2 shows the high-level idea of constrained joint embedding.

The remainder of this section provides a more detailed overview of the problem of alignment and the algorithm of manifold alignment. Following these informal descriptions, Section 5.2 develops the formal loss functions for manifold alignment and proves the optimality of the manifold alignment algorithm. Section 5.3 describes four variants of the basic manifold alignment framework. Then, Section 5.4 explores three applications of manifold alignment that illustrate how manifold alignment and its extensions are useful for identify-

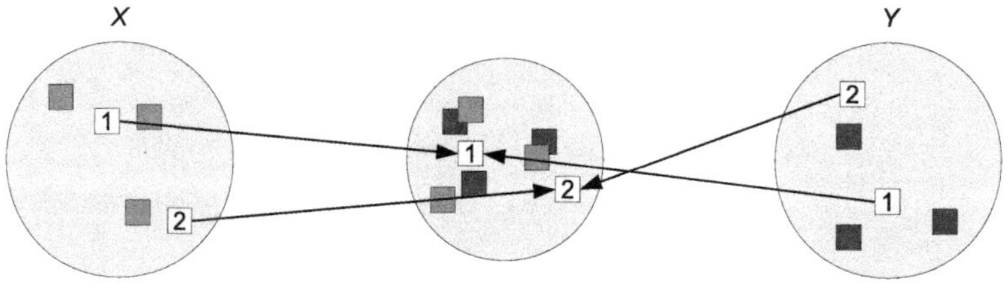

Figure 5.2: Given two datasets X and Y with two instances from both datasets that are known to be in correspondence, manifold alignment embeds all of the instances from each dataset in a new space where the corresponding instances are constrained to be equal and the internal structures of each dataset are preserved.

ing new correspondences between datasets, performing cross-dataset information retrieval, and performing exploratory data analysis; though, of course, manifold alignment's utility is not limited to these situations. Finally, Section 5.5 summarizes the chapter and discusses some limitations of manifold alignment and Section 5.6 reviews various approaches related to manifold alignment.

5.1.1 Problem Statement

The problem of alignment is to identify a transformation of one dataset that "matches it up" with a transformation of another dataset. That is, given two datasets,[1] X and Y, whose instances lie on the same manifold, $\mathcal{Z}$, but who may be represented by different features, the problem of alignment is to find two functions f and g, such that $f(x_i)$ is close to $g(y_j)$ in terms of Euclidean distance if x_i and y_j are close with respect to geodesic distance along $\mathcal{Z}$. Here, X is an $n \times p$ matrix containing n data instances in p-dimensional space, Y is an $m \times q$ matrix containing m data instances in q-dimensional space, $f : \mathbb{R}^p \to \mathbb{R}^k$, and $g : \mathbb{R}^q \to \mathbb{R}^k$ for some k called the latent dimensionality.

The instances x_i and y_j are in exact correspondence if and only if $f(x_i) = g(y_j)$. On the other hand, prior correspondence information includes any information about the similarity of the instances in X and Y, not just exact correspondence information. The union of the range of f and the range of g is the joint latent space, and the concatenation of the new coordinates $\begin{pmatrix} f(X) \\ g(Y) \end{pmatrix}$ is the unified representation of X and Y, where $f(X)$ is an $n \times k$ matrix containing the result f applied to each row of X, and $g(Y)$ an $m \times k$ matrix containing the result of g applied to each row of Y. $f(X)$ and $g(Y)$ are the new coordinates of X and Y in the joint latent space.

5.1.2 Overview of the Algorithm

Manifold alignment is one solution to the problem of alignment. There are two key ideas to manifold alignment: considering local geometry as as well as correspondence information and viewing multiple datasets as being samples on the same manifold. First, instead of only preserving correspondences across datasets, manifold alignment also preserves the individual structures within each dataset by mapping similar instances in each dataset to similar locations in Euclidean space. In other words, manifold alignment maps each dataset to a new joint latent space where locally similar instances within each dataset and given corresponding instances across datasets are close or identical in that space (see Figure 5.3).

[1]There is an analogous definition for alignment of multiple datasets. This statement only considers two datasets for simplicity of notation.

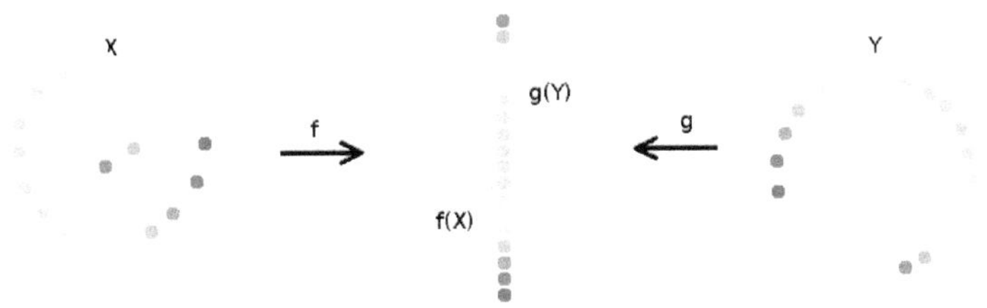

Figure 5.3: (See Color Insert.) An illustration of the problem of manifold alignment. The two datasets X and Y are embedded into a single space where the corresponding instances are equal and local similarities within each dataset are preserved.

This algorithm can be supervised, semi-supervised, or unsupervised. With complete correspondence information, the algorithm is supervised, and it simply finds a unifying representation of all the instances. With incomplete correspondence information, the algorithm is semi-supervised, and it relies only on the known correspondences and the datasets' intrinsic structures to form the embedding. With no correspondence information, the algorithm is unsupervised, and some correspondences must be inferred. Section 5.3.4 discusses one way to infer correspondences.

Second, manifold alignment views each individual dataset as belonging to one larger dataset. Since all the datasets have the same manifold structure, the graph Laplacians associated with each dataset are all discrete approximations of the same manifold, and thus, the diagonal concatenation of these Laplacians, along with the off-diagonal matrices filled with correspondence information, is still an approximation of that manifold. The idea is simple but elegant — by viewing two or more samples as actually being one large sample, making inferences about multiple datasets reduces to making inferences about a single dataset.

Embedding this joint Laplacian combines these ideas. By using a graph embedding algorithm, manifold alignment preserves the local similarities and correspondence information encoded by the joint Laplacian. Thus, by combining these ideas together, the problem of manifold alignment can be reduced to a variant of the standard manifold learning problem.

5.2 Formalization and Analysis

Figure 5.4 summarizes the notation used in this section.

5.2.1 Loss Functions

This section develops the intuition behind the loss function for manifold alignment in two ways, each analogous to one of the two key ideas from Section 5.1.2. The first way illustrates that the loss function captures the idea that manifold alignment should preserve local similarity and correspondence information. The second way illustrates the idea that after forming the joint Laplacian, manifold alignment is equivalent to Laplacian eigenmaps [4]. Subsequent sections use the second approach because it greatly simplifies the notation and the proofs of optimality.

The First Derivation of Manifold Alignment: Preserving Similarities

The first loss function has two parts: one part to preserve local similarity within each dataset and one part to preserve correspondence information about instances across datasets. With

For any $n \times p$ matrix, M, $M(i,j)$ is the i,jth entry of M, $M(i,\cdot)$ is ith row, and $M(\cdot,j)$ is the jth column. $(M)^+$ denotes the Moore-Penrose pseudoinverse. $\|M(i,\cdot)\|$ denotes the l_2 norm. M' denotes the transpose of M.

$X^{(a)}$ is an $n_a \times p_a$ data matrix with n_a observations and p_a features.
$W^{(a)}$ is an $n_a \times n_a$ matrix, where $W^{(a)}(i,j)$ is the similarity of $X^{(a)}(i,\cdot)$ and $X^{(a)}(j,\cdot)$ (could be defined by heat kernel).
$D^{(a)}$ is an $n_a \times n_a$ diagonal matrix: $D^{(a)}(i,i) = \sum_j W^{(a)}(i,j)$.
$L^{(a)} = D^{(a)} - W^{(a)}$ is the Laplacian associated with $X^{(a)}$.

$W_{(a,b)}$ is an $n_a \times n_b$ matrix, where $W^{(a,b)}(i,j) \neq 0$, when $X^{(a)}(i,\cdot)$ and $X^{(b)}(j,\cdot)$ are in correspondence and 0 otherwise. $W^{(a,b)}(i,j)$ is the similarity, or the strength of correspondence, of the two instances. Typically, $W^{(a,b)}(i,j) = 1$ if the instances $X^{(a)}(i,\cdot)$ and $X^{(b)}(j,\cdot)$ are in correspondence.

If c is the number of manifolds being aligned, $\mathbf{X}$ is the joint dataset, a $(\sum_i n_i) \times (\sum_i p_i)$ matrix, and $\mathbf{W}$ is the $(\sum_i n_i) \times (\sum_i n_i)$ joint adjacency matrix,

$$\mathbf{X} = \begin{pmatrix} X^{(1)} & \cdots & 0 \\ & \cdots & \\ 0 & \cdots & X^{(c)} \end{pmatrix}, \quad \mathbf{W} = \begin{pmatrix} \nu W^{(1)} & \mu W^{(1,2)} & \cdots & \mu W^{(1,c)} \\ & & \cdots & \\ \mu W^{(c,1)} & \mu W^{(c,2)} & \cdots & \nu W^{(c)} \end{pmatrix}.$$

ν and μ are scalars that control how much the alignment should try to respect local similarity versus correspondence information. Typically, $\nu = \mu = 1$. Equivalently, $\mathbf{W}$ is a $(\sum_i n_i) \times (\sum_i n_i)$ matrix with zeros on the diagonal and for all i and j,

$$\mathbf{W}(i,j) = \begin{cases} \nu W^{(a)}(i,j) & \text{if } \mathbf{X}(i,\cdot) \text{ and } \mathbf{X}(j,\cdot) \text{ are both from } X^{(a)} \\ \mu W^{(a,b)}(i,j) & \text{if } \mathbf{X}(i,\cdot) \text{ and } \mathbf{X}(j,\cdot) \text{ are corresponding instances from} \\ & X^{(a)} \text{ and } X^{(b)}, \text{ respectively} \\ 0 & \text{otherwise,} \end{cases}$$

where the $W^{(a)}(i,j)$ and $W^{(a,b)}(i,j)$ here are an abuse of notation with i and j being the row and column that $\mathbf{W}(i,j)$ came from. The precise notation would be $W^{(a)}(i_a, j_a)$ and $W^{(a,b)}(i_a, j_b)$, where k_g is the index such that $\mathbf{X}(k,\cdot) = [0 \ldots 0 \; X^{(g)}(k_g,\cdot) \; 0 \ldots 0]$, $k_g = k - \sum_{l=0}^{g-1} n_l$, $n_0 = 0$.

$\mathbf{D}$ is an $\sum_i n_i \times \sum_i n_i$ diagonal matrix with $\mathbf{D}(i,i) = \sum_j \mathbf{W}(i,j)$.

$\mathbf{L} = \mathbf{D} - \mathbf{W}$ is the joint graph Laplacian.

If the dimension of the new space is d, the embedded coordinates are given by

1. in the nonlinear case, $\mathbf{F}$, a $\left(\sum_i n_i\right) \times d$ matrix representing the new coordinates.
2. in the linear case, $\mathbf{F}$, a $\left(\sum_i p_i\right) \times d$ matrix, where $\mathbf{XF}$ represents the new coordinates.

$F^{(a)}$ or $X^{(a)}F^{(a)}$ are the new coordinates of the dataset $X^{(a)}$.

Figure 5.4: Notation used in this chapter.

c datasets, $X^{(1)}, \ldots, X^{(c)}$, for each dataset the loss function includes a term of the following form:

$$C_\lambda(F^{(a)}) = \sum_{i,j} \|F^{(a)}(i,\cdot) - F^{(a)}(j,\cdot)\|^2 W^{(a)}(i,j),$$

where $F^{(a)}$ is the embedding of the ath dataset and the sum is taken over all pairs of instances in that dataset. $C_\lambda(F^{(a)})$ is the cost of preserving the local similarities within $X^{(a)}$. This equation says that if two data instances from $X^{(a)}$, $X^{(a)}(i,\cdot)$, and $X^{(a)}(j,\cdot)$ are similar, which happens when $W^{(a)}(i,j)$ is larger, their locations in the latent space, $F^{(a)}(i,\cdot)$ and $F^{(a)}(j,\cdot)$, should be closer together.

Additionally, to preserve correspondence information, for each pair of datasets the loss function includes

$$C_\kappa(F^{(a)}, F^{(b)}) = \sum_{i,j} \|F^{(a)}(i,\cdot) - F^{(b)}(j,\cdot)\|^2 W^{(a,b)}(i,j).$$

$C_\kappa(F^{(a)}, F^{(b)})$ is the cost of preserving correspondence information between $F^{(a)}$ and $F^{(b)}$. This equation says that if two data points, $X^{(a)}(i,\cdot)$ and $X^{(b)}(j,\cdot)$, are in stronger correspondence, which happens when $W^{(a,b)}(i,j)$ is larger, their locations in the latent space, $F^{(a)}(i,\cdot)$ and $F^{(b)}(j,\cdot)$, should be closer together.

The complete loss function is thus

$$C_1(F^{(1)}, \ldots, F^{(c)}) = \nu \sum_a C_\lambda(F^{(a)}) + \mu \sum_{a \neq b} C_\kappa(F^{(a)}, F^{(b)})$$

$$= \nu \sum_a \sum_{i,j} \|F^{(a)}(i,\cdot) - F^{(a)}(j,\cdot)\|^2 W^{(a)}(i,j) \; + \mu \sum_{a \neq b} \sum_{i,j} \|F^{(a)}(i,\cdot) - F^{(b)}(j,\cdot)\|^2 W^{(a,b)}(i,j)$$

The Second Derivation of Manifold Alignment: Embedding the Joint Laplacian

The second loss function is simply the loss function for Laplacian eigenmaps using the joint adjacency matrix:

$$C_2(\mathbf{F}) = \sum_{i,j} \|\mathbf{F}(i,\cdot) - \mathbf{F}(j,\cdot)\|^2 \mathbf{W}(i,j),$$

where the sum is taken over all pairs of instances from all datasets. Here $\mathbf{F}$ is the unified representation of all the datasets and $\mathbf{W}$ is the joint adjacency matrix. This equation says that if two data instances, $X^{(a)}(i',\cdot)$ and $X^{(b)}(j',\cdot)$, are similar, regardless of whether they are in the same dataset ($a = b$) or from different datasets ($a \neq b$), which happens when $\mathbf{W}(i,j)$ is larger in either case, their locations in the latent space, $\mathbf{F}(i,\cdot)$ and $\mathbf{F}(j,\cdot)$, should be closer together.

Equivalently, making use of the facts that $\|M(i,\cdot)\|^2 = \sum_k M(i,k)^2$ and that the Laplacian is a quadratic difference operator,

$$\begin{aligned}
C_2(\mathbf{F}) &= \sum_{i,j} \sum_k [\mathbf{F}(i,k) - \mathbf{F}(j,k)]^2 \mathbf{W}(i,j) \\
&= \sum_k \sum_{i,j} [\mathbf{F}(i,k) - \mathbf{F}(j,k)]^2 \mathbf{W}(i,j) \\
&= \sum_k \operatorname{tr}(\mathbf{F}(\cdot,k)' \mathbf{L} \mathbf{F}(\cdot,k)) \\
&= \operatorname{tr}(\mathbf{F}' \mathbf{L} \mathbf{F}),
\end{aligned}$$

where $\mathbf{L}$ is the joint Laplacian of all the datasets.

Overall, this formulation of the loss function says that, given the joint Laplacian, aligning all the datasets of interest is equivalent to embedding the joint dataset according to the Laplacian eigenmap loss function.

Equivalence of the Two Loss Functions

The equivalence of the two loss functions follows directly from the definition of the joint Laplacian, **L**. Note that for all i and j,

$$\mathbf{W}(i,j) = \begin{cases} \nu W^{(a)}(i,j) & \text{if } \mathbf{X}(i,\cdot) \text{ and } \mathbf{X}(j,\cdot) \text{ are both from } X^{(a)} \\ \mu W^{(a,b)}(i,j) & \text{if } \mathbf{X}(i,\cdot) \text{ and } \mathbf{X}(j,\cdot) \text{ are corresponding instances from} \\ & X^{(a)} \text{ and } X^{(b)}\text{, respectively} \\ 0 & \text{otherwise.} \end{cases}$$

Then, the terms of $C_2(\mathbf{F})$ containing instances from the same dataset are exactly the $C_\lambda(F^{(a)})$ terms, and the terms of containing instances from different datasets are exactly the $C_\kappa(F^{(a)}, F^{(b)})$ terms. Since all other terms are 0,

$$C_1(F^{(1)}, \ldots, F^{(c)}) = C_2(\mathbf{F}) = C(\mathbf{F}).$$

This equivalence means that embedding the joint Laplacian is equivalent to preserving local similarity within each dataset and correspondence information between all pairs of datasets.

The Final Optimization Problem

Although this loss function captures the intuition of manifold alignment, for it to work mathematically it needs an additional constraint,

$$\mathbf{F}'\mathbf{D}\mathbf{F} = I,$$

where I is the $d \times d$ identity matrix. Without this constraint, the trivial solution of mapping all instances to zero would minimize the loss function.

Note, however, that two other constraints are commonly used instead. The first is

$$\mathbf{F}'\mathbf{F} = I$$

and the second is

$$\mathbf{F}' \begin{pmatrix} D^{(1)} & \cdots & 0 \\ & \cdots & \\ 0 & \cdots & D^{(c)} \end{pmatrix} \mathbf{F} = I.$$

Each of these gives slightly different results and interpretations, and in fact in Section 5.4, we use the third constraint. The best constraint to use in any given application depends on how correspondence information is given, how large the weight on correspondence information, μ, is, and whether the researcher would like the disparate datasets to have the same scale in the joint representation. For simplicity of notation in our exposition, we elect to use the first constraint, which views the correspondence information as being edges in the joint weight matrix.

Thus, the final optimization equation for manifold alignment is

$$\underset{\mathbf{F}:\mathbf{F}'\mathbf{D}\mathbf{F}=I}{\arg\min}\ C(\mathbf{F}) = \underset{\mathbf{F}:\mathbf{F}'\mathbf{D}\mathbf{F}=I}{\arg\min}\ \mathrm{tr}(\mathbf{F}'\mathbf{L}\mathbf{F}).$$

5.2.2 Optimal Solutions

This section derives the optimal solution to the optimization problem posed in the previous section using the method of Lagrange multipliers. The technique is well-known in the literature (see for example Bishop's derivation of PCA [12]), and the solution is equivalent to that of Laplacian eigenmaps.

Consider the case when $d = 1$. Then $\mathbf{F}$ is just a vector, f, and

$$\arg\min_{f:f'\mathbf{D}f=1} C(f) = \arg\min_{f} \ f'\mathbf{L}f + \lambda(1 - f'\mathbf{D}f)$$

Differentiating with respect to f and λ and setting equal to zero gives

$$\mathbf{L}f = \lambda \mathbf{D}f$$

and

$$f'\mathbf{D}f = 1.$$

The first equation shows that the optimal f is a solution of the generalized eigenvector problem, $\mathbf{L}f = \lambda\mathbf{D}f$. Multiplying both sides of this equation by f' and using $f'\mathbf{D}f = 1$ gives $f'\mathbf{L}f = \lambda$, which means that minimizing $f'\mathbf{L}f$ requires the smallest nonzero eigenvector.

For $d > 1$, $\mathbf{F} = [f_1, f_2, \ldots, f_d]$, and the optimization problem becomes

$$\arg\min_{\mathbf{F}:\mathbf{F}'\mathbf{D}\mathbf{F}=1} C(\mathbf{F}) = \arg\min_{f_1,\ldots,f_d} \sum_i f_i'\mathbf{L}f_i + \lambda_i(1 - f_i'\mathbf{D}f_i),$$

and the solution is the d smallest nonzero eigenvectors. In this case, the total cost is $\sum_{i=1}^{d} \lambda_i$ if the eigenvalues, $\lambda_1, \ldots, \lambda_n$, are sorted in ascending order and exclude the zero eigenvalues.

5.2.3 The Joint Laplacian Manifold Alignment Algorithm

Using the optimal solution derived in the last section, this section describes the algorithm for manifold alignment using Laplacian eigenmaps on the joint Laplacian.

Given c datasets, $X^{(1)}, \ldots, X^{(c)}$, all lying on the same manifold, a similarity function (or a distance function), S, that returns the similarity of any two instances from the same dataset with respect to geodesic distance along the manifold (perhaps $S = e^{-\|x-y\|}$), and some given correspondence information in the form of pairs of similarities of instances from different datasets, the algorithm is as follows:

1. **Find the adjacency matrices, $W^{(1)}, \ldots, W^{(c)}$, of each dataset using S, possibly only including a weight between two instances if one is in the k-nearest neighbors of the other.**

2. **Construct the joint Laplacian, $\mathbf{L}$.**

3. **Compute the d smallest nonzero eigenvectors of $\mathbf{L}f = \lambda\mathbf{D}f$.**

4. **The rows $1 + \sum_{l=0}^{g-1} n_l, 2 + \sum_{l=0}^{g-1} n_l, \ldots, n_g + \sum_{l=0}^{g-1} n_l$ of $\mathbf{F}$ are the new coordinates of $X^{(g)}$.**

5.3 Variants of Manifold Alignment

There are a number of extensions to the basic joint Laplacian manifold alignment framework. This section explores four important variants: restricting the embedding functions to be linear, enforcing hard constraints on corresponding pairs of instances, finding alignments at multiple scales, and performing alignment with no given correspondence information.

5.3.1 Linear Restriction

In nonlinear manifold alignment, the eigenvectors of the Laplacian are exactly the new coordinates of the embedded instances — there is no simple closed form for the mapping function from the original data to the latent space. Linear manifold alignment, however, enforces an explicit, linear functional form for the embedding function. Besides being useful for making out-of-sample estimates, linear alignment helps diminish the problem of missing correspondence information, finds relationships between the features of multiple datasets instead of just between instances, and attempts to identify a common linear subspace of the original datasets.

This section develops linear alignment in steps analogous to the development of nonlinear alignment with many of the details omitted because of the similarity between the two approaches.

Problem Statement

The problem of linear alignment is slightly different from the general problem of alignment; it is to identify a linear transformation, instead of an arbitrary transformation, of one dataset that best "matches that dataset up" with a linear transformation of another dataset. That is, given two datasets,[2] X and Y, whose instances lie on the same manifold, $\mathcal{Z}$, but who may be represented by different features, the problem of linear alignment is to find two matrices F and G, such that x_iF is close to y_jG in terms of Euclidean distance if x_i and y_j are close with respect to geodesic distance along $\mathcal{Z}$.

The Linear Loss Function

The motivation and intuition for linear manifold alignment are the same as for nonlinear alignment. The loss function is thus similar, only requiring an additional term for the linear constraint on the mapping function. The new loss function is

$$C(\mathbf{F}) = \sum_{i \neq j} \|\mathbf{X}(i,\cdot)\mathbf{F} - \mathbf{X}(j,\cdot)\mathbf{F}\|^2 \mathbf{W}(i,j) = \mathrm{tr}(\mathbf{F}'\mathbf{X}'\mathbf{L}\mathbf{X}\mathbf{F}),$$

where the sum is taken over all pairs of instances from all datasets. Once again, the constraint $\mathbf{F}'\mathbf{X}'\mathbf{D}\mathbf{X}\mathbf{F} = I$ allows for nontrivial solutions to the optimization problem. This equation captures the same intuitions as the nonlinear loss function, namely that if $\mathbf{X}(i,\cdot)$ is similar to $\mathbf{X}(j,\cdot)$, which occurs when $\mathbf{W}(i,j)$ is large, the embedded coordinates, $\mathbf{X}(i,\cdot)\mathbf{F}$ and $\mathbf{X}(j,\cdot)\mathbf{F}$, will be closer together, but it restricts the embedding of the $\mathbf{X}$ to being a linear embedding.

Optimal Solutions

Much like nonlinear alignment reduces to Laplacian eigenmaps on the joint Laplacian of the datasets, linear alignment reduces to locality preserving projections [6] on the joint Laplacian of the datasets. The solution to the optimization problem is the minimum eigenvectors of the generalized eigenvector problem:

$$\mathbf{X}'\mathbf{L}\mathbf{X}f = \lambda\mathbf{X}'\mathbf{D}\mathbf{X}f.$$

The proof of this fact is similar to the nonlinear case (just replace the matrix $\mathbf{L}$ in that proof with the matrix $\mathbf{X}'\mathbf{L}\mathbf{X}$, and the matrix $\mathbf{D}$ with $\mathbf{X}'\mathbf{D}\mathbf{X}$).

[2]Once again this definition could include more than two datasets.

Comparison to Nonlinear Alignment

The most immediate practical benefit of using linear alignment is that the explicit functional forms of the alignment functions allow for embedding new instances from any of the datasets into the latent space without having to use an interpolation method. This functional form is also useful for mapping instances from one dataset directly to the space of another dataset. Given some point $X^{(g)}(i,\cdot)$, the function $F^{(g)}(F^{(h)})^{+}$ maps that point to the coordinate system of $X^{(h)}$. This direct mapping function is useful for transfer. Given some function, f trained on $X^{(h)}$ but which is inconsistent with the coordinate system of $X^{(g)}$ (perhaps f takes input from $\mathbb{R}^3$ but the instances from $X^{(g)}$ are in $\mathbb{R}^4$), $f(X^{(g)}(i,\cdot)F^{(g)}(F^{(h)})^{+})$ is an estimate of the value of what $f(X^{(g)}(i,\cdot))$ would be.

Linear alignment is also often more efficient than nonlinear alignment. If $\sum_i p_i \ll \sum_i n_i$, linear alignment will be much faster than nonlinear alignment, since the matrices $\mathbf{X}'\mathbf{L}\mathbf{X}$ and $\mathbf{X}'\mathbf{D}\mathbf{X}$ are $(\sum_i p_i)\times(\sum_i p_i)$ instead of $(\sum_i n_i)\times(\sum_i n_i)$. Of course, these benefits come at a heavy cost if the manifold structure of the original datasets cannot be expressed by a linear function of the original dataset. Linear alignment sacrifices the ability to align arbtitarily warped manifolds. However, as in any linear regression, including nonlinear transformation of the original features of the datasets is one way to circumvent this problem in some cases.

At the theoretical level, linear alignment is interesting because, letting $\mathbf{X}$ be a variable, the linear loss function is a generalization of the simpler, nonlinear loss function. Setting $\mathbf{X}$ to the identity matrix, linear alignment reduces to the nonlinear formulation. This observation highlights the fact that even in the nonlinear case the embedded coordinates, $\mathbf{F}$, are functions—they are functions of the indices of the original datasets.

Other Interpretations

Since each mapping function is linear, the features of the embedded datasets (the projected coordinate systems) are linear combinations of the features of the original datasets, which means that another way to view linear alignment and its associated loss function is as a joint feature selection algorithm. Linear alignment thus tries to select the features of the original datasets that are shared across datasets; it tries to select a combination of the original features that best respects similarities within and between each dataset. Because of this, examining the coefficients in $\mathbf{F}$ is informative about which features of the original datasets are most important for respecting local similarity and correspondence information. In applications where the underlying manifold of each dataset has a semantic interpretation, linear alignment attempts to filter out the features that are dataset-specific and defines a set of invariant features of the datasets.

Another related interpretation of linear alignment is as feature-level alignment. The function $F^{(g)}(F^{(h)})^{+}$ that maps the instances of one dataset to the coordinate frame of another dataset also represents the relationship between the features of each of those datasets. For example, if $X^{(2)}$ is a rotation of $X^{(1)}$, $F^{(2)}(F^{(1)})^{+}$ should ideally be that rotation matrix (it may not be if there is not enough correspondence information or if the dimensionality of the latent space is different from that of the datasets, for example). From a more abstract perspective, each column of the embedded coordinates $\mathbf{XF}$ is composed of a set of linear combinations of the columns from each of the original datasets. That is, the columns $F^{(g)}(i,\cdot)$ and $F^{(h)}(i,\cdot)$, which define the ith feature of the latent space, combine some number of the columns from $X^{(g)}(i,\cdot)$ and $X^{(h)}(i,\cdot)$. They unify the features from $X^{(g)}(i,\cdot)$ and $X^{(h)}(i,\cdot)$ into a feature in the latent space. Thus linear alignment defines an alignment of the features of each dataset.

5.3.2 Hard Constraints

In nonlinear alignment, hard constraints can replace some of the soft constraints specified by the loss function. Two instances that should be exactly equal in the latent space can be constrained to be the same by merging them in the joint Laplacian. This merging action forms a new row by combining the individual edge weights in each row of the instances that are equal and removing the original rows of those instances from the joint Laplacian [9]. The eigenvectors of the joint Laplacian will then have one less entry. To recover the full embedding, the final coordinates must include two copies of the embedded merged row, each in the appropriate locations.

5.3.3 Multiscale Alignment

Many real-world datasets exhibit non-trivial regularities at *multiple* levels. For example, for the dataset involving abstracts of NIPS conference papers,[3] at the most abstract level, the set of all papers can be categorized into two main topics: machine learning and neuroscience. At the next level, the papers can be categorized into a number of areas, such as cognitive science, computer vision, dimensionality reduction, reinforcement learning, etc. To transfer knowledge across domains taking consideration of their intrinsic multilevel structures, we need to develop algorithms for multiscale manifold alignment. All previously studied approaches to manifold alignment are restricted to a single scale. In this section, we discuss how to extend multiscale algorithms such as diffusion wavelets [13] to yield hierarchical solutions to the alignment problem. The goal of this multiscale approach is to produce alignment results that preserve local geometry of each manifold and match instances in correspondence at every scale. Compared to "flat" methods, multiscale alignment automatically generates alignment results at different scales by exploring the intrinsic structures (in common) of the two data sets, avoiding the need to specify the dimensionality of the new space.

Finding multiscale alignments using diffusion wavelets enables a natural multiscale interpretation and gives a sparse solution. Multiscale alignment offers additional advantages in transfer learning and in exploratory data analysis. Most manifold alignment methods must be modified to deal with asymmetric similarity relations, which occur when constructing graphs using k-nearest neighbor relationships, in directed citation and web graphs, in Markov decision processes, and in many other applications. In contrast to most manifold alignment methods, multiscale alignment using diffusion wavelets can be used without modification, although there is no optimality guarantee in that case. Furthermore, multiscale alignment is useful to exploratory data analysis because it generates a hierarchy of alignments that reflects a hierarchical structure common to the datasets of interest.

Intuitively, multiscale alignment is appealing because many datasets show regularity at multiple scales. For example, in the NIPS conference paper dataset,[4] there are two main topics at the most abstract level: machine learning and neuroscience. At the next level, the papers fall into a number of categories, such as dimensionality reduction or reinforcement learning. Another dataset with a similar topic structure should be able to be aligned at each of these scales. Multiscale manifold alignment simultaneously extracts this type of structure across all datasets of interest.

This section formulates the problem of multiscale alignment using the framework of multiresolution wavelet analysis [13]. In contrast to "flat" alignment methods which result in a single latent space for alignment in a pre-selected dimension, multiscale alignment using diffusion wavelets automatically generates alignment results at different levels by discovering the shared intrinsic multilevel structures of the given datasets. This multilevel approach

[3]Available at www.cs.toronto.edu/~roweis/data.html.

[4]www.cs.toronto.edu/~roweis/data.html

results in multiple alignments in spaces of different dimension, where the dimensions are automatically decided according to a precision term.

Problem Statement

Given a fixed sequence of dimensions, $d_1 > d_2 > \ldots > d_h$, as well as two datasets, X and Y, and some partial correspondence information, $x_i \in X_l \longleftrightarrow y_i \in Y_l$, the multiscale manifold alignment problem is to compute mapping functions, $\mathcal{A}_k$ and $\mathcal{B}_k$, at each level k ($k = 1, 2, \ldots, h$) that project X and Y to a new space, preserving local geometry of each dataset and matching instances in correspondence. Furthermore, the associated sequence of mapping functions should satisfy $span(\mathcal{A}_1) \supseteq span(\mathcal{A}_2) \supseteq \ldots \supseteq span(\mathcal{A}_h)$ and $span(\mathcal{B}_1) \supseteq span(\mathcal{B}_2) \supseteq \ldots \supseteq span(\mathcal{B}_h)$, where $span(\mathcal{A}_i)$ (or $span(\mathcal{B}_i)$) represents the subspace spanned by the columns of $\mathcal{A}_i$ (or $\mathcal{B}_i$).

This view of multiscale manifold alignment consists of two parts: (1) determining a hierarchy in terms of number of levels and the dimensionality at each level and (2) finding alignments to minimize the cost function at each level. Our approach solves both of these problems simultaneously while satisfying the subspace hierarchy constraint.

Optimal Solutions

There is one key property of diffusion wavelets that needs to be emphasized. Given a diffusion operator T, such as a random walk on a graph or manifold, the diffusion wavelet (DWT) algorithm produces a subspace hierarchy associated with the eigenvectors of T (if T is symmetric). Letting λ_i be the eigenvalue associated with the ith eigenvector of T, the kth level of the DWT hiearchy is spanned by the eigenvectors of T with $\lambda_i^{2^k} \geq \epsilon$, for some precision parameter, ϵ. Although each level of the hierarchy is spanned by a certain set of eigenvectors, the DWT algorithm returns a set of scaling functions, ϕ_k, at each level, which span the same space as the eigenvectors but have some desirable properties.

To apply diffusion wavelets to a multiscale alignment problem, the algorithm must address the following challenge: the regular diffusion wavelets algorithm can only handle regular eigenvalue decomposition in the form of $A\gamma = \lambda\gamma$, where A is the given matrix, γ is an eigenvector, and λ is the corresponding eigenvalue. However, the problem we are interested in is a generalized eigenvalue decomposition, $A\gamma = \lambda B\gamma$, where we have two input matrices A and B. This multiple manifold alignment algorithm overcomes this challenge.

The Main Algorithm

Using the notation defined in Figure 5.4, the algorithm is as follows:

1. **Construct a matrix representing the joint manifold, L.**

2. **Find an $(\sum p_i) \times r$ matrix, G, such that $G'G = \mathbf{X}'\mathbf{X}$ using SVD.**

3. **Define $T = ((G')^{+}\mathbf{X}'\mathbf{L}\mathbf{X}G^{+})^{+}$.**

4. **Use diffusion wavelets to explore the intrinsic structure of the joint manifold:**

   $[\phi_k]_{\phi_0} = \mathcal{DWT}(T^{+}, \epsilon)$, where $\mathcal{DWT}()$ is the diffusion wavelets implementation described in [13] with extraneous parameters omitted. $[\phi_k]_{\phi_0}$ are the scaling function bases at level k represented as an $r \times d_k$ matrix, $k = 1, \ldots, h$.

5. **Compute mapping functions for manifold alignment (at level k): $\mathbf{F}_k = (G)^{+}[\phi_k]_{\phi_0}$.**

Benefits

As discussed in [13], the benefits of using diffusion wavelets are:

- Wavelet analysis generalizes to asymmetric matrices.
- Diffusion wavelets result in sets of mapping functions that capture different spectral bands of the relevant operator.
- The basis vectors in ϕ_k are localized (sparse).

5.3.4 Unsupervised Alignment

Performing unsupervised alignment requires generating the portions of the joint Laplacian that represent the between-dataset similarities. One way to do this is by local pattern matching. With no given correspondence information, if the datasets $X^{(a)}$ and $X^{(b)}$ are represented by different features, there is no easy way to directly compare $X^{(a)}(i,\cdot)$ and $X^{(b)}(j,\cdot)$. One way to build connections between them is to use the relations between $X^{(a)}(i,\cdot)$ and its neighbors to characterize $X^{(a)}(i,\cdot)$'s local geometry. Using relations rather than features to represent local geometry makes the direct comparison of $X^{(a)}(i,\cdot)$ and $X^{(b)}(j,\cdot)$ possible. After generating correspondence information using this approach, any of the previous manifold alignment algorithms work. This section shows how to compute local patterns representing local geometry, and shows that these patterns are valid for comparison across datasets.

Given $X^{(a)}$, pattern matching first constructs an $n_a \times n_a$ distance matrix $Distance_a$, where $Distance_a(i,j)$ is Euclidean distance between $X^{(a)}(i,\cdot)$ and $X^{(a)}(j,\cdot)$. The algorithm then decomposes this matrix into elementary contact patterns of fixed size $k+1$. $R_{X^{(a)}(i,\cdot)}$ is a $(k+1)\times(k+1)$ matrix representing the local geometry of $X^{(a)}(i,\cdot)$.

$$R_{X^{(a)}(i,\cdot)}(u,v) = distance(z_u, z_v),$$

where $z_1 = X^{(a)}(i,\cdot)$ and $z_2,\ldots,z_{k+1}$ are $X^{(a)}(i,\cdot)$'s k nearest neighbors. Similarly, $R_{X^{(b)}(j,\cdot)}$ is a $(k+1)\times(k+1)$ matrix representing the local geometry of $X^{(b)}(j,\cdot)$. The order of $X^{(b)}(j,\cdot)$'s k nearest neighbors have $k!$ permutations, so $R_{X^{(b)}(j,\cdot)}$ has $k!$ variants. Let $\{R_{X^{(b)}(j,\cdot)}\}_h$ denote its hth variant.

Each local contact pattern $R_{X^{(a)}(i,\cdot)}$ is represented by a submatrix, which contains all pairwise distances between local neighbors around $X^{(a)}(i,\cdot)$. Such a submatrix is a two-dimensional representation of a high dimensional substructure. It is independent of the coordinate frame and contains enough information to reconstruct the whole manifold. $X^{(b)}$ is processed similarly and distance between $R_{X^{(a)}(i,\cdot)}$ and $R_{X^{(b)}(j,\cdot)}$ is defined as follows:

$$dist(R_{X^{(a)}(i,\cdot)}, R_{X^{(b)}(j,\cdot)}) = \min_{1\le h\le k!} \min(dist_1(h), dist_2(h)),$$

where

$$\begin{aligned}
dist_1(h) &= \|\{R_{X^{(b)}(j,\cdot)}\}_h - k_1 R_{X^{(a)}(i,\cdot)}\|_F,\\
dist_2(h) &= \|R_{X^{(a)}(i,\cdot)} - k_2\{R_{X^{(b)}(j,\cdot)}\}_h\|_F,\\
k_1 &= \mathrm{tr}(R'_{X^{(a)}(i,\cdot)}\{R_{X^{(b)}(j,\cdot)}\}_h)/\mathrm{tr}(R'_{X^{(a)}(i,\cdot)}R_{X^{(a)}(i,\cdot)}),\\
k_2 &= \mathrm{tr}(\{R_{X^{(b)}(j,\cdot)}\}'_h R_{X^{(a)}(i,\cdot)})/\mathrm{tr}(\{R_{X^{(b)}(j,\cdot)}\}'_h\{R_{X^{(b)}(j,\cdot)}\}_h).
\end{aligned}$$

Finally, $W^{(a,b)}$ is computed as follows:

$$W^{(a,b)}(i,j) = e^{-dist(R_{X^{(a)}(i,\cdot)}, R_{X^{(b)}(j,\cdot)})/\delta^2}.$$

Theorem: Given two $(k+1) \times (k+1)$ distance matrices R_1 and R_2, $k_2 = \mathbf{tr}(R_2' R_1)/\mathbf{tr}(R_2' R_2)$ minimizes $\|R_1 - k_2 R_2\|_F$ and $k_1 = \mathbf{tr}(R_1' R_2)/\mathbf{tr}(R_1' R_1)$ minimizes $\|R_2 - k_1 R_1\|_F$.

Proof:

Finding k_2 is formalized as

$$k_2 = \arg\min_{k_2} \|R_1 - k_2 R_2\|_F,$$

where $\|\cdot\|_F$ represents Frobenius norm.

It is easy to verify that

$$\|R_1 - k_2 R_2\|_F = \mathrm{tr}(R_1' R_1) - 2k_2 \mathrm{tr}(R_2' R_1) + k_2^2 \mathrm{tr}(R_2' R_2).$$

Since $\mathrm{tr}(R_1' R_1)$ is a constant, the minimization problem is equal to

$$k_2 = \arg\min_{k_2} k_2^2 \mathrm{tr}(R_2' R_2) - 2k_2 \mathrm{tr}(R_2' R_1).$$

Differentiating with respect to k_2 gives

$$2k_2 \mathrm{tr}(R_2' R_2) = 2\mathrm{tr}(R_2' R_1),$$

which implies

$$k_2 = \mathrm{tr}(R_2' R_1)/\mathrm{tr}(R_2' R_2).$$

Similarly,

$$k_1 = \mathrm{tr}(R_1' R_2)/\mathrm{tr}(R_1' R_1).$$

To compute matrix $W^{(a,b)}$, the algorithm needs to compare all pairs of local patterns. When comparing local pattern $R_{X^{(a)}(i,\cdot)}$ and $R_{X^{(b)}(j,\cdot)}$, the algorithm assumes $X^{(a)}(i,\cdot)$ matches $X^{(b)}(j,\cdot)$. However, the algorithm does not know how $X^{(a)}(i,\cdot)$'s k neighbors match $X^{(b)}(j,\cdot)$'s k neighbors. To find the best possible match, it considers all $k!$ possible permutations, which is tractable since k is always small.

$R_{X^{(a)}(i,\cdot)}$ and $R_{X^{(b)}(j,\cdot)}$ are from different manifolds, so their sizes could be quite different. The previous theorem shows how to find the best re-scaler to enlarge or shrink one of them to match the other. Showing that $dist(R_{X^{(a)}(i,\cdot)}, R_{X^{(b)}(j,\cdot)})$ considers all the possible matches between two local patterns and returns the distance computed from the best possible match is straightforward.

5.4 Application Examples

5.4.1 Protein Alignment

One simple application of alignment is aligning the three-dimensional structures of proteins. This example shows how alignment can identify the corresponding parts of datasets.

Protein three-dimensional structure reconstruction is an important step in Nuclear Magnetic Resonance (NMR) protein structure determination. Basically, it finds a map from distances to coordinates. A protein three-dimensional structure is a chain of amino acids. Let n be the number of amino acids in a given protein and $C(1,\cdot), \cdots, C(n,\cdot)$ be the coordinate vectors for the amino acids, where $C(i,\cdot) = (C(i,1), C(i,2), C(i,3))$ and $Ci,1, Ci,2$, and $Ci,3$ are the x, y, z coordinates of amino acid i (in biology, one usually uses atoms but not amino acids as the basic elements in determining protein structure. Since the number of atoms is huge, for simplicity, we use amino acids as the basic elements). Then the distance

$d(i,j)$ between amino acids i and j can be defined as $d(i,j) = \|C(i,\cdot) - C(j,\cdot)\|$. Define $A = \{d(i,j) \mid i,j = 1,\cdots,n\}$, and $C = \{C(i,\cdot) \mid i = 1,\cdots,n\}$. It is easy to see that if C is given, then we can immediately compute A. However, if A is given, it is non-trivial to compute C. The latter problem is called protein structure reconstruction. In fact, the problem is even more tricky, since only the distances between neighbors are reliable, and A is an incomplete distance matrix. The problem has been proved to be NP-complete for general sparse distance matrices [14]. In the real world, other techniques such as angle constraints and human experience are used together with the partial distance matrix to determine protein structures. With the information available to us, NMR techniques might find multiple estimations (models), since more than one configuration can be consistent with the distance matrix and the constraints. Thus, the result is an ensemble of models, rather than a single structure. Most usually, the ensemble of structures, with perhaps 10 to 50 members, all of which fit the NMR data and retain good stereochemistry, is deposited with the Protein Data Bank (PDB) [15]. Models related to the same protein should be similar and comparisons between the models in this ensemble provides some information about how well the protein conformation was determined by NMR. In this application, we study a Glutaredoxin protein PDB-1G7O (this protein has 215 amino acids in total), whose 3D structure has 21 models. We pick up Model 1, Model 21, and Model 10 for test. These models are related to the same protein, so it makes sense to treat them as manifolds to display our techniques. We denote the 3 data matrices $X^{(1)}$, $X^{(2)}$, and $X^{(3)}$, all 215×3 matrices. To evaluate how manifold alignment can re-scale manifolds, we multiply two of the datasets by a constant, $X^{(1)} = 4X^{(1)}$ and $X^{(3)} = 2X^{(3)}$. The comparison of $X^{(1)}$ and $X^{(2)}$ (row vectors of $X^{(1)}$ and $X^{(2)}$ represent points in the three-dimensional space) is shown in Figure 5.5(a). The comparison of all three manifolds are shown in Figure 5.6(a). In biology, such chains are called protein backbones. These pictures show that the rescaled protein represented by $X^{(1)}$ is larger than that of $X^{(3)}$, which is larger than that of $X^{(2)}$. The orientations of these proteins are also different. To simulate pairwise correspondences information, we uniformly selected a fourth of the amino acids as correspondence resulting in three 54×3 matrices. We compare the results of five alignment approaches on these datasets.

Procrustes Manifold Alignment

One of the simplest alignment algorithms is Procrustes alignment [10]. Since such models are already low dimensional (3D) embeddings of the distance matrices, we skip Step 1 and 2 in Procrustes alignment algorithm, which are normally used to get an initial low dimension embedding of the datasets. We run the algorithm from Step 3, which attempts to find a rotation matrix that best aligns two datasets $X^{(1)}$ and $X^{(2)}$. Procrustes alignment removes the translational, rotational, and scaling components so that the optimal alignment between the instances in correspondence is achieved. The algorithm identifies the re-scale factor k as 4.2971, and the rotation matrix Q as

$$Q = \begin{pmatrix} 0.56151 & -0.53218 & 0.63363 \\ 0.65793 & 0.75154 & 0.048172 \\ -0.50183 & 0.38983 & 0.77214 \end{pmatrix}.$$

$Y^{(2)}$, the new representation of $X^{(2)}$, is computed as $Y^{(2)} = kX^{(2)}Q$. We plot $Y^{(2)}$ and $X^{(1)}$ in the same graph (Figure 5.5(B)). The plot shows that after the second protein is rotated and rescaled to be similar in size to the first protein, the two proteins are aligned well.

Semi-supervised Manifold Alignment

Next we compare the result of nonlinear alignment, also called semi-supervised alignment [9], using the same data and correspondence. The alignment result is shown in Figure 5.5(c). From the figure, we can see that semi-supervised alignment can map data instances in correspondence to similar locations in the new space, but the instances outside of the correspondence are not aligned well.

Manifold Projections

Next we show the results for linear alignment, also called manifold projections. The three-dimensional (Figure 5.5(c)), two-dimensional (Figure 5.5(d)) and one-dimensional (Figure 5.5(e)) alignment results are shown in Figure 5.5. These figures clearly show that the alignment of two different manifolds is achieved by projecting the data (represented by the original features) onto a new space using our carefully generated mapping functions. Compared to the three-dimensional alignment result of Procrustes alignment, three-dimensional alignment from manifold projection changes the topologies of both manifolds to make them match. Recall that Procrustes alignment does not change the shapes of the given manifolds. The real mapping functions $F^{(1)}$ and $F^{(2)}$ to compute the alignment are

$$F^{(1)} = \begin{pmatrix} -0.1589 & -0.0181 & -0.2178 \\ 0.1471 & 0.0398 & -0.1073 \\ 0.0398 & -0.2368 & -0.0126 \end{pmatrix}, \quad F^{(2)} = \begin{pmatrix} -0.6555 & -0.7379 & -0.3007 \\ 0.0329 & 0.0011 & -0.8933 \\ 0.7216 & -0.6305 & 0.2289 \end{pmatrix}.$$

Manifold Alignment without Correspondence

We also show the unsupervised manifold alignment approach assuming no given pairwise correspondence information. We plot three-dimensional (Figure 5.5(g)), two-dimensional (Figure 5.5h) and one-dimensional (Figure 5.5(i)) alignment results in Figure 5.5. These figures show that alignment can still be achieved using local geometry matching algorithm when no pairwise correspondence information is given.

Multiple Manifold Alignment

Finally, we show the algorithm with all three datasets (using feature-level alignment, $c = 3$). The alignment results are shown in Figure 5.6. From these figures, we can see that all three manifolds are projected to one space, where alignment is achieved. The mapping functions $F^{(1)}$, $F^{(2)}$, and $F^{(3)}$ to compute alignment are as follows:

$$F^{(1)} = \begin{pmatrix} -0.0518 & 0.2133 & 0.0810 \\ -0.2098 & 0.0816 & 0.0046 \\ -0.0073 & -0.0175 & 0.2093 \end{pmatrix}, \quad F^{(2)} = \begin{pmatrix} 0.3808 & 0.2649 & 0.6860 \\ -0.7349 & 0.7547 & 0.2871 \\ -0.2862 & -0.3352 & 0.4509 \end{pmatrix},$$

$$F^{(3)} = \begin{pmatrix} 0.1733 & 0.2354 & -0.0043 \\ -0.3785 & 0.3301 & -0.0787 \\ -0.1136 & 0.1763 & 0.4325 \end{pmatrix}.$$

5.4.2 Parallel Corpora

Another simple example of manifold alignment is in aligning parallel corpora for cross-lingual document retrieval. The data we use in this example is a collection of the proceedings of the European Parliament [11], dating from 04/1996 to 10/2006. The corpus includes versions in 11 European languages: French, Italian, Spanish, Portuguese, English, Dutch,

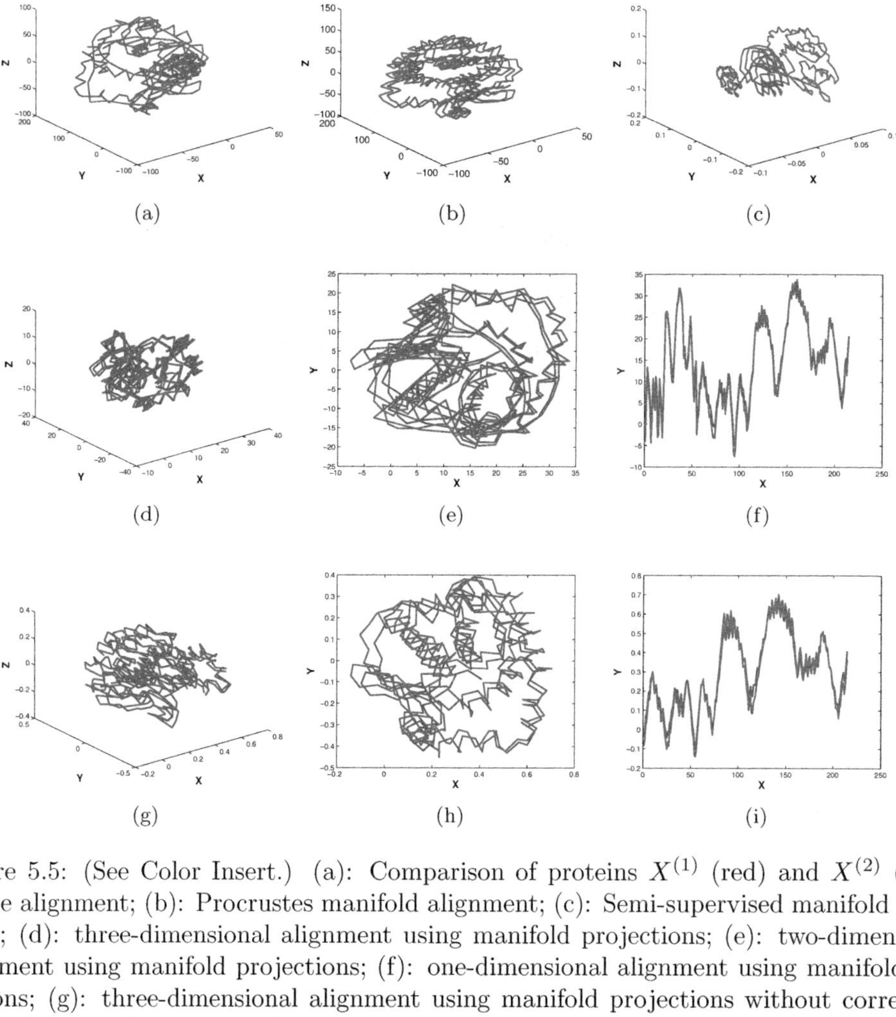

Figure 5.5: (See Color Insert.) (a): Comparison of proteins $X^{(1)}$ (red) and $X^{(2)}$ (blue) before alignment; (b): Procrustes manifold alignment; (c): Semi-supervised manifold alignment; (d): three-dimensional alignment using manifold projections; (e): two-dimensional alignment using manifold projections; (f): one-dimensional alignment using manifold projections; (g): three-dimensional alignment using manifold projections without correspondence; (h): two-dimensional alignment using manifold projections without correspondence; (i): one-dimensional alignment using manifold projections without correspondence.

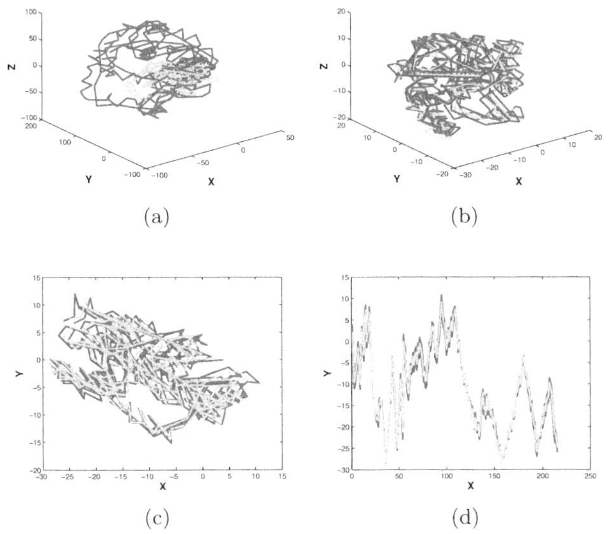

Figure 5.6: (See Color Insert.) (a): Comparison of the proteins $X^{(1)}$ (red), $X^{(2)}$ (blue), and $X^{(3)}$ (green) before alignment; (b): three-dimensional alignment using multiple manifold alignment; (c): two-dimensional alignment using multiple manifold alignment; (d): one-dimensional alignment using multiple manifold alignment.

German, Danish, Swedish, Greek and Finnish. Altogether, the corpus comprises about 30 million words for each language. Assuming that similar documents have similar word usage within each language, we can generate eleven graphs, one for each language, each of which reflects the semantic similarity of the documents written in that language.

The data for these experiments came from the English–Italian parallel corpora, each of which has more than 36,000,000 words. The dataset has many files, and each file contains the utterances of one speaker in turn. We treat an utterance as a document. We first extracted English–Italian document pairs where both documents have at least 100 words. This resulted in 59,708 document pairs. We then represented each English document with the most commonly used 4,000 English words, and each Italian document with the most commonly used 4,000 Italian words. The documents are represented as bags of words, and no tag information is included. 10,000 resulting document pairs are used for training and the remaining 49,708 document pairs are held for testing.

We first show our algorithmic framework using this dataset. In this application, the only parameter we need to set is $d = 200$, i.e., we map two manifolds to the same 200-dimensional space. The other parameters directly come with the input datasets $X^{(1)}$ (for English) and $X^{(2)}$ (for Italian): $p_1 = p_2 = 4000$; $n_1 = n_2 = 10,000$; $c = 2$; $W^{(1)}$ and $W^{(2)}$ are constructed using heat kernels, where $\delta = 1$; $W^{(1,2)}$ is given by the training correspondence information. Since the number of documents is huge, we only do feature-level alignment, which results in mapping functions $F^{(1)}$ (for English) and $F^{(2)}$ (for Italian). These two mapping functions map documents from the original English language/Italian language spaces to the new latent 200-dimensional space. The procedure for the experiment is as follows: for each given English document, we retrieve its top k most similar Italian documents in the new latent space. The probability that the true match is among the top k documents is used to show the goodness of the method. The results are summarized

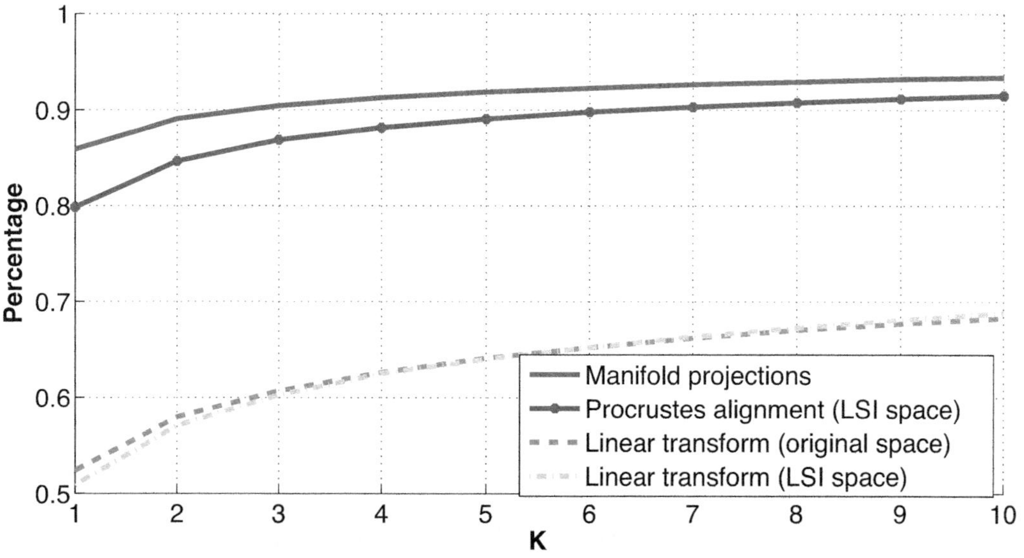

Figure 5.7: EU parallel corpus alignment example.

in Figure 5.7. If we retrieve the most relevant Italian document, then the true match has an 86% probability of being retrieved. If we retrieve 10, this probability jumps to 90%. Different from most approaches in cross-lingual knowledge transfer, we are not using any method from informational retrieval area to tune our framework to this task. For the purpose of comparison, we also used a linear transformation $\mathcal{F}$ to directly align two corpora, where $X^{(1)}\mathcal{F}$ is used to approximate $X^{(2)}$. This is a regular least square problem and the solution is given by $\mathcal{F} = (X^{(1)})^{+}X^{(2)}$, which is a $4,000 \times 4,000$ matrix for our case. The result of this approach is roughly 35% worse than the manifold alignment approach. The true match has a 52% probability of being the first retrieved document. We also applied LSI [16] to preprocess the data and mapped each document to a 200-dimensional LSI space. Procrustes alignment and Linear transform were then applied to align the corpora in these 200-dimensional spaces. The result of Procrustes alignment (Figure 5.7) is roughly 6% worse than manifold projections. Performance of linear transform in LSI space is almost the same as the linear transform result in the original space. There are two reasons why manifold alignment approaches perform much better than the regular linear transform approaches: (1) Manifold alignment approach preserves the topologies of the given manifolds in the computation of alignment. This lowers the chance of getting into "overfitting" problems. (2) Manifold alignment maps the data to a lower dimensional space, getting rid of the information that does not model the common underlying structure of the given manifolds. In manifold projections, each column of $F^{(1)}$ is a $4,000 \times 1$ vector. Each entry on this vector corresponds to a word. To illustrate how the alignment is achieved using our approach, we show five selected corresponding columns of $F^{(1)}$ and $F^{(2)}$ in Table 5.1 and Table 5.2. From these tables, we can see that our approach can automatically map the words with similar meanings from different language spaces to similar locations in the new space.

5.4.3 Aligning Topic Models

Next, we applied the diffusion wavelet-based multiscale alignment algorithm to align corpora represented in different topic spaces. We show that the alignment is useful for finding topics shared by the different topic extraction methods, which suggests that alignment may be

Table 5.1: Five selected mapping functions for English corpus.

	Top 10 Terms
1	ahern tuberculosis eta watts dublin wogau october september yielded structural
2	lakes vienna a4 wednesday chirac lebanon fischler ahern vaccines keys
3	scotland oostlander london tuberculosis finns chirac vaccines finland lisbon prosper
4	hiv jarzembowski tuberculosis mergers virus adjourned march chirac merger parents
5	corruption jarzembowski wednesday mayor parents thursday rio oostlander ruijten vienna

Table 5.2: Five selected mapping functions for Italian corpus.

	Top 10 Terms
1	ahern tubercolosi eta watts dublino ottobre settembre wogau carbonica dicembre
2	laghi vienna mercoledi a4 chirac ahern vaccini libano fischler svedese
3	tubercolosi scozia oostlander londra finlandesi finlandia chirac lisbona vaccini svezia
4	hiv jarzembowski fusioni tubercolosi marzo chirac latina genitori vizioso venerdi
5	corruzione mercoledi jarzembowski statistici sindaco rio oostlander limitiamo concentrati vienna

useful for integrating multiple topic spaces. Since we align the topic spaces at multiple levels, the alignment results are also useful for exploring the hierarchical topic structure of the data.

Given two collections, $X^{(1)}$ (a $n_1 \times p_1$ matrix) and $X^{(2)}$ (a $n_2 \times p_2$ matrix), where p_i is the size of the vocabulary set and n_i is the number of the documents in collection $X^{(i)}$, assume the topics learned from the two collections are given by S_1 and S_2, where S_i is a $p_i \times r_i$ matrix and r_i is the number of the topics in $X^{(i)}$. Then the representations of $X^{(i)}$ in the topic space is $X^{(i)}S_i$. Following our main algorithm, $X^{(1)}S_1$ and $X^{(2)}S_2$ can be aligned in the latent space at level k by using mapping functions $F_k^{(1)}$ and $F_k^{(2)}$. The representations of $X^{(1)}$ and $X^{(2)}$ after alignment become $X^{(1)}S_1F_k^{(1)}$ and $X^{(2)}S_2F_k^{(2)}$. The document contents ($X^{(1)}$ and $X^{(2)}$) are not changed. The only thing that has been changed is S_i, the topic matrix. Recall that the columns of S_i are topics of $X^{(i)}$. The alignment algorithm changes S_1 to $S_1F_k^{(1)}$ and S_2 to $S_2F_k^{(2)}$. The columns of $S_1F_k^{(1)}$ and $S_2F_k^{(2)}$ are still of length p_i. Such columns are in fact the new "aligned" topics.

In this application, we used the NIPS (1-12) full paper dataset, which includes 1,740 papers and 2,301,375 tokens in total. We first represented this dataset using two different topic spaces: LSI space [16] and LDA space [17]. In other words, $X^{(1)} = X^{(2)}$, but $S_1 \neq S_2$ for this set. The reasons for aligning these two datasets is that while they define different features, they are constructed from the same data, and hence admit a correspondence under which the resulting datasets should be aligned well. Also, LSI and LDA topics can be mapped back to the English words, so the mapping functions are semantically interpretable. This helps us understand how the alignment of two collections is achieved (by aligning their underlying topics). We extracted 400 topics from the dataset with both LDA and LSI models ($r_1 = r_2 = 400$). The top eight words of the first five topics from each model are shown in Figure 5.8a and Figure 5.8b. It is clear that none of those topics are similar across the two sets. We ran the main algorithm ($\mu = \nu = 1$) using 20% uniformly selected documents as correspondences. This identified a three-level hierarchy of mapping functions. The number of basis functions spanning each level was: 800, 91, and 2. These numbers correspond to the structure of the latent space at each scale. At the finest scale, the space is spanned by 800 vectors because the joint manifold is spanned by 400 LSI topics plus 400 LDA topics. At the second level the joint manifold is spanned by 91 vectors, which we now examine more closely. Looking at how the original topics were changed can help us better

Top 8 Terms
generalization function generalize shown performance theory size shepard
hebbian hebb plasticity activity neuronal synaptic anti hippocampal
grid moore methods atkeson steps weighted start interpolation
measure standard data dataset datasets results experiments measures
energy minimum yuille minima shown local university physics

(a) Topic 1-5 (LDA) before alignment.

Top 8 Terms
fish terminals gaps arbor magnetic die insect cone
learning algorithm data model state function models distribution
model cells neurons cell visual figure time neuron
data training set model recognition image models gaussian
state neural network model time networks control system

(b) Topic 1-5 (LSI) before alignment.

Top 8 Terms
road car vehicle autonomous lane driving range unit
processor processors brain ring computation update parallel activation
hopfield epochs learned synapses category modulation initial pulse
brain loop constraints color scene fig conditions transfer
speech capacity peak adaptive device transition type connections

(c) 5 LDA topics at level 2 after alignment.

Top 8 Terms
road autonomous vehicle range navigation driving unit video
processors processor parallel approach connection update brain activation
hopfield pulse firing learned synapses stable states network
brain color visible maps fig loop elements constrained
speech connections capacity charge type matching depth signal

(d) 5 LSI topics at level 2 after alignment.

Top 8 Terms
recurrent direct events pages oscillator user hmm oscillators
false chain protein region mouse human proteins roc

(e) 2 LDA topics at level 3 after alignment.

Top 8 Terms
recurrent belief hmm filter user head obs routing
chain mouse region human receptor domains proteins heavy

(f) 2 LSI topics at level 3 after alignment.

Figure 5.8: The eight most probable terms in corresponding pairs of LSI and LDA topics before alignment and at two different scales after alignment.

understand the alignment algorithm. In Figures 5.8c and 5.8d, we show five corresponding topics (corresponding columns of $S_1\alpha_2$ and $S_2\beta_2$) at the second level. From these figures, we can see that the new topics in correspondence are very similar to each other across the datasets, and interestingly the new aligned topics are semantically meaningful — they represent some areas in either machine learning or neuroscience. At the third level, there are only two aligned topics (Figure 5.8e and 5.8f). Clearly, one of them is about machine learning and another is about neuroscience, which are the most abstract topics of the papers submitted to the NIPS conference. From these results, we can see that our algorithm can automatically align the given data sets at different scales following the intrinsic structure of the datasets. Also, the multiscale alignment algorithm was useful for finding the common topics shared by the given collections, and thus it is useful for finding more robust topic spaces.

5.5 Summary

Manifold alignment is useful in applications where the utility of a dataset depends only on the relative geodesic distances between its instances, which lie on some manifold. In these cases, embedding the instances in a space of the same dimensionality as the original manifold while preserving the geodesic similarity maintains the utility of the dataset. Alignment of multiple such datasets allows for simple a simple framework for transfer learning between the datasets.

The fundamental idea of manifold alignment is to view all datasets of interest as lying on the same manifold. To capture this idea mathematically, the alignment algorithm concatenates the graph Laplacians of each dataset, forming a joint Laplacian. A within-dataset similarity function gives all of the edge weights of this joint Laplacian between the instances within each dataset, and correspondence information fills in the edge weights between the instances in separate datasets. The manifold alignment algorithm then embeds this joint Laplacian in a new latent space.

A corollary of this algorithm is that any embedding technique that depends on the similarities (or distances) between data instances can also find a unifying representation of disparate datasets. To perform this extension, the embedding algorithm must use both the regular similarities within each datasets and must treat correspondence information as an additional set of similarities for instances from different datasets, thus viewing multiple datasets as all belonging to one joint dataset. Running the embedding algorithm on this joint dataset results in a unified set of features for the initially disparate datasets.

In practice, the difficulties of manifold alignment are identifying whether the datasets are actually sampled from a single underlying manifold, defining a similarity function that captures the appropriate structures of the datasets, inferring any reliable correspondence information, and finding the true dimensionality of this underlying manifold. Nevertheless, once an appropriate representation and an effective similarity metric are available, manifold alignment is optimal with respect to its loss function and efficient, requiring only the order of complexity of an eigenvalue decomposition.

5.6 Bibliographical and Historical Remarks

The problem of alignment occurs in a variety of fields. Often the alignment methods used in these fields are specialized for particular applications. Some notable field-specific problems are image alignment, protein sequence alignment, and protein structure alignment. Researchers also study the more general problem of alignment under the name informa-

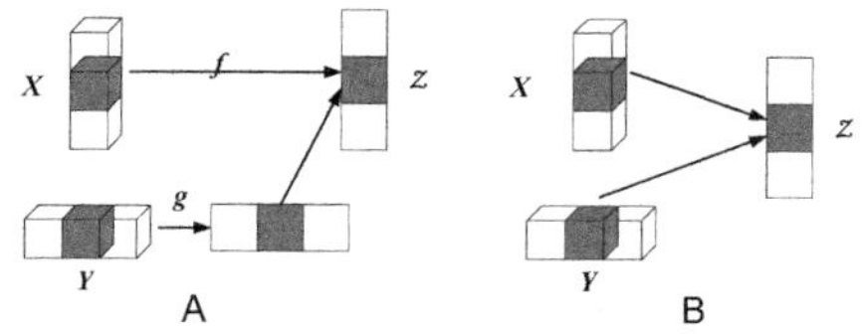

Figure 5.9: Two types of manifold alignment (this figure only shows two manifolds, but the same idea also applies to multiple manifold alignment). X and Y are both sampled from the manifold $\mathcal{Z}$, which the latent space estimates. The red regions represent the subsets that are in correspondence. f and g are functions to compute lower dimensional embedding of X and Y. Type **A** is two-step alignment, which includes diffusion map-based alignment and procrustes alignment; Type **B** is one-step alignment, which includes semi-supervised alignment, manifold projections, and semi-definite alignment.

tion fusion or data fusion. Canonical correlation analysis [18], which has many of its own extensions, is a well-known method for alignment from the statistics community.

Manifold alignment is essentially a graph-based algorithm, but there is also a vast literature on graph-based methods for alignment that are unrelated to manifold learning. The graph theoretic formulation of alignment is typically called graph matching, graph isomorphism, or approximate graph isomorphism.

This section focuses on the smaller but still substantial body of literature on methods for manifold alignment. There are two general types of manifold alignment algorithm. The first type (illustrated in Figure 5.9(A)) includes diffusion map-based alignment [19] and Procrustes alignment [10]. These approaches first map the original datasets to low dimensional spaces reflecting their intrinsic geometries using a standard manifold learning algorithm for dimensionality reduction (linear like LPP [20] or nonlinear like Laplacian eigenmaps [4]). After this initial embedding, the algorithms rotate or scale one of the embedded datatsets to achieve alignment with the other dataset. In this type of alignment, the computation of the initial embedding is unrelated to the actual alignment, so the algorithms do not guarantee that corresponding instances will be close to one another in the final alignment. Even if the second step includes some consideration of correspondence information, the embeddings are independent of this new constraint, so they may not be suited for optimal alignment of corresponding instances.

The second type of manifold alignment algorithm (illustrated in Figure 5.9(B)) includes semi-supervised alignment [9], manifold projections [21] and semi-definite alignment [22]. Semi-supervised alignment first creates a joint manifold representing the union of the given manifolds, then maps that joint manifold to a lower dimensional latent space preserving local geometry of each manifold, and matching instances in correspondence. Semi-supervised alignment is based on eigenvalue decomposition. Semi-definite alignment solves a similar problem using a semi-definite programming framework. Manifold projections is a linear approximation of semi-supervised alignment that directly builds connections between features rather than instances and can naturally handle new test instances. The manifold alignment algorithm discussed in this chapter is a one-step approach.

5.7 Acknowledgments

This research is funded in part by the National Science Foundation under Grant Nos. NSF CCF-1025120, IIS-0534999, and IIS-0803288.

Bibliography

[1] S. Mahadevan. *Representation Discovery Using Harmonic Analysis.* Morgan and Claypool Publishers, 2008.

[2] R. Coifman, S. Lafon, A. Lee, M. Maggioni, B. Nadler, F. Warner, and S. Zucker. Geometric diffusions as a tool for harmonic analysis and structure definition of data: Diffusion maps. *Proceedings of the National Academy of Sciences*, 102(21):7426–7431, 2005.

[3] J. Tenenbaum, V. de Silva, and J. Langford. A global geometric framework for nonlinear dimensionality reduction. *Science*, 290(5500):2319–2323, 2000.

[4] M. Belkin and P. Niyogi. Laplacian eigenmaps for dimensionality reduction and data representation. *Neural Computation*, 6(15):1373–1396, 2003.

[5] S. Roweis and L. Saul. Nonlinear dimensionality reduction by local linear embedding. *Science*, 290(5500):2323–2326, 2000.

[6] X. He and P. Niyogi. Locality preserving projections. In *Proceedings of the Advances in Neural Information Processing Systems*, 2003.

[7] K. Weinberger and L. Saul. Unsupervised learning of image manifolds by semidefinite programming. In *IEEE Computer Society Conference on Computer Vision and Pattern Recognition*, 2004.

[8] T. Jolliffe. *Principal Components Analysis.* Springer-Verlag, 1986.

[9] J. Ham, D. Lee, and L. Saul. Semisupervised alignment of manifolds. In *Proceedings of the International Workshop on Artificial Intelligence and Statistics*, 2005.

[10] C. Wang and S. Mahadevan. Manifold alignment using procrustes analysis. In *Proceedings of the 25th International Conference on Machine Learning*, 2008.

[11] P. Koehn. Europarl: A parallel corpus for statistical machine translation. In *MT Summit*, 2005.

[12] C. Bishop. *Pattern Recognition and Machine Learning.* Springer, 2007.

[13] R. Coifman and M. Maggioni. Diffusion wavelets. *Applied and Computational Harmonic Analysis*, 21(1):53–94, 2006.

[14] L. Hogben. *Handbook of linear algebra.* Chapman/Hall CRC Press, 2006.

[15] H. M. Berman, J. Westbrook, Z. Feng, G. Gillilandand, T. N. Bhat, H. Weissig, I. N. Shindyalov, and P. E. Bourne. The protein data bank. *Nucleic Acids Research*, 28(1):235–242, 2000.

[16] S. Deerwester, S. T. Dumais, G. W. Furnas, T. K. Landauer, and R. Harshman. Indexing by latent semantic analysis. *Journal of the American Society for Information Science*, 41(6):391–407, 1990.

[17] D. Blei, A. Ng, and M. Jordan. Latent dirichlet allocation. *Journal of Machine Learning Research*, 3(4/5):993–1022, 2003.

[18] H. Hotelling. Relations between two sets of variates. *Biometrika*, 28(3/4):321–377, 1936.

[19] S. Lafon, Y. Keller, and R. Coifman. Data fusion and multicue data matching by diffusion maps. *IEEE Transactions on Pattern Analysis and Machine Intelligence*, 28(11):1784–1797, 2006.

[20] X. He and P. Niyogi. Locality preserving projections. In *Proceedings of the Advances in Neural Information Processing Systems*, 2003.

[21] C. Wang and S. Mahadevan. Manifold alignment without correspondence. In *Proceedings of the 21st International Joint Conference on Artificial Intelligence*, 2009.

[22] L. Xiong, F. Wang, and C. Zhang. Semi-definite manifold alignment. In *Proceedings of the 18th European Conference on Machine Learning*, 2007.

Chapter 6

Large-Scale Manifold Learning

Ameet Talwalkar, Sanjiv Kumar, Mehryar Mohri, Henry Rowley

6.1 Introduction

The problem of dimensionality reduction arises in many computer vision applications, where it is natural to represent images as vectors in a high-dimensional space. Manifold learning techniques extract low-dimensional structure from high-dimensional data in an unsupervised manner. These techniques typically try to unfold the underlying manifold so that some quantity, e.g., pairwise geodesic distances, is maintained invariant in the new space. This makes certain applications such as K-means clustering more effective in the transformed space.

In contrast to linear dimensionality reduction techniques such as Principal Component Analysis (PCA), manifold learning methods provide more powerful non-linear dimensionality reduction by preserving the local structure of the input data. Instead of assuming global linearity, these methods typically make a weaker local-linearity assumption, i.e., for nearby points in high-dimensional input space, l_2 distance is assumed to be a good measure of geodesic distance, or distance along the manifold. Good sampling of the underlying manifold is essential for this assumption to hold. In fact, many manifold learning techniques provide guarantees that the accuracy of the recovered manifold increases as the number of data samples increases. In the limit of infinite samples, one can recover the true underlying manifold for certain classes of manifolds [88, 5, 11]. However, there is a trade-off between improved sampling of the manifold and the computational cost of manifold learning algorithms. In this chapter, we address the computational challenges involved in learning manifolds given millions of face images extracted from the Web.

Several manifold learning techniques have been proposed, e.g., Semidefinite Embedding (SDE) [35], Isomap [34], Laplacian Eigenmaps [4], and Local Linear Embedding (LLE) [31]. SDE aims to preserve distances and angles between all neighboring points. It is formulated as an instance of semidefinite programming, and is thus prohibitively expensive for large-scale problems. Isomap constructs a dense matrix of approximate geodesic distances between *all* pairs of inputs, and aims to find a low dimensional space that best preserves these distances. Other algorithms, e.g., Laplacian Eigenmaps and LLE, focus only on preserving local neighborhood relationships in the input space. They generate low-dimensional representations via manipulation of the graph Laplacian or other sparse matrices related to the graph Laplacian [6]. In this chapter, we focus mainly on Isomap and Laplacian Eigenmaps, as both methods have good theoretical properties and the differences in their approaches allow us to make interesting comparisons between dense and sparse methods.

All of the manifold learning methods described above can be viewed as specific instances of Kernel PCA [19]. These kernel-based algorithms require SVD of matrices of size $n \times n$, where n is the number of samples. This generally takes $O(n^3)$ time. When only a few singular values and singular vectors are required, there exist less computationally intensive techniques such as Jacobi, Arnoldi, Hebbian, and more recent randomized methods [17, 18, 30]. These iterative methods require computation of matrix-vector products at each step and involve multiple passes through the data. When the matrix is sparse, these techniques can be implemented relatively efficiently. However, when dealing with a large, dense matrix, as in the case of Isomap, these products become expensive to compute. Moreover, when working with 18M data points, it is not possible even to store the full 18M$\times$18M matrix ($\sim$1300TB), rendering the iterative methods infeasible. Random sampling techniques provide a powerful alternative for approximate SVD and only operate on a subset of the matrix.

In this chapter, we examine both the Nyström and Column sampling methods (defined in Section 6.3), providing the first direct comparison between their performances on practical applications. The Nyström approximation has been studied in the machine learning community [36] [13]. In parallel, Column sampling techniques have been analyzed in the theoretical Computer Science community [16, 12, 10]. However, prior to initial work in [33, 22], the relationship between these approximations had not been well studied. We provide an extensive analysis of these algorithms, show connections between these approximations, and provide a direct comparison between their performances.

Apart from singular value decomposition, the other main computational hurdle associated with Isomap and Laplacian Eigenmaps is large-scale graph construction and manipulation. These algorithms first need to construct a local neighborhood graph in the input space, which is an $O(n^2)$ problem given n data points. Moreover, Isomap requires shortest paths between every pair of points requiring $O(n^2 \log n)$ computation. Both of these steps become intractable when n is as large as 18M. In this study, we use approximate nearest neighbor methods, and explore random sampling based SVD that requires the computation of shortest paths only for a subset of points. Furthermore, these approximations allow for an efficient distributed implementation of the algorithms.

We now summarize our main contributions. First, we present the largest scale study so far on manifold learning, using 18M data points. To date, the largest manifold learning study involves the analysis of music data using 267K points [29]. In vision, the largest study is limited to less than 10K images [20]. Second, we show connections between two random sampling based singular value decomposition algorithms and provide the first direct comparison of their performances. Finally, we provide a quantitative comparison of Isomap and Laplacian Eigenmaps for large-scale face manifold construction on clustering and classification tasks.

6.2 Background

In this section, we introduce notation (summarized in Table 6.1) and present basic definitions of two of the most common sampling-based techniques for matrix approximation.

Table 6.1: Summary of notation used throughout this chapter.

$\mathbf{T}$	arbitrary matrix in $\mathbb{R}^{a\times b}$
$\mathbf{T}^{(j)}$	jth column vector of $\mathbf{T}$ for $j = 1\dots b$
$\mathbf{T}_{(i)}$	ith row vector of $\mathbf{T}$ for $i = 1\dots a$
$\mathbf{T}^{(i:j)}, \mathbf{T}_{(i:j)}$	ith through jth columns / rows of $\mathbf{T}$
$\mathbf{T}_k$	'best' rank-k approximation to $\mathbf{T}$
$\|\cdot\|_2, \|\cdot\|_F$	Spectral, Frobenius norms of $\mathbf{T}$
$\mathbf{v}$	arbitrary vector in $\mathbb{R}^a$
$\|\cdot\|$	l_2 norm of a vector
$\mathbf{T} = \mathbf{U}_T\mathbf{\Sigma}_T\mathbf{V}_T^\top$	Singular Value Decomposition (SVD) of $\mathbf{T}$
$\mathbf{K}$	SPSD kernel matrix in $\mathbb{R}^{n\times n}$ with $\text{rank}(\mathbf{K}) = r \le n$
$\mathbf{K} = \mathbf{U\Sigma V}^\top$	SVD of $\mathbf{K}$
$\mathbf{K}^+$	Pseudo-inverse of $\mathbf{K}$
$\widetilde{\mathbf{K}}$	approximation to $\mathbf{K}$ derived from $l \ll n$ of its columns

6.2.1 Notation

Let $\mathbf{T} \in \mathbb{R}^{a\times b}$ be an arbitrary matrix. We define $\mathbf{T}^{(j)}$, $j = 1\dots b$, as the jth column vector of $\mathbf{T}$, $\mathbf{T}_{(i)}$, $i = 1\dots a$, as the ith row vector of $\mathbf{T}$, and $\|\cdot\|$ the l_2 norm of a vector. Furthermore, $\mathbf{T}^{(i:j)}$ refers to the ith through jth columns of $\mathbf{T}$ and $\mathbf{T}_{(i:j)}$ refers to the ith through jth rows of $\mathbf{T}$. We denote by $\mathbf{T}_k$ the 'best' rank-k approximation to $\mathbf{T}$, i.e., $\mathbf{T}_k = \text{argmin}_{\mathbf{V}\in\mathbb{R}^{a\times b},\text{rank}(\mathbf{V})=k}\|\mathbf{T}-\mathbf{V}\|_\xi$, where $\xi \in \{2, F\}$ and $\|\cdot\|_2$ denotes the spectral norm and $\|\cdot\|_F$ the Frobenius norm of a matrix. Assuming that $\text{rank}(\mathbf{T}) = r$, we can write the compact Singular Value Decomposition (SVD) of this matrix as $\mathbf{T} = \mathbf{U}_T\mathbf{\Sigma}_T\mathbf{V}_T^\top$ where $\mathbf{\Sigma}_T$ is diagonal and contains the singular values of $\mathbf{T}$ sorted in decreasing order and $\mathbf{U}_T \in \mathbb{R}^{a\times r}$ and $\mathbf{V}_T \in \mathbb{R}^{b\times r}$ have orthogonal columns that contain the left and right singular vectors of $\mathbf{T}$ corresponding to its singular values. We can then describe $\mathbf{T}_k$ in terms of its SVD as $\mathbf{T}_k = \mathbf{U}_{T,k}\mathbf{\Sigma}_{T,k}\mathbf{V}_{T,k}^\top$ where $\mathbf{\Sigma}_{T,k}$ is a diagonal matrix of the top k singular values of $\mathbf{T}$ and $\mathbf{U}_{T,k}$ and $\mathbf{V}_{T,k}$ are the associated left and right singular vectors.

Now let $\mathbf{K} \in \mathbb{R}^{n\times n}$ be a symmetric positive semidefinite (SPSD) kernel or Gram matrix with $\text{rank}(\mathbf{K}) = r \le n$, i.e. a symmetric matrix for which there exists an $\mathbf{X} \in \mathbb{R}^{N\times n}$ such that $\mathbf{K} = \mathbf{X}^\top\mathbf{X}$. We will write the SVD of $\mathbf{K}$ as $\mathbf{K} = \mathbf{U\Sigma U}^\top$, where the columns of $\mathbf{U}$ are orthogonal and $\mathbf{\Sigma} = \text{diag}(\sigma_1, \dots, \sigma_r)$ is diagonal. The pseudo-inverse of $\mathbf{K}$ is defined as $\mathbf{K}^+ = \sum_{t=1}^r \sigma_t^{-1}\mathbf{U}^{(t)}{\mathbf{U}^{(t)}}^\top$, and $\mathbf{K}^+ = \mathbf{K}^{-1}$ when $\mathbf{K}$ is full rank. For $k < r$, $\mathbf{K}_k = \sum_{t=1}^k \sigma_t\mathbf{U}^{(t)}{\mathbf{U}^{(t)}}^\top = \mathbf{U}_k\mathbf{\Sigma}_k\mathbf{U}_k^\top$ is the 'best' rank-k approximation to $\mathbf{K}$, i.e., $\mathbf{K}_k = \text{argmin}_{\mathbf{K}'\in\mathbb{R}^{n\times n},\text{rank}(\mathbf{K}')=k}\|\mathbf{K} - \mathbf{K}'\|_{\xi\in\{2,F\}}$, with $\|\mathbf{K} - \mathbf{K}_k\|_2 = \sigma_{k+1}$ and $\|\mathbf{K} - \mathbf{K}_k\|_F = \sqrt{\sum_{t=k+1}^r \sigma_t^2}$ [17].

We will be focusing on generating an approximation $\widetilde{\mathbf{K}}$ of $\mathbf{K}$ based on a sample of $l \ll n$ of its columns. We assume that we sample columns uniformly without replacement as suggested by [23], though various methods have been proposed to select columns (see Chapter 4 of 22 for more details on various sampling schemes). Let $\mathbf{C}$ denote the $n \times l$ matrix formed by these columns and $\mathbf{W}$ the $l \times l$ matrix consisting of the intersection of these l columns with the corresponding l rows of $\mathbf{K}$. Note that $\mathbf{W}$ is SPSD since $\mathbf{K}$ is SPSD. Without loss of generality, the columns and rows of $\mathbf{K}$ can be rearranged based on this sampling so that $\mathbf{K}$ and $\mathbf{C}$ be written as follows:

$$\mathbf{K} = \begin{bmatrix} \mathbf{W} & \mathbf{K}_{21}^\top \\ \mathbf{K}_{21} & \mathbf{K}_{22} \end{bmatrix} \quad \text{and} \quad \mathbf{C} = \begin{bmatrix} \mathbf{W} \\ \mathbf{K}_{21} \end{bmatrix}. \tag{6.1}$$

The approximation techniques discussed next use the SVD of $\mathbf{W}$ and $\mathbf{C}$ to generate approximations for $\mathbf{K}$.

6.2.2 Nyström Method

The Nyström method was first introduced as a quadrature method for numerical integration, used to approximate eigenfunction solutions [27, 2]. More recently, it was presented in [36] to speed up kernel algorithms and has been used in applications ranging from manifold learning to image segmentation [29, 15, 33]. The Nyström method uses $\mathbf{W}$ and $\mathbf{C}$ from (6.1) to approximate $\mathbf{K}$. Assuming a uniform sampling of the columns, the Nyström method generates a rank-k approximation $\widetilde{\mathbf{K}}$ of $\mathbf{K}$ for $k < n$ defined by:

$$\widetilde{\mathbf{K}}_k^{nys} = \mathbf{C}\mathbf{W}_k^+\mathbf{C}^\top \approx \mathbf{K}, \tag{6.2}$$

where $\mathbf{W}_k$ is the best k-rank approximation of $\mathbf{W}$ with respect to the spectral or Frobenius norm and $\mathbf{W}_k^+$ denotes the pseudo-inverse of $\mathbf{W}_k$. If we write the SVD of $\mathbf{W}$ as $\mathbf{W} = \mathbf{U}_W\mathbf{\Sigma}_W\mathbf{U}_W^\top$, then from (6.2) we can write

$$\begin{aligned} \widetilde{\mathbf{K}}_k^{nys} &= \mathbf{C}\mathbf{U}_{W,k}\mathbf{\Sigma}_{W,k}^+\mathbf{U}_{W,k}^\top\mathbf{C}^\top \\ &= \left(\sqrt{\frac{l}{n}}\mathbf{C}\mathbf{U}_{W,k}\mathbf{\Sigma}_{W,k}^+\right)\left(\frac{n}{l}\mathbf{\Sigma}_{W,k}\right)\left(\sqrt{\frac{l}{n}}\mathbf{C}\mathbf{U}_{W,k}\mathbf{\Sigma}_{W,k}^+\right)^\top, \end{aligned}$$

and hence the Nyström method approximates the top k singular values ($\mathbf{\Sigma}_k$) and singular vectors ($\mathbf{U}_k$) of $\mathbf{K}$ as:

$$\widetilde{\mathbf{\Sigma}}_{nys} = \left(\frac{n}{l}\right)\mathbf{\Sigma}_{W,k} \quad \text{and} \quad \widetilde{\mathbf{U}}_{nys} = \sqrt{\frac{l}{n}}\mathbf{C}\mathbf{U}_{W,k}\mathbf{\Sigma}_{W,k}^+. \tag{6.3}$$

Since the running time complexity of compact SVD on $\mathbf{W}$ is in $O(l^2k)$ and matrix multiplication with $\mathbf{C}$ takes $O(nlk)$, the total complexity of the Nyström approximation computation is in $O(nlk)$.

6.2.3 Column Sampling Method

The Column sampling method was introduced to approximate the SVD of any rectangular matrix [16]. It generates approximations of $\mathbf{K}$ by using the SVD of $\mathbf{C}$.[1] If we write the SVD of $\mathbf{C}$ as $\mathbf{C} = \mathbf{U}_C\mathbf{\Sigma}_C\mathbf{V}_C^\top$ then the Column sampling method approximates the top k singular values ($\mathbf{\Sigma}_k$) and singular vectors ($\mathbf{U}_k$) of $\mathbf{K}$ as:

$$\widetilde{\mathbf{\Sigma}}_{col} = \sqrt{\frac{n}{l}}\mathbf{\Sigma}_{C,k} \quad \text{and} \quad \widetilde{\mathbf{U}}_{col} = \mathbf{U}_C = \mathbf{C}\mathbf{V}_{C,k}\mathbf{\Sigma}_{C,k}^+. \tag{6.4}$$

The runtime of the Column sampling method is dominated by the SVD of $\mathbf{C}$. The algorithm takes $O(nlk)$ time to perform compact SVD on $\mathbf{C}$, but is still more expensive than the Nyström method as the constants for SVD are greater than those for the $O(nlk)$ matrix multiplication step in the Nyström method.

[1]The Nyström method also uses sampled columns of $\mathbf{K}$, but the Column sampling method is named so because it uses direct decomposition of $\mathbf{C}$, while the Nyström method decomposes its submatrix, $\mathbf{W}$.

6.3 Comparison of Sampling Methods

Given that two sampling-based techniques exist to approximate the SVD of SPSD matrices, we pose a natural question: which method should one use to approximate singular values, singular vectors, and low-rank approximations? We first analyze the form of these approximations and then empirically evaluate their performance in Section 6.3.3 on a variety of datasets.

6.3.1 Singular Values and Singular Vectors

As shown in (6.3) and (6.4), the singular values of $\mathbf{K}$ are approximated as the scaled singular values of $\mathbf{W}$ and $\mathbf{C}$, respectively. The scaling terms are quite rudimentary and are primarily meant to *compensate* for the 'small sample size' effect for both approximations. Formally, these scaling terms make the approximations in (6.3) and (6.4) unbiased estimators of the true singular values. The form of singular vectors is more interesting. The Column sampling singular vectors ($\widetilde{\mathbf{U}}_{col}$) are orthonormal since they are the singular vectors of $\mathbf{C}$. In contrast, the Nyström singular vectors ($\widetilde{\mathbf{U}}_{nys}$) are approximated by *extrapolating* the singular vectors of $\mathbf{W}$ as shown in (6.3), and are *not* orthonormal. It is easy to verify that $\widetilde{\mathbf{U}}_{nys}^\top \widetilde{\mathbf{U}}_{nys} \neq \mathbf{I}_l$, where $\mathbf{I}_l$ is the identity matrix of size l. As we show in Section 6.3.3, this adversely affects the accuracy of singular vector approximation from the Nyström method.

It is possible to orthonormalize the Nyström singular vectors by using QR decomposition. Since $\widetilde{\mathbf{U}}_{nys} \propto \mathbf{C}\mathbf{U}_W \mathbf{\Sigma}_W^+$, where $\mathbf{U}_W$ is orthogonal and $\mathbf{\Sigma}_W$ is diagonal, this simply implies that QR decomposition creates an orthonormal span of $\mathbf{C}$ rotated by $\mathbf{U}_W$. However, the complexity of QR decomposition of $\widetilde{\mathbf{U}}_{nys}$ is the same as that of the SVD of $\mathbf{C}$. Thus, the computational cost of orthogonalizing $\widetilde{\mathbf{U}}_{nys}$ would nullify the computational benefit of the Nyström method over Column sampling.

6.3.2 Low-Rank Approximation

Several studies have empirically shown that the accuracy of low-rank approximations of kernel matrices is tied to the performance of kernel-based learning algorithms [36, 33, 37]. Furthermore, the effect of an approximation in the kernel matrix on the *hypothesis* generated by several widely used kernel-based learning algorithms has been theoretically analyzed [7]. Hence, accurate low-rank approximations are of great practical interest in machine learning. As discussed in Section 6.2.1, the optimal $\mathbf{K}_k$ is given by,

$$\mathbf{K}_k = \mathbf{U}_k \mathbf{\Sigma}_k \mathbf{U}_k^\top = \mathbf{U}_k \mathbf{U}_k^\top \mathbf{K} = \mathbf{K}\mathbf{U}_k \mathbf{U}_k^\top \tag{6.5}$$

where the columns of $\mathbf{U}_k$ are the k singular vectors of $\mathbf{K}$ corresponding to the top k singular values of $\mathbf{K}$. We refer to $\mathbf{U}_k \mathbf{\Sigma}_k \mathbf{U}_k^\top$ as *Spectral Reconstruction*, since it uses both the singular values and vectors of $\mathbf{K}$, and $\mathbf{U}_k \mathbf{U}_k^\top \mathbf{K}$ as *Matrix Projection*, since it uses only singular vectors to compute the projection of $\mathbf{K}$ onto the space spanned by vectors $\mathbf{U}_k$. These two low-rank approximations are equal only if $\mathbf{\Sigma}_k$ and $\mathbf{U}_k$ contain the true singular values and singular vectors of $\mathbf{K}$. Since this is not the case for approximate methods such as Nyström and Column sampling these two measures generally give different errors. From an application point of view, matrix projection approximations, although they can be quite accurate, are not necessarily symmetric and require storage of and multiplication with $\mathbf{K}$. Hence, although matrix projection is often analyzed theoretically, for large-scale problems, the storage and computational requirements may be inefficient or even infeasible. As such, in the context of large-scale manifold learning, we focus on spectral reconstructions in this chapter (for further discussion on matrix projection, see 22).

Using (6.3), the Nyström spectral reconstruction is:

$$\widetilde{\mathbf{K}}_k^{nys} = \widetilde{\mathbf{U}}_{nys,k}\widetilde{\mathbf{\Sigma}}_{nys,k}\widetilde{\mathbf{U}}_{nys,k}^\top = \mathbf{C}\mathbf{W}_k^+\mathbf{C}^\top. \tag{6.6}$$

When $k = l$, this approximation perfectly reconstructs three blocks of $\mathbf{K}$, and $\mathbf{K}_{22}$ is approximated by the Schur Complement of $\mathbf{W}$ in $\mathbf{K}$:

$$\widetilde{\mathbf{K}}_l^{nys} = \mathbf{C}\mathbf{W}^+\mathbf{C}^\top = \begin{bmatrix} \mathbf{W} & \mathbf{K}_{21}^\top \\ \mathbf{K}_{21} & \mathbf{K}_{21}\mathbf{W}^+\mathbf{K}_{21} \end{bmatrix}. \tag{6.7}$$

The Column sampling spectral reconstruction has a similar form as (6.6):

$$\widetilde{\mathbf{K}}_k^{col} = \widetilde{\mathbf{U}}_{col,k}\widetilde{\mathbf{\Sigma}}_{col,k}\widetilde{\mathbf{U}}_{col,k}^\top = \sqrt{n/l}\,\mathbf{C}\big((\mathbf{C}^\top\mathbf{C})_k^{\frac{1}{2}}\big)^+\mathbf{C}^\top. \tag{6.8}$$

Note that a scaling term appears in the Column sampling reconstruction. To analyze the two approximations, we consider an alternative characterization using the fact that $\mathbf{K} = \mathbf{X}^\top\mathbf{X}$ for some $\mathbf{X} \in \mathbb{R}^{N\times n}$. Similar to [13], we define a zero-one sampling matrix, $\mathbf{S} \in \mathbb{R}^{n\times l}$, that selects l columns from $\mathbf{K}$, i.e., $\mathbf{C} = \mathbf{KS}$. Each column of $\mathbf{S}$ has exactly one non-zero entry per column. Further, $\mathbf{W} = \mathbf{S}^\top\mathbf{KS} = (\mathbf{XS})^\top\mathbf{XS} = \mathbf{X}'^\top\mathbf{X}'$, where $\mathbf{X}' \in \mathbb{R}^{N\times l}$ contains l sampled columns of $\mathbf{X}$ and $\mathbf{X}' = \mathbf{U}_{X'}\mathbf{\Sigma}_{X'}\mathbf{V}_{X'}^\top$ is the SVD of $\mathbf{X}'$. We use these definitions to present Theorems 9 and 10.

Theorem 9 *Column sampling and Nyström spectral reconstructions of rank k are of the form*

$$\mathbf{X}^\top\mathbf{U}_{X',k}\mathbf{Z}\mathbf{U}_{X',k}^\top\mathbf{X},$$

where $\mathbf{Z} \in \mathbb{R}^{k\times k}$ *is SPSD. Further, among all approximations of this form, neither the Column sampling nor the Nyström approximation is optimal (in* $\|\cdot\|_F$*).*

Proof 11 *If* $\alpha = \sqrt{n/l}$*, then starting from (6.8) and expressing* $\mathbf{C}$ *and* $\mathbf{W}$ *in terms of* $\mathbf{X}$ *and* $\mathbf{S}$*, we have*

$$\begin{aligned}\widetilde{\mathbf{K}}_k^{col} &= \alpha\mathbf{KS}((\mathbf{S}^\top\mathbf{K}^2\mathbf{S})_k^{1/2})^+\mathbf{S}^\top\mathbf{K}^\top \\ &= \alpha\mathbf{X}^\top\mathbf{X}'\big((\mathbf{V}_{C,k}\mathbf{\Sigma}_{C,k}^2\mathbf{V}_{C,k}^\top)^{1/2}\big)^+\mathbf{X}'^\top\mathbf{X} \\ &= \mathbf{X}^\top\mathbf{U}_{X',k}\mathbf{Z}_{col}\mathbf{U}_{X',k}^\top\mathbf{X},\end{aligned} \tag{6.9}$$

where $\mathbf{Z}_{col} = \alpha\mathbf{\Sigma}_{X'}\mathbf{V}_{X'}^\top\mathbf{V}_{C,k}\mathbf{\Sigma}_{C,k}^+\mathbf{V}_{C,k}^\top\mathbf{V}_{X'}\mathbf{\Sigma}_{X'}$*. Similarly, from (6.6) we have:*

$$\begin{aligned}\widetilde{\mathbf{K}}_k^{nys} &= \mathbf{KS}(\mathbf{S}^\top\mathbf{KS})_k^+\mathbf{S}^\top\mathbf{K}^\top \\ &= \mathbf{X}^\top\mathbf{X}'\big(\mathbf{X}'^\top\mathbf{X}'\big)_k^+\mathbf{X}'^\top\mathbf{X} \\ &= \mathbf{X}^\top\mathbf{U}_{X',k}\mathbf{U}_{X',k}^\top\mathbf{X}.\end{aligned} \tag{6.10}$$

Clearly, $\mathbf{Z}_{nys} = \mathbf{I}_k$*. Next, we analyze the error,* $\mathbf{E}$*, for an arbitrary* $\mathbf{Z}$*, which yields the approximation* $\widetilde{\mathbf{K}}_k^Z$*:*

$$\mathbb{E} = \|\mathbf{K} - \widetilde{\mathbf{K}}_k^Z\|_F^2 = \|\mathbf{X}^\top(\mathbf{I}_N - \mathbf{U}_{X',k}\mathbf{Z}\mathbf{U}_{X',k}^\top)\mathbf{X}\|_F^2. \tag{6.11}$$

Let $\mathbf{X} = \mathbf{U}_X\mathbf{\Sigma}_X\mathbf{V}_X^\top$ *and* $\mathbf{Y} = \mathbf{U}_X^\top\mathbf{U}_{X',k}$*. Then,*

$$\begin{aligned}\mathbf{E} &= \mathit{Trace}\big[\big((\mathbf{I}_N - \mathbf{U}_{X',k}\mathbf{Z}\mathbf{U}_{X',k}^\top)\mathbf{U}_X\mathbf{\Sigma}_X^2\mathbf{U}_X^\top\big)^2\big] \\ &= \mathit{Trace}\big[\big(\mathbf{U}_X\mathbf{\Sigma}_X\mathbf{U}_X^\top(\mathbf{I}_N - \mathbf{U}_{X',k}\mathbf{Z}\mathbf{U}_{X',k}^\top)\mathbf{U}_X\mathbf{\Sigma}_X\mathbf{U}_X^\top\big)^2\big] \\ &= \mathit{Trace}\big[\big(\mathbf{U}_X\mathbf{\Sigma}_X(\mathbf{I}_N - \mathbf{YZY}^\top)\mathbf{\Sigma}_X\mathbf{U}_X^\top\big)^2\big] \\ &= \mathit{Trace}\big[\mathbf{\Sigma}_X(\mathbf{I}_N - \mathbf{YZY}^\top)\mathbf{\Sigma}_X^2(\mathbf{I}_N - \mathbf{YZY}^\top)\mathbf{\Sigma}_X)\big] \\ &= \mathit{Trace}\big[\mathbf{\Sigma}_X^4 - 2\mathbf{\Sigma}_X^2\mathbf{YZY}^\top\mathbf{\Sigma}_X^2 + \mathbf{\Sigma}_X\mathbf{YZY}^\top\mathbf{\Sigma}_X^2\mathbf{YZY}^\top\mathbf{\Sigma}_X)\big].\end{aligned} \tag{6.12}$$

To find $\mathbf{Z}^*$*, the* $\mathbf{Z}$ *that minimizes (6.12), we use the convexity of (6.12) and set:*

$$\partial \mathbf{E}/\partial \mathbf{Z} = -2\mathbf{Y}^\top \mathbf{\Sigma}_X^4 \mathbf{Y} + 2(\mathbf{Y}^\top \mathbf{\Sigma}_X^2 \mathbf{Y})\mathbf{Z}^*(\mathbf{Y}^\top \mathbf{\Sigma}_X^2 \mathbf{Y}) = 0$$

and solve for $\mathbf{Z}^*$*, which gives us:*

$$\mathbf{Z}^* = (\mathbf{Y}^\top \mathbf{\Sigma}_X^2 \mathbf{Y})^+(\mathbf{Y}^\top \mathbf{\Sigma}_X^4 \mathbf{Y})(\mathbf{Y}^\top \mathbf{\Sigma}_X^2 \mathbf{Y})^+.$$

$\mathbf{Z}^* = \mathbf{Z}_{nys} = \mathbf{I}_k$ *if* $\mathbf{Y} = \mathbf{I}_k$*, though* $\mathbf{Z}^*$ *does not in general equal either* $\mathbf{Z}_{col}$ *or* $\mathbf{Z}_{nys}$*, which is clear by comparing the expressions of these three matrices.*[2] *Furthermore, since* $\mathbf{\Sigma}_X^2 = \mathbf{\Sigma}_K$*,* $\mathbf{Z}^*$ *depends on the spectrum of* $\mathbf{K}$*.*

While Theorem 9 shows that the optimal approximation is data dependent and may differ from the Nyström and Column sampling approximations, Theorem 10 presented below reveals that in certain instances the Nyström method is optimal. In contrast, the Column sampling method enjoys no such guarantee.

Theorem 10 *Let* $r = \text{rank}(\mathbf{K}) \leq k \leq l$ *and* $\text{rank}(\mathbf{W}) = r$*. Then, the Nyström approximation is exact for spectral reconstruction. In contrast, Column sampling is exact iff* $\mathbf{W} = \big((l/n)\mathbf{C}^\top\mathbf{C}\big)^{1/2}$*. When this specific condition holds, Column-Sampling trivially reduces to the Nyström method.*

Proof 12 *Since* $\mathbf{K} = \mathbf{X}^\top\mathbf{X}$, $\text{rank}(\mathbf{K}) = \text{rank}(\mathbf{X}) = r$*. Similarly,* $\mathbf{W} = \mathbf{X}'^\top\mathbf{X}'$ *implies* $\text{rank}(\mathbf{X}') = r$*. Thus the columns of* $\mathbf{X}'$ *span the columns of* $\mathbf{X}$ *and* $\mathbf{U}_{X',r}$ *is an orthonormal basis for* $\mathbf{X}$*, i.e.,* $\mathbf{I}_N - \mathbf{U}_{X',r}\mathbf{U}_{X',r}^\top \in \text{Null}(\mathbf{X})$*. Since* $k \geq r$*, from (6.10) we have*

$$\|\mathbf{K} - \widetilde{\mathbf{K}}_k^{nys}\|_F = \|\mathbf{X}^\top(\mathbf{I}_N - \mathbf{U}_{X',r}\mathbf{U}_{X',r}^\top)\mathbf{X}\|_F = 0, \tag{6.13}$$

which proves the first statement of the theorem. To prove the second statement, we note that $\text{rank}(C) = r$*. Thus,* $\mathbf{C} = \mathbf{U}_{C,r}\mathbf{\Sigma}_{C,r}\mathbf{V}_{C,r}^\top$ *and* $(\mathbf{C}^\top\mathbf{C})_k^{1/2} = (\mathbf{C}^\top\mathbf{C})^{1/2} = \mathbf{V}_{C,r}\mathbf{\Sigma}_{C,r}\mathbf{V}_{C,r}^\top$ *since* $k \geq r$*. If* $\mathbf{W} = (1/\alpha)(\mathbf{C}^\top\mathbf{C})^{1/2}$*, then the Column sampling and Nyström approximations are identical and hence exact. Conversely, to exactly reconstruct* $\mathbf{K}$*, Column sampling necessarily reconstructs* $\mathbf{C}$ *exactly. Using* $\mathbf{C}^\top = [\mathbf{W} \quad \mathbf{K}_{21}^\top]$ *in (6.8) we have:*

$$\begin{aligned}
\widetilde{\mathbf{K}}_k^{col} = \mathbf{K} \quad &\Longrightarrow \quad \alpha\mathbf{C}\big((\mathbf{C}^\top\mathbf{C})_k^{\frac{1}{2}}\big)^+\mathbf{W} = \mathbf{C} && (6.14)\\
&\Longrightarrow \quad \alpha\mathbf{U}_{C,r}\mathbf{V}_{C,r}^\top\mathbf{W} = \mathbf{U}_{C,r}\mathbf{\Sigma}_{C,r}\mathbf{V}_{C,r}^\top && (6.15)\\
&\Longrightarrow \quad \alpha\mathbf{V}_{C,r}\mathbf{V}_{C,r}^\top\mathbf{W} = \mathbf{V}_{C,r}\mathbf{\Sigma}_{C,r}\mathbf{V}_{C,r}^\top && (6.16)\\
&\Longrightarrow \quad \mathbf{W} = \frac{1}{\alpha}(\mathbf{C}^\top\mathbf{C})^{1/2}. && (6.17)
\end{aligned}$$

In (6.16) we use $\mathbf{U}_{C,r}^\top\mathbf{U}_{C,r} = \mathbf{I}_r$*, while (6.17) follows since* $\mathbf{V}_{C,r}\mathbf{V}_{C,r}^\top$ *is an orthogonal projection onto the span of the rows of* $\mathbf{C}$ *and the columns of* $\mathbf{W}$ *lie within this span implying* $\mathbf{V}_{C,r}\mathbf{V}_{C,r}^\top\mathbf{W} = \mathbf{W}$*.*

6.3.3 Experiments

To test the accuracy of singular values/vectors and low-rank approximations for different methods, we used several kernel matrices arising in different applications, as described in Table 6.2. We worked with datasets containing less than ten thousand points to be able to

[2] This fact is illustrated in our experimental results for the 'DEXT' dataset in Figure 6.2(a).

Table 6.2: Description of the datasets used in our experiments comparing sampling-based matrix approximations.

Dataset	Data	n	d	Kernel
PIE-2.7K	faces	2731	2304	linear
PIE-7K	faces	7412	2304	linear
MNIST	digits	4000	784	linear
ESS	proteins	4728	16	RBF
ABN	abalones	4177	8	RBF

compare with exact SVD. We fixed k to be 100 in all the experiments, which captures more than 90% of the spectral energy for each dataset.

For singular values, we measured percentage accuracy of the approximate singular values with respect to the exact ones. For a fixed l, we performed 10 trials by selecting columns uniformly at random from $\mathbf{K}$. We show in Figure 6.1(a) the difference in mean percentage accuracy for the two methods for $l = n/10$, with results bucketed by groups of singular values, i.e., we sorted the singular values in descending order, grouped them as indicated in the figure, and report the average percentage accuracy for each group. The empirical results show that the Column sampling method generates more accurate singular values than the Nyström method. A similar trend was observed for other values of l.

For singular vectors, the accuracy was measured by the dot product, i.e., cosine of principal angles between the exact and the approximate singular vectors. Figure 6.1(b) shows the difference in mean accuracy between Nyström and Column sampling methods, once again bucketed by groups of singular vectors sorted in descending order based on their corresponding singular values. The top 100 singular vectors were all better approximated by Column sampling for all datasets. This trend was observed for other values of l as well. Furthermore, even when the Nyström singular vectors are orthogonalized, the Column sampling approximations are superior, as shown in Figure 6.1(c).

Next we compared the low-rank approximations generated by the two methods using spectral reconstruction as described in Section 6.3.2. We measured the accuracy of reconstruction relative to the optimal rank-k approximation, $\mathbf{K}_k$, as:

$$\text{relative accuracy} = \frac{\|\mathbf{K} - \mathbf{K}_k\|_F}{\|\mathbf{K} - \widetilde{\mathbf{K}}_k^{nys/col}\|_F}. \tag{6.18}$$

The relative accuracy will approach one for good approximations. Results are shown in Figure 6.2(a). The Nyström method produces superior results for spectral reconstruction. These results are somewhat surprising given the relatively poor quality of the singular values/vectors for the Nyström method, but they are in agreement with the consequences of Theorem 10. Furthermore, as stated in Theorem 9, the optimal spectral reconstruction approximation is tied to the spectrum of $\mathbf{K}$. Our results suggest that the relative accuracies of Nyström and Column sampling spectral reconstructions are also tied to this spectrum. When we analyzed spectral reconstruction performance on a sparse kernel matrix with a slowly decaying spectrum, we found that Nyström and Column sampling approximations were roughly equivalent ('DEXT' in Figure 6.2(a)). This result contrasts the results for dense kernel matrices with exponentially decaying spectra arising from the other datasets used in the experiments.

One factor that impacts the accuracy of the Nyström method for some tasks is the non-orthonormality of its singular vectors (Section 6.3.1). Although orthonormalization is computationally costly and typically avoided in practice, we nonetheless evaluated the effect

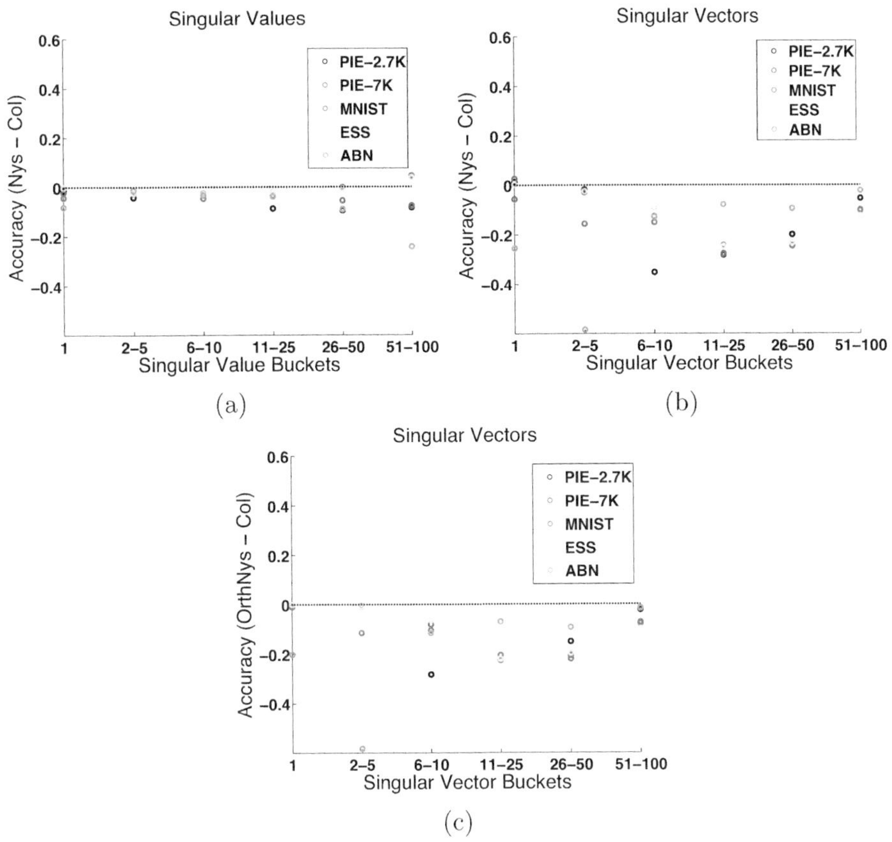

Figure 6.1: (See Color Insert.) Differences in accuracy between Nyström and column sampling. Values above zero indicate better performance of Nyström and vice versa. (a) Top 100 singular values with $l = n/10$. (b) Top 100 singular vectors with $l = n/10$. (c) Comparison using orthogonalized Nyström singular vectors.

of such orthonormalization. Empirically, the accuracy of Orthonormal Nyström spectral reconstruction is actually worse relative to the standard Nyström approximation, as shown in Figure 6.2(b). This surprising result can be attributed to the fact that orthonormalization of the singular vectors leads to the loss of some of the unique properties described in Section 6.3.2. For instance, Theorem 10 no longer holds and the scaling terms do not cancel out, i.e., $\widetilde{\mathbf{K}}_k^{nys} \neq \mathbf{C}\mathbf{W}_k^+\mathbf{C}^\top$.

6.4 Large-Scale Manifold Learning

In the previous section, we discussed two sampling-based techniques that generate approximations for kernel matrices. Although we analyzed the effectiveness of these techniques for approximating singular values, singular vectors and low-rank matrix reconstruction, we have yet to discuss the effectiveness of these techniques in the context of actual machine learning tasks. In fact, the Nyström method has been shown to be successful on a variety of learning tasks including Support Vector Machines [14], Gaussian Processes [36], Spectral Clustering [15], manifold learning [33], Kernel Logistic Regression [21], Kernel Ridge Regression [7] and more generally to approximate regularized matrix inverses via the Woodbury approximation

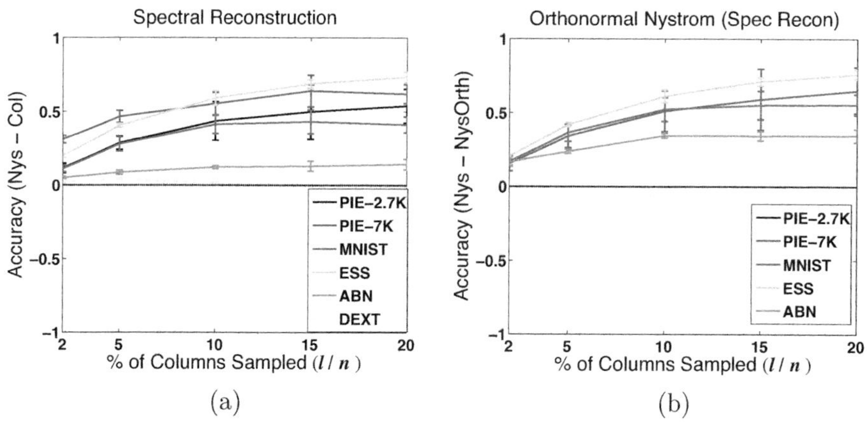

Figure 6.2: (See Color Insert.) Performance accuracy of spectral reconstruction approximations for different methods with $k = 100$. Values above zero indicate better performance of the Nyström method. (a) Nyström versus column sampling. (b) Nyström versus orthonormal Nyström.

[36]. In this section, we will discuss in detail how approximate embeddings can be used in the context of manifold learning, relying on the sampling based algorithms from the previous section to generate an approximate SVD. In particular, we present the largest study to date for manifold learning, and provide a quantitative comparison of Isomap and Laplacian Eigenmaps for large scale face manifold construction on clustering and classification tasks.

6.4.1 Manifold Learning

Manifold learning considers the problem of extracting low-dimensional structure from high-dimensional data. Given n input points, $\mathbf{X} = \{\mathbf{x}_i\}_{i=1}^n$ and $\mathbf{x}_i \in \mathbb{R}^d$, the goal is to find corresponding outputs $\mathbf{Y} = \{\mathbf{y}_i\}_{i=1}^n$, where $\mathbf{y}_i \in \mathbb{R}^k$, $k \ll d$, such that $\mathbf{Y}$ 'faithfully' represents $\mathbf{X}$. We now briefly review the Isomap and Laplacian Eigenmaps techniques to discuss their computational complexity.

Isomap

Isomap aims to extract a low-dimensional data representation that best preserves all pairwise distances between input points, as measured by their geodesic distances along the manifold [34]. It approximates the geodesic distance assuming that input space distance provides good approximations for nearby points, and for faraway points it estimates distance as a series of hops between neighboring points. This approximation becomes exact in the limit of infinite data. Isomap can be viewed as an adaptation of Classical Multidimensional Scaling [8], in which geodesic distances replace Euclidean distances.

Computationally, Isomap requires three steps:

1. Find the t nearest neighbors for each point in input space and construct an undirected neighborhood graph, denoted by $\mathcal{G}$, with points as nodes and links between neighbors as edges. This requires $O(n^2)$ time.

2. Compute the approximate geodesic distances, Δ_{ij}, between all pairs of nodes (i, j) by finding shortest paths in $\mathcal{G}$ using Dijkstra's algorithm at each node. Perform double centering, which converts the squared distance matrix into a dense $n \times n$

similarity matrix, i.e., compute $\mathbf{K} = -\frac{1}{2}\mathbf{HDH}$, where $\mathbf{D}$ is the squared distance matrix, $\mathbf{H} = \mathbf{I}_n - \frac{1}{n}\mathbf{1}\mathbf{1}^\top$ is the centering matrix, $\mathbf{I}_n$ is the $n \times n$ identity matrix and $\mathbf{1}$ is a column vector of all ones. This step takes $O(n^2 \log n)$ time, dominated by the calculation of geodesic distances.

3. Find the optimal k dimensional representation, $\mathbf{Y} = \{\mathbf{y}_i\}_{i=1}^n$, such that $\mathbf{Y} = \operatorname{argmin}_{\mathbf{Y}'} \sum_{i,j} \left(\|\mathbf{y}'_i - \mathbf{y}'_j\|_2^2 - \Delta_{ij}^2\right)$. The solution is given by,

$$\mathbf{Y} = (\mathbf{\Sigma}_k)^{1/2}\mathbf{U}_k^\top \tag{6.19}$$

where $\mathbf{\Sigma}_k$ is the diagonal matrix of the top k singular values of $\mathbf{K}$ and $\mathbf{U}_k$ are the associated singular vectors. This step requires $O(n^2)$ space for storing $\mathbf{K}$, and $O(n^3)$ time for its SVD.

The time and space complexities for all three steps are intractable for $n = 18$M.

Laplacian Eigenmaps

Laplacian Eigenmaps aims to find a low-dimensional representation that best preserves neighborhood relations as measured by a weight matrix $\mathbf{W}$ [4].[3] The algorithm works as follows:

1. Similar to Isomap, first find t nearest neighbors for each point. Then construct $\mathbf{W}$, a sparse, symmetric $n \times n$ matrix, where $\mathbf{W}_{ij} = \exp\left(-\|\mathbf{x}_i - \mathbf{x}_j\|_2^2/\sigma^2\right)$ if $(\mathbf{x}_i, \mathbf{x}_j)$ are neighbors, 0 otherwise, and σ is a scaling parameter.

2. Construct the diagonal matrix $\mathbf{D}$, such that $\mathbf{D}_{ii} = \sum_j \mathbf{W}_{ij}$, in $O(tn)$ time.

3. Find the k dimensional representation by minimizing the normalized, weighted distance between neighbors as,

$$\mathbf{Y} = \operatorname*{argmin}_{\mathbf{Y}'} \sum_{i,j} \left(\frac{\mathbf{W}_{ij}\|\mathbf{y}'_i - \mathbf{y}'_j\|_2^2}{\sqrt{\mathbf{D}_{ii}\mathbf{D}_{jj}}} \right). \tag{6.20}$$

This objective function penalizes nearby inputs for being mapped to faraway outputs, with 'nearness' measured by the weight matrix $\mathbf{W}$ [6]. To find $\mathbf{Y}$, we define $\mathcal{L} = \mathbf{I}_n - \mathbf{D}^{-1/2}\mathbf{W}\mathbf{D}^{-1/2}$ where $\mathcal{L} \in \mathbb{R}^{n \times n}$ is the symmetrized, normalized form of the graph Laplacian, given by $\mathbf{D} - \mathbf{W}$. Then, the solution to the minimization in (6.20) is

$$\mathbf{Y} = \mathbf{U}_{\mathcal{L},k}^\top \tag{6.21}$$

where $\mathbf{U}_{\mathcal{L},k}^\top$ are the bottom k singular vectors of $\mathcal{L}$, excluding the last singular vector corresponding to the singular value 0. Since $\mathcal{L}$ is sparse, it can be stored in $O(tn)$ space, and iterative methods, such as Lanczos, can be used to find these k singular vectors relatively quickly.

To summarize, in both the Isomap and Laplacian Eigenmaps methods, the two main computational efforts required are neighborhood graph construction and manipulation and SVD of a symmetric positive semidefinite (SPSD) matrix. In the next section, we further discuss the Nyström and Column sampling methods in the context of manifold learning, and describe the graph operations in Section 6.4.3.

[3] The weight matrix should not be confused with the subsampled SPSD matrix, $\mathbf{W}$, associated with the Nyström method. Since sampling-based approximation techniques will not be used with Laplacian Eigenmaps, the notation should be clear from the context.

6.4.2 Approximation Experiments

Since we use sampling-based SVD approximation to scale Isomap, we first examined how well the Nyström and Column sampling methods approximated our desired low-dimensional embeddings, i.e., $\mathbf{Y} = (\boldsymbol{\Sigma}_k)^{1/2}\mathbf{U}_k^\top$. Using (6.3), the Nyström low-dimensional embeddings are:

$$\widetilde{\mathbf{Y}}^{nys} = \widetilde{\boldsymbol{\Sigma}}_{nys,k}^{1/2}\widetilde{\mathbf{U}}_{nys,k}^\top = \big((\boldsymbol{\Sigma}_W)_k^{1/2}\big)^+\mathbf{U}_{W,k}^\top\mathbf{C}^\top. \tag{6.22}$$

Similarly, from (6.4) we can express the Column sampling low-dimensional embeddings as:

$$\widetilde{\mathbf{Y}}^{col} = \widetilde{\boldsymbol{\Sigma}}_{col,k}^{1/2}\widetilde{\mathbf{U}}_{col,k}^\top = \sqrt[4]{\frac{n}{l}}\big((\boldsymbol{\Sigma}_C)_k^{1/2}\big)^+\mathbf{V}_{C,k}^\top\mathbf{C}^\top. \tag{6.23}$$

Both approximations are of a similar form. Further, notice that the optimal low-dimensional embeddings are in fact the square root of the optimal rank k approximation to the associated SPSD matrix, i.e., $\mathbf{Y}^\top\mathbf{Y} = \mathbf{K}_k$, for Isomap. As such, there is a connection between the task of approximating low-dimensional embeddings and the task of generating low-rank approximate spectral reconstructions, as discussed in Section 6.3.2. Recall that the theoretical analysis in Section 6.3.2 as well as the empirical results in Section 6.3.3 both suggested that the Nyström method was superior in its spectral reconstruction accuracy. Hence, we performed an empirical study using the datasets from Table 6.2 to measure the quality of the low-dimensional embeddings generated by the two techniques and see if the same trend exists.

We measured the quality of the low-dimensional embeddings by calculating the extent to which they preserve distances, which is the appropriate criterion in the context of manifold learning. For each dataset, we started with a kernel matrix, $\mathbf{K}$, from which we computed the associated $n \times n$ squared distance matrix, $\mathbf{D}$, using the fact that $\|\mathbf{x}_i - \mathbf{x}_j\|^2 = \mathbf{K}_{ii} + \mathbf{K}_{jj} - 2\mathbf{K}_{ij}$. We then computed the approximate low-dimensional embeddings using the Nyström and Column sampling methods, and then used these embeddings to compute the associated approximate squared distance matrix, $\widetilde{\mathbf{D}}$. We measured accuracy using the notion of relative accuracy defined in (6.18), which can be expressed in terms of distance matrices as:

$$\text{relative accuracy} = \frac{\|\mathbf{D} - \mathbf{D}_k\|_F}{\|\mathbf{D} - \widetilde{\mathbf{D}}\|_F},$$

where $\mathbf{D}_k$ corresponds to the distance matrix computed from the optimal k dimensional embeddings obtained using the singular values and singular vectors of $\mathbf{K}$. In our experiments, we set $k = 100$ and used various numbers of sampled columns, ranging from $l = n/50$ to $l = n/5$. Figure 6.3 presents the results of our experiments. Surprisingly, we do not see the same trend in our empirical results for embeddings as we previously observed for spectral reconstruction, as the two techniques exhibit roughly similar behavior across datasets. As a result, we decided to use both the Nyström and Column sampling methods for our subsequent manifold learning study.

6.4.3 Large-Scale Learning

In this section, we outline the process of learning a manifold of faces. We first describe the datasets used in our experiments. We then explain how to extract nearest neighbors, a common step between Laplacian Eigenmaps and Isomap. The remaining steps of Laplacian Eigenmaps are straightforward, so the subsequent sections focus on Isomap, and specifically on the computational efforts required to generate a manifold using Webfaces-18M.

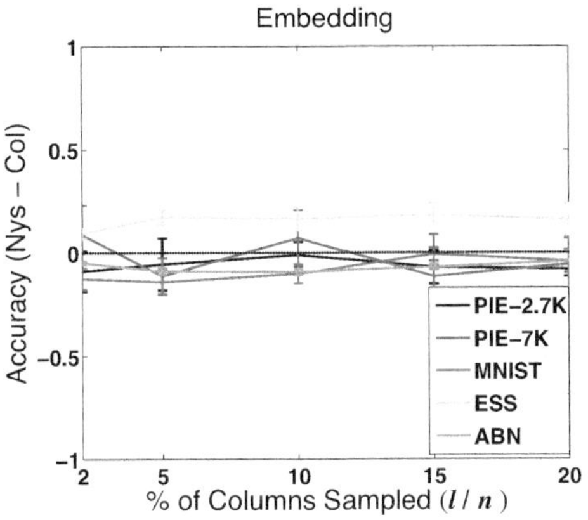

Figure 6.3: (See Color Insert.) Embedding accuracy of Nyström and column sampling. Values above zero indicate better performance of Nyström and vice versa.

Datasets

We used two faces datasets consisting of 35K and 18M images. The CMU PIE face dataset [32] contains 41,368 images of 68 subjects under 13 different poses and various illumination conditions. A standard face detector extracted 35,247 faces (each 48×48 pixels), which comprised our 35K set (PIE-35K). We used this set because, being labeled, it allowed us to perform quantitative comparisons. The second dataset, named Webfaces-18M, contains 18.2 million images of faces extracted from the Web using the same face detector. For both datasets, face images were represented as 2304 dimensional pixel vectors which were globally normalized to have zero mean and unit variance. No other pre-processing, e.g., face alignment, was performed. In contrast, [20] used well-aligned faces (as well as much smaller data sets) to learn face manifolds. Constructing Webfaces-18M, including face detection and duplicate removal, took 15 hours using a cluster of several hundred machines. We used this cluster for all experiments requiring distributed processing and data storage.

Nearest neighbors and neighborhood graph

The cost of naive nearest neighbor computation is $O(n^2)$, where n is the size of the dataset. It is possible to compute exact neighbors for PIE-35K, but for Webfaces-18M this computation is prohibitively expensive. So, for this set, we used a combination of random projections and spill trees [26] to get approximate neighbors. Computing 5 nearest neighbors in parallel with spill trees took $\sim$2 days on the cluster. Figure 6.4 shows the top 5 neighbors for a few randomly chosen images in Webfaces-18M. In addition to this visualization, comparison of exact neighbors and spill tree approximations for smaller subsets suggested good performance of spill trees.

We next constructed the neighborhood graph by representing each image as a node and connecting all neighboring nodes. Since Isomap and Laplacian Eigenmaps require this graph to be connected, we used depth-first search to find its largest connected component. These steps required $O(tn)$ space and time. Constructing the neighborhood graph for Webfaces-18M and finding the largest connected component took 10 minutes on a single machine using the OpenFST library [1].

For neighborhood graph construction, an 'appropriate' choice of number of neighbors, t, is crucial. A small t may give too many disconnected components, while a large t may introduce unwanted edges. These edges stem from inadequately sampled regions of the

Figure 6.4: Visualization of neighbors for Webfaces-18M. The first image in each row is the input, and the next five are its neighbors.

Table 6.3: Number of components in the Webfaces-18M neighbor graph and the percentage of images within the largest connected component for varying numbers of neighbors with and without an upper limit on neighbor distances.

	No Upper Limit		Upper Limit Enforced	
t	# Comp	% Largest	# Comp	% Largest
1	1.7M	0.05 %	4.3M	0.03 %
2	97K	97.2 %	285K	80.1 %
3	18K	99.3 %	277K	82.2 %
5	1.9K	99.9 %	275K	83.1 %

manifold and false positives introduced by the face detector. Since Isomap needs to compute shortest paths in the neighborhood graph, the presence of bad edges can adversely impact these computations. This is known as the problem of leakage or 'short-circuits' [3]. Here, we chose $t = 5$ and also enforced an upper limit on neighbor distance to alleviate the problem of leakage. We used a distance limit corresponding to the 95th percentile of neighbor distances in the PIE-35K dataset.

Table 6.3 shows the effect of choosing different values for t with and without enforcing the upper distance limit. As expected, the size of the largest connected component increases as t increases. Also, enforcing the distance limit reduces the size of the largest component. Figure 6.5 shows a few random samples from the largest component. Images not within the largest component are either part of a strongly connected set of images (Figure 6.6) or do not have any neighbors within the upper distance limit (Figure 6.7). There are significantly more false positives in Figure 6.7 than in Figure 6.5, although some of the images in Figure 6.7 are actually faces. Clearly, the distance limit introduces a trade-off between filtering out non-faces and excluding actual faces from the largest component.[4]

Approximating geodesics

To construct the similarity matrix $\mathbf{K}$ in Isomap, one approximates geodesic distance by shortest-path lengths between every pair of nodes in the neighborhood graph. This requires $O(n^2 \log n)$ time and $O(n^2)$ space, both of which are prohibitive for 18M nodes. However,

[4]To construct embeddings with Laplacian Eigenmaps, we generated $\mathbf{W}$ and $\mathbf{D}$ from nearest neighbor data for images within the largest component of the neighborhood graph and solved (6.21) using a sparse eigensolver.

Figure 6.5: A few random samples from the largest connected component of the Webfaces-18M neighborhood graph.

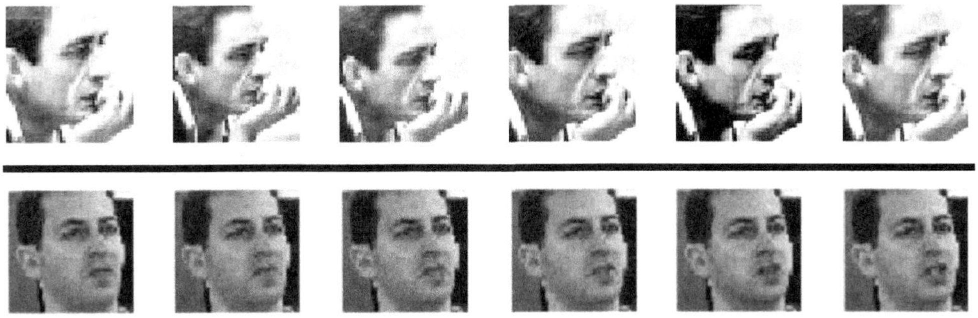

Figure 6.6: Visualization of disconnected components of the neighborhood graphs from Webfaces-18M (top row) and from PIE-35K (bottom row). The neighbors for each of these images are all within this set, thus making the entire set disconnected from the rest of the graph. Note that these images are not exactly the same.

Figure 6.7: Visualization of disconnected components containing exactly one image. Although several of the images above are not faces, some are actual faces, suggesting that certain areas of the face manifold are not adequately sampled by Webfaces-18M.

since we use sampling-based approximate decomposition, we need only $l \ll n$ columns of $\mathbf{K}$, which form the submatrix $\mathbf{C}$. We thus computed geodesic distance between l randomly selected nodes (called landmark points) and the rest of the nodes, which required $O(ln \log n)$ time and $O(ln)$ space. Since this computation can easily be parallelized, we performed geodesic computation on the cluster and stored the output in a distributed fashion. The overall procedure took 60 minutes for Webfaces-18M using l = 10K. The bottom four rows in Figure 6.9 show sample shortest paths for images within the largest component for Webfaces-18M, illustrating smooth transitions between images along each path.[5]

Generating low-dimensional embeddings

Before generating low-dimensional embeddings using Isomap, one needs to convert distances into similarities using a process called centering [8]. For the Nyström approximation, we computed $\mathbf{W}$ by double centering $\mathbf{D}$, the $l \times l$ matrix of squared geodesic distances between all landmark nodes, as $\mathbf{W} = -\frac{1}{2}\mathbf{HDH}$, where $\mathbf{H} = \mathbf{I}_l - \frac{1}{l}\mathbf{1}\mathbf{1}^\top$ is the centering matrix, $\mathbf{I}_l$ is the $l \times l$ identity matrix, and $\mathbf{1}$ is a column vector of all ones. Similarly, the matrix $\mathbf{C}$ was obtained from squared geodesic distances between the landmark nodes and all other nodes using single-centering as described in [9].

For the Column sampling approximation, we decomposed $\mathbf{C}^\top\mathbf{C}$, which we constructed by performing matrix multiplication in parallel on $\mathbf{C}$. For both approximations, decomposition on an $l \times l$ matrix ($\mathbf{C}^\top\mathbf{C}$ or $\mathbf{W}$) took about one hour. Finally, we computed low-dimensional embeddings by multiplying the scaled singular vectors from approximate decomposition with $\mathbf{C}$. For Webfaces-18M, generating low dimensional embeddings took 1.5 hours for the Nyström method and 6 hours for the Column sampling method.

6.4.4 Manifold Evaluation

Manifold learning techniques typically transform the data such that Euclidean distance in the transformed space between *any* pair of points is meaningful, under the assumption that in the original space Euclidean distance is meaningful only in local neighborhoods. Since K-means clustering computes Euclidean distances between all pairs of points, it is a natural choice for evaluating these techniques. We also compared the performance of various techniques using nearest neighbor classification. Since CMU-PIE is a labeled dataset, we first focused on quantitative evaluation of different embeddings using face pose as class labels. The PIE set contains faces in 13 poses, and such a fine sampling of the pose space makes clustering and classification tasks very challenging. In all the experiments we fixed the dimension of the reduced space, k, to be 100.

The first set of experiments was aimed at finding how well different Isomap approximations perform in comparison to exact Isomap. We used a subset of PIE with 10K images (PIE-10K) since, for this size, exact SVD could be done on a single machine within reasonable time and memory limits. We fixed the number of clusters in our experiments to equal the number of pose classes, and measured clustering performance using two measures, *Purity* and *Accuracy*. Purity measures the frequency of data belonging to the same cluster sharing the same class label, while Accuracy measures the frequency of data from the same class appearing in a single cluster. Thus, ideal clustering will have 100% Purity and 100% Accuracy.

Table 6.4 shows that clustering with Nyström Isomap with just l=1K performs almost

[5]In fact, the techniques we described in the context of approximating geodesic distances via shortest path are currently used by Google for its "People Hopper" application, which runs on the social networking site Orkut [24].

Table 6.4: K-means clustering of face poses applied to PIE-10K for different algorithms. Results are averaged over 10 random K-means initializations.

Methods	Purity (%)	Accuracy (%)
PCA	54.3 ($\pm$0.8)	46.1 ($\pm$1.4)
Exact Isomap	58.4 ($\pm$1.1)	53.3 ($\pm$4.3)
Nyström Isomap	59.1 ($\pm$0.9)	53.3 ($\pm$2.7)
Col-Sampling Isomap	56.5 ($\pm$0.7)	49.4 ($\pm$3.8)
Laplacian Eigenmaps	35.8 ($\pm$5.0)	69.2 ($\pm$10.8)

Table 6.5: K-means clustering of face poses applied to PIE-35K for different algorithms. Results are averaged over 10 random K-means initializations.

Methods	Purity (%)	Accuracy (%)
PCA	54.6 ($\pm$1.3)	46.8 ($\pm$1.3)
Nyström Isomap	59.9 ($\pm$1.5)	53.7 ($\pm$4.4)
Col-Sampling Isomap	56.1 ($\pm$1.0)	50.7 ($\pm$3.3)
Laplacian Eigenmaps	39.3 ($\pm$4.9)	74.7 ($\pm$5.1)

as well as exact Isomap on this dataset.[6] This matches with the observation made in [36], where the Nyström approximation was used to speed up kernel machines. Also, Column sampling Isomap performs slightly worse than Nyström Isomap. The clustering results on the full PIE-35K set (Table 6.5) with $l = 10$K also affirm this observation. Figure 6.8 shows the optimal 2D projections from different methods for PIE-35K. The Nyström method separates the pose clusters better than Column sampling verifying the quantitative results.

The fact that Nyström outperforms Column sampling is somewhat surprising given the experimental evaluations in Section 6.4.2, where we found the two approximation techniques to achieve similar performance. One possible reason for the poor performance of Column sampling Isomap is due to the form of the similarity matrix $\mathbf{K}$. When using a finite number of data points for Isomap, $\mathbf{K}$ is not guaranteed to be SPSD. We verified that $\mathbf{K}$ was not SPSD in our experiments, and a significant number of top eigenvalues, i.e., those with largest magnitudes, were negative. The two approximation techniques differ in their treatment of negative eigenvalues and the corresponding eigenvectors. The Nyström method allows one to use eigenvalue decomposition (EVD) of $\mathbf{W}$ to yield signed eigenvalues, making it possible to discard the negative eigenvalues and the corresponding eigenvectors. On the contrary, it is not possible to discard these in the Column-based method, since the signs of eigenvalues are lost in the SVD of the rectangular matrix $\mathbf{C}$ (or EVD of $\mathbf{C}^\top\mathbf{C}$). Thus, the presence of negative eigenvalues deteriorates the performance of Column sampling method more than the Nyström method.

Table 6.4 and Table 6.5 also show a significant difference in the Isomap and Laplacian Eigenmaps results. The 2D embeddings of PIE-35K (Figure 6.8) reveal that Laplacian Eigenmaps projects data points into a small compact region, consistent with its objective function defined in (6.20), as it tends to map neighboring inputs as nearby as possible in the low-dimensional space. When used for clustering, these compact embeddings lead to a few large clusters and several tiny clusters, thus explaining the high accuracy and low purity of the clusters. This indicates poor clustering performance of Laplacian Eigenmaps, since one can achieve even 100% accuracy simply by grouping all points into a single cluster. However, the purity of such clustering would be very low. Finally, the improved clustering results of Isomap over PCA for both datasets verify that the manifold of faces is not linear

[6] The differences are statistically insignificant.

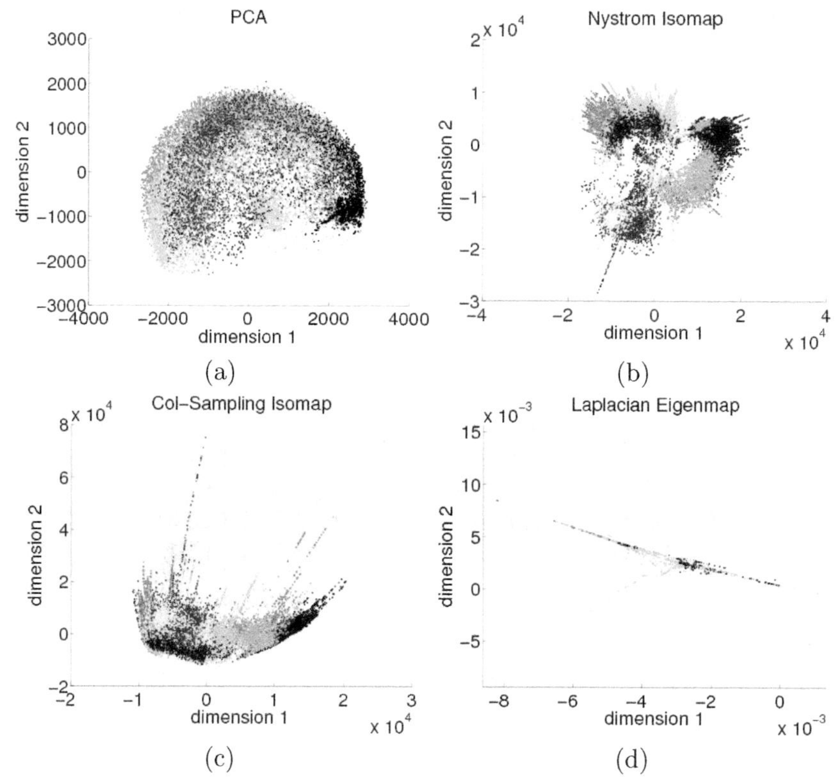

Figure 6.8: (See Color Insert.) Optimal 2D projections of PIE-35K where each point is color coded according to its pose label. (a) PCA projections tend to spread the data to capture maximum variance. (b) Isomap projections with Nyström approximation tend to separate the clusters of different poses while keeping the cluster of each pose compact. (c) Isomap projections with column sampling approximation have more overlap than with Nyström approximation. (d) Laplacian eigenmaps project the data into a very compact range.

in the input space.

Moreover, we compared the performance of Laplacian Eigenmaps and Isomap embeddings on pose classification.[7] The data was randomly split into a training and a test set, and K-Nearest Neighbor (KNN) was used for classification. $K = 1$ gives lower error than higher K as shown in Table 6.6. Also, the classification error is lower for both exact and approximate Isomap than for Laplacian Eigenmaps, suggesting that neighborhood information is better preserved by Isomap (Table 6.6 and Table 6.7). Note that, similar to clustering, the Nyström approximation performs as well as Exact Isomap (Table 6.6). Better clustering and classification results, combined with 2D visualizations, imply that approximate Isomap outperforms exact Laplacian Eigenmaps. Moreover, the Nyström approximation is computationally cheaper and empirically more effective than the Column sampling approximation. Thus, we used Nyström Isomap to generate embeddings for Webfaces-18M.

After learning a face manifold from Webfaces-18M, we analyzed the results with various visualizations. The top row of Figure 6.9 shows the 2D embeddings from Nyström Isomap.

[7]KNN only uses nearest neighbor information for classification. Since neighborhoods are considered to be locally linear in the input space, we expect KNN to perform well in the input space. Hence, using KNN to compare low-level embeddings indirectly measures how well nearest neighbor information is preserved.

Table 6.6: K-nearest neighbor face pose classification error (%) on PIE-10K subset for different algorithms.

Methods	$K = 1$	$K = 3$	$K = 5$
Isomap	10.9 (±0.5)	14.1 (±0.7)	15.8 (±0.3)
Nyström Isomap	11.0 (±0.5)	14.0 (±0.6)	15.8 (±0.6)
Col-Sampling Isomap	12.0 (±0.4)	15.3 (±0.6)	16.6 (±0.5)
Laplacian Eigenmaps	12.7 (±0.7)	16.6 (±0.5)	18.9 (±0.9)

Table 6.7: 1-nearest neighbor face pose classification error on PIE-35K for different algorithms.

Nyström Isomap	Col-Sampling Isomap	Laplacian Eigenmaps
9.8 (±0.2)	10.3 (±0.3)	11.1 (±0.3)

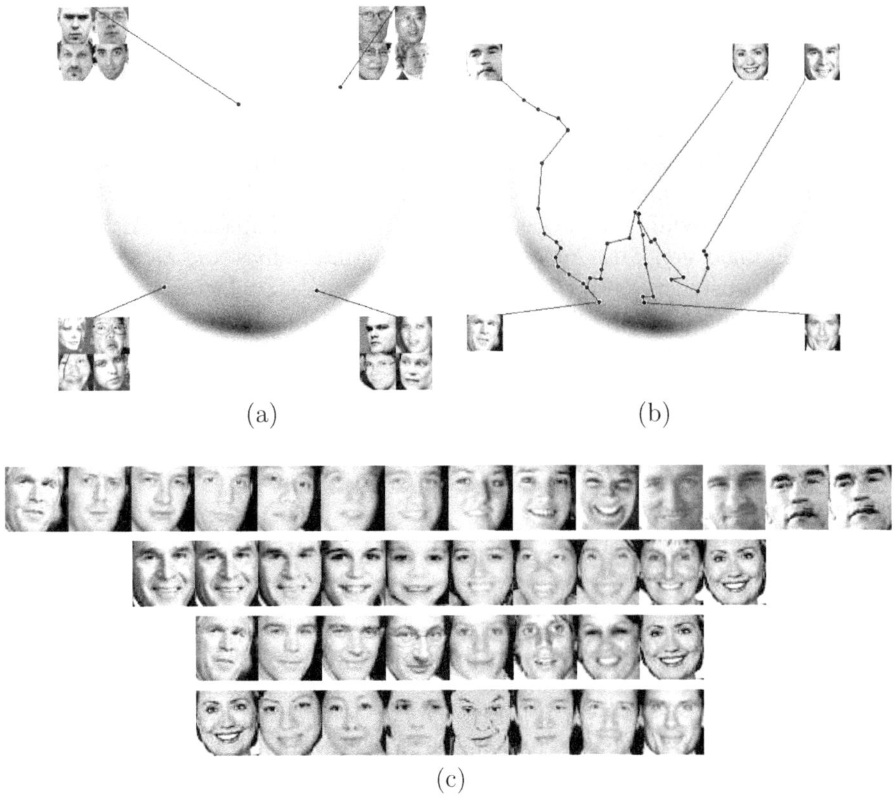

Figure 6.9: 2D embedding of Webfaces-18M using Nyström isomap (top row). Darker areas indicate denser manifold regions. (a) Face samples at different locations on the manifold. (b) Approximate geodesic paths between celebrities. (c) Visualization of paths shown in (b).

The top left figure shows the face samples from various locations in the manifold. It is interesting to see that embeddings tend to cluster the faces by pose. These results support the good clustering performance observed using Isomap on PIE data. Also, two groups (bottom left and top right) with similar poses but different illuminations are projected at different locations. Additionally, since 2D projections are very condensed for 18M points, one can expect more discrimination for higher k, e.g., $k = 100$.

In Figure 6.9, the top right figure shows the shortest paths on the manifold between different public figures. The images along the corresponding paths have smooth transitions as shown in the bottom of the figure. In the limit of infinite samples, Isomap guarantees that the distance along the shortest path between any pair of points will be preserved as Euclidean distance in the embedded space. Even though the paths in the figure are reasonable approximations of straight lines in the embedded space, these results suggest that either (i) 18M faces are perhaps not enough samples to learn the face manifold exactly, or (ii) a low-dimensional manifold of faces may not actually exist (perhaps the data clusters into multiple low dimensional manifolds). It remains an open question as to how we can measure and evaluate these hypotheses, since even very large-scale testing has not provided conclusive evidence.

6.5 Summary

We have presented large-scale nonlinear dimensionality reduction using unsupervised manifold learning. In order to work on a such a large scale, we first studied sampling based algorithms, presenting an analysis of two techniques for approximating SVD on large dense SPSD matrices and providing a theoretical and empirical comparison. Although the Column sampling method generates more accurate singular values and singular vectors, the Nyström method constructs better low-rank approximations, which are of great practical interest as they do not use the full matrix. Furthermore, our large-scale manifold learning studies reveal that Isomap coupled with the Nyström approximation can effectively extract low-dimensional structure from datasets containing millions of images. Nonetheless, the existence of an underlying manifold of faces remains an open question.

6.6 Bibliography and Historical Remarks

Manifold learning algorithms are extensions of classical linear dimensionality reduction techniques introduced over a century ago, e.g., Principal Component Analysis (PCA) and Classical Multidimensional Scaling [28, 8]. Pioneering work on non-linear dimensionality reduction was introduced by [34, 31] which led to the development of several related algorithms for manifold learning [4, 11, 35]. The connection between manifold learning algorithms and Kernel PCA was noted by [19]. The Nyström method was initially introduced as a quadrature method for numerical integration, used to approximate eigenfunction solutions [27, 2]. More recently it has been studied for a variety of kernel-based algorithms and other algorithms involving symmetric positive semidefinite matrices [36, 14, 15, 13, 21, 37, 23, 7], and in particular for large-scale manifold learning [9, 29, 33]. Column sampling techniques have also been analyzed for approximating general rectangular matrices, including notable work by [16, 12, 10]. Initial comparisons between the Nyström method and these more general Column sampling methods were first discussed in [33, 22].

Bibliography

[1] C. Allauzen, M. Riley, J. Schalkwyk, W. Skut, and M. Mohri. OpenFST: A general and efficient weighted finite-state transducer library. In *Conference on Implementation and Application of Automata*, 2007.

[2] Christopher T. Baker. *The numerical treatment of integral equations*. Clarendon Press, Oxford, 1977.

[3] M. Balasubramanian and E. L. Schwartz. The Isomap algorithm and topological stability. *Science*, 295, 2002.

[4] M. Belkin and P. Niyogi. Laplacian Eigenmaps and spectral techniques for embedding and clustering. In *Neural Information Processing Systems*, 2001.

[5] M. Belkin and P. Niyogi. Convergence of Laplacian Eigenmaps. In *Neural Information Processing Systems*, 2006.

[6] O. Chapelle, B. Schölkopf, and A. Zien, editors. *Semi-Supervised Learning*. MIT Press, Cambridge, MA, 2006.

[7] Corinna Cortes, Mehryar Mohri, and Ameet Talwalkar. On the impact of kernel approximation on learning accuracy. In *Conference on Artificial Intelligence and Statistics*, 2010.

[8] T. F. Cox, M. A. A. Cox, and T. F. Cox. *Multidimensional Scaling*. Chapman & Hall/CRC, 2nd edition, 2000.

[9] Vin de Silva and Joshua Tenenbaum. Global versus local methods in nonlinear dimensionality reduction. In *Neural Information Processing Systems*, 2003.

[10] Amit Deshpande, Luis Rademacher, Santosh Vempala, and Grant Wang. Matrix approximation and projective clustering via volume sampling. In *Symposium on Discrete Algorithms*, 2006.

[11] David L. Donoho and Carrie Grimes. Hessian Eigenmaps: locally linear embedding techniques for high dimensional data. *Proceedings of the National Academy of Sciences of the United States of America*, 100(10):5591–5596, 2003.

[12] Petros Drineas, Ravi Kannan, and Michael W. Mahoney. Fast Monte Carlo algorithms for matrices II: Computing a low-rank approximation to a matrix. *SIAM Journal on Computing*, 36(1), 2006.

[13] Petros Drineas and Michael W. Mahoney. On the Nyström method for approximating a Gram matrix for improved kernel-based learning. *Journal of Machine Learning Research*, 6:2153–2175, 2005.

[14] Shai Fine and Katya Scheinberg. Efficient SVM training using low-rank kernel representations. *Journal of Machine Learning Research*, 2:243–264, 2002.

[15] Charless Fowlkes, Serge Belongie, Fan Chung, and Jitendra Malik. Spectral grouping using the Nyström method. *Transactions on Pattern Analysis and Machine Intelligence*, 26(2):214–225, 2004.

[16] Alan Frieze, Ravi Kannan, and Santosh Vempala. Fast Monte-Carlo algorithms for finding low-rank approximations. In *Foundation of Computer Science*, 1998.

[17] Gene Golub and Charles Van Loan. *Matrix Computations*. Johns Hopkins University Press, Baltimore, 2nd edition, 1983.

[18] G. Gorrell. Generalized Hebbian algorithm for incremental Singular Value Decomposition in natural language processing. In *European Chapter of the Association for Computational Linguistics*, 2006.

[19] J. Ham, D. D. Lee, S. Mika, and B. Schölkopf. A kernel view of the dimensionality reduction of manifolds. In *International Conference on Machine Learning*, 2004.

[20] X. He, S. Yan, Y. Hu, and P. Niyogi. Face recognition using Laplacianfaces. *IEEE Transactions on Pattern Analysis and Machine Intelligence*, 27(3):328–340, 2005.

[21] Peter Karsmakers, Kristiaan Pelckmans, Johan Suykens, and Jugo Van Hamme. Fixed-size Kernel Logistic Regression for phoneme classification. In *Interspeech*, 2007.

[22] Sanjiv Kumar, Mehryar Mohri, and Ameet Talwalker. On sampling-based approximate spectral decomposition. In *International Conference on Machine Learning*, 2009.

[23] Sanjiv Kumar, Mehryar Mohri, and Ameet Talwalkar. Sampling techniques for the Nyström method. In *Conference on Artificial Intelligence and Statistics*, 2009.

[24] Sanjiv Kumar and Henry Rowley. People Hopper. `http://googleresearch.blogspot.com/2010/03/hopping-on-face-manifold-via-people.html`, 2010.

[25] Yann LeCun and Corinna Cortes. The MNIST database of handwritten digits. `http://yann.lecun.com/exdb/mnist/`, 1998.

[26] T. Liu, A. W. Moore, A. G. Gray, and K. Yang. An investigation of practical approximate nearest neighbor algorithms. In *Neural Information Processing Systems*, 2004.

[27] E.J. Nyström. Über die praktische auflösung von linearen integralgleichungen mit anwendungen auf randwertaufgaben der potentialtheorie. *Commentationes Physico-Mathematicae*, 4(15):1–52, 1928.

[28] Karl Pearson. On lines and planes of closest fit to systems of points in space. *Philosophical Magazine*, 2(6):559–572, 1901.

[29] John C. Platt. Fast embedding of sparse similarity graphs. In *Neural Information Processing Systems*, 2004.

[30] Vladimir Rokhlin, Arthur Szlam, and Mark Tygert. A randomized algorithm for Principal Component Analysis. *SIAM Journal on Matrix Analysis and Applications*, 31(3):1100–1124, 2009.

[31] Sam Roweis and Lawrence Saul. Nonlinear dimensionality reduction by Locally Linear Embedding. *Science*, 290(5500), 2000.

[32] Terence Sim, Simon Baker, and Maan Bsat. The CMU pose, illumination, and expression database. In *Conference on Automatic Face and Gesture Recognition*, 2002.

[33] Ameet Talwalkar, Sanjiv Kumar, and Henry Rowley. Large-scale manifold learning. In *Conference on Vision and Pattern Recognition*, 2008.

[34] J. Tenenbaum, V. de Silva, and J.C. Langford. A global geometric framework for nonlinear dimensionality reduction. *Science*, 290(5500), 2000.

[35] Kilian Q. Weinberger and Lawrence K. Saul. An introduction to nonlinear dimensionality reduction by maximum variance unfolding. In *AAAI Conference on Artificial Intelligence*, 2006.

[36] Christopher K. I. Williams and Matthias Seeger. Using the Nyström method to speed up kernel machines. In *Neural Information Processing Systems*, 2000.

[37] Kai Zhang, Ivor Tsang, and James Kwok. Improved Nyström low-rank approximation and error analysis. In *International Conference on Machine Learning*, 2008.

Chapter 7

Metric and Heat Kernel

Wei Zeng, Jian Sun, Ren Guo, Feng Luo, and Xianfeng Gu

7.1 Introduction

Jay Jorgenson and Serge Lang [33] called the heat kernel "... a universal gadget which is a dominant factor practically everywhere in mathematics, also in physics, and has very simple and powerful properties." In the past few decades, heat kernel has been studied and used in various sections of mathematics [24]. Recently, researchers from applied fields have witnessed the rise in usage of heat kernel in various areas in science and engineering. In machine leading, heat kernel has been used for ranking, dimensionality reduction, and date representation [10, 2, 11, 32]. In geometry processing, it has been used for shape signature, fining correspondence and shape segmentation [54, 41, 15], to name a few.

In this book chapter, we will consider the heat kernel of the Laplace–Beltrami operator on a Riemannian manifold and focus on the relation between metric and heat kernel. Specifically, it is well-known that the Laplace–Beltrami operator Δ of a smooth Riemannian manifold is determined by its Riemannian metric; so is the heat kernel as it is the kernel of the integral operator $e^{-t\Delta}$. Conversely, one can recover the Riemannian metric from heat kernel. We will consider the following two problems: (i) In practice, we are often given a discrete approximation of a Riemannian manifold. In such a discrete setting, can heat kernel and metric be recovered from one another? (ii) In many applications, it is desirable to have heat kernel to represent the metric as it organizes the metric information in a nice multi-scale way and thus is more robust in the presence of noise. Can we further simplify the heat kernel representation without losing the metric information? We will partially answer those two questions based on the work by the authors and their coauthors and finally present a few applications of heat kernel in the field of geometry processing.

The Laplace–Beltrami operator of a smooth Riemannian manifold is determined by the Riemannian metric. Conversely, the heat kernel constructed from its eigenvalues and eigenfunctions determines the Riemannian metric. This work proves the analogy on Euclidean polyhedral surfaces (triangle meshes), that the discrete Laplace–Beltrami operator and the discrete Riemannian metric (uniquely up to a scaling) are mutually determined by each other.

Given a Euclidean polyhedral surface, its Riemannian metric is represented as edge lengths, satisfying triangle inequalities on all faces. The Laplace–Beltrami operator is formulated using the cotangent formula, where the edge weight is defined as the sum of the cotangent of angles against the edge. We prove that the edge lengths can be determined by the edge weights uniquely up to a scaling using the variational approach.

First, we show that the space of all possible metrics of a polyhedral surface is convex. Then, we construct a special energy defined on the metric space, such that the gradient of the energy equals to the edge weights. Third, we show the Hessian matrix of the energy is positive definite, restricted on the tangent space of the metric space; therefore the energy is convex. Finally, by the fact that the parameter on a convex domain and the gradient of a convex function defined on the domain have one-to-one correspondence, we show the edge weights determine the polyhedral metric uniquely up to a scaling.

The constructive proof leads to a computational algorithm that finds the unique metric on a topological triangle mesh from a discrete Laplace–Beltrami operator matrix.

Laplace–Beltrami operator plays a fundamental role in Riemannian geometry [52]. Discrete Laplace–Beltrami operators on triangulated surface meshes span the entire spectrum of geometry processing applications, including mesh parameterization, segmentation, reconstruction, compression, re-meshing and so on [37, 50, 60]. Laplace–Beltrami operator is determined by the Riemannian metric. The heat kernel can be constructed from the eigenvalues and eigenfunctions of the Laplace–Beltrami operator; conversely, it fully determines the Riemannian metric (uniquely up to a scaling). In this work, we prove the discrete analogy to this fundamental fact for surface case, that the discrete Laplace–Beltrami operator and the discrete Riemannian metric are mutually determined by each other.

In real applications, a smooth metric surface is usually represented as a triangulated mesh. The manifold heat kernel is estimated from the discrete Laplace operator. There are many ways to discretize the Laplace–Beltrami operator.

Discretizations of Laplace–Beltrami Operator

The most well-known and widely-used discrete formulation of Laplace operator over triangulated meshes is the so-called *cotangent scheme*, which was originally introduced in [17, 43]. Xu [59] proposed several simple discretization schemes of Laplace operators over triangulated surfaces, and established the theoretical analysis on convergence. Wardetzky et al. [58] proved the theoretical limitation that the discrete Laplacians cannot satisfy all natural properties, and thus explained the diversity of existing discrete Laplace operators. A family of operations were presented by extending more natural properties into the existing operators. Reuter et al. [45] computed a discrete Laplace operator using the finite element method, and exploited the isometry invariance of the Laplace operator as shape fingerprint for object comparison. Belkin et al. [3] proposed the first discrete Laplacian that pointwise converges to the true Laplacian as the input mesh approximates a smooth manifold better. Dey et al. [16] employed this mesh Laplacian and provided the first convergence to relate the discrete spectrum with the true spectrum, and studied the stability and robustness of the discrete approximation of Laplace spectra.

Discrete Curvature Flow

Laplace–Beltrami operator has been closely related to discrete curvature flow method. In general, discrete curvature flow is the gradient flow of special energy forms, and the Hessians of the energies are Laplace–Beltrami operators.

One way to a discretization of conformality is the circle packing metric introduced by Thurston [55]. The notion of circle packing has appeared in the work of Koebe [35]. Thurston conjectured in [56] that for a discretization of the Jordan domain in the plane, the sequence of circle packings converge to the Riemann mapping. This was proved by Rodin and Sullivan [46]. Colin de Verdiere [13] established the first variational principle for circle packing and proved Thurston's existence of circle packing metrics. This paved a

Table 7.1: Symbol List

S	smooth surface	Σ	triangular mesh
$\mathbf{g}$	Riemannian metric	v_i	ith vertex
Δ_M	Laplace–Beltrami operator	$[v_i, v_j]$	edge connecting v_i and v_j
H_t	heat operator	g^{ij}	inverse of the Riemannian metric tensor
K_t	heat kernel	θ_i	corner angle at v_i
λ_i	eigen value of Δ_M	d_k	edge length of $[v_i, v_j]$
ϕ_i	eigen function of Δ_M	Ψ_p^M	heat kernal map
d_t	diffusion distance	H	Hessian matrix
ecc_t	eccentricity	d_{GW}^p	Gromov–Wasserstein distance

way for a fast algorithmic implementation of finding the circle packing metrics, such as the one by Collins and Stephenson [12]. In [19], Chow and Luo generalized Colin de Verdiere's work and introduced the discrete Ricci flow and discrete Ricci energy on surfaces. The algorithmic was later implemented and applied for surface parameterization [31, 30].

Another related discretization method is called circle pattern. Circle pattern was proposed by Bowers and Hurdal [7], and has been proven to be a minimizer of a convex energy by Bobenko and Springborn [6]. An efficient circle pattern algorithm was developed by Kharevych et al. [34] Discrete Yamabe flow was introduced by Luo in [39]. In a recent work of Springborn et al. [51], the Yamabe energy is explicitly given by using the Milnor–Lobachevsky function.

In Glickenstein's work on the monotonicity property of weighted Delaunay triangles in [21], most above Hessians are unified.

The symbols used for presentation are listed in Table 7.1.

7.2 Theoretic Background

This section briefly introduces elementary theories for the heat kernel of the Laplace-Beltrami Operator.

7.2.1 Laplace–Beltrami Operator

Laplace–Beltrami Operator is closely related to the heat diffusion process. Assume $(M, \mathbf{g})$ is a compact Riemannian manifold, $\mathbf{g}$ is the Riemannian metric. Denote Δ_M as the Laplace-Beltrami operator of M, which maps a function on M to another one by (see e.g. [47] for a good introduction)

$$\Delta_M f = \frac{1}{\sqrt{\det g}} \sum_{i,j} \frac{\partial}{\partial x^j} \left(g^{ij} \sqrt{\det g} \frac{\partial f}{\partial x^i} \right).$$

The Laplace–Beltrami operator over a compact manifold is bounded and symmetric negative semi-definite, and hence has a countable set of eigenfunctions $\phi_i : O \to \mathbb{R}$ and eigenvalues $\lambda_i \in \mathbb{R}$, such that

$$\Delta_M \phi_i = \lambda_i \phi_i.$$

The set of eigenfunctions forms a complete orthonormal basis for the space of L_2 functions on the manifold. That is, for any square integrable function f:

$$f(\mathbf{p}) = \sum_i a_i \phi_i(\mathbf{p}), \text{ where } a_i = \langle f, \phi_i \rangle \ \forall\ \mathbf{p} \in O$$

$$\langle \phi_i, \phi_j \rangle = \delta_{ij} = 1 \text{ if } i = j, \text{ and } 0 \text{ otherwise.}$$

7.2.2 Heat Kernel

The heat diffusion process over M is governed by the heat equation

$$\Delta_M u(x,t) = -\frac{\partial u(x,t)}{\partial t}, \tag{7.1}$$

where Δ_M is the Laplace-Beltrami operator of M. If M has boundaries, we additionally require u to satisfy certain boundary conditions (e.g. the Dirichlet boundary condition: $u(x,t) = 0$ for all $x \in \partial M$ and all t) in order to solve this partial differential equation. Given an initial heat distribution $f : M \rightarrow$, let $H_t(f)$ denote the heat distribution at time t, namely $H_t(f)$ satisfies the heat equation for all t, and $\lim_{t\to 0} H_t(f) = f$. H_t is called the *heat operator*. Both Δ_M and H_t are operators that map one real-valued function defined on M to another such function. It is easy to verify that they satisfy the following relation $H_t = e^{-t\Delta_M}$. Thus both operators share the same eigenfunctions and if λ is an eigenvalue of Δ_M, then $e^{-\lambda t}$ is an eigenvalue of H_t corresponding to the same eigenfunction.

It is well-known (see, e.g., [29]) that for any M, there exists a function $k_t(x,y) :^+ \times M \times M \rightarrow$ such that

$$H_t f(x) = \int_M k_t(x,y) f(y) dy, \tag{7.2}$$

where dy is the volume form at $y \in M$. The minimum function $k_t(x,y)$ that satisfies Eqn. 7.2), is called the *heat kernel*, and can be thought of as the amount of heat that is transferred from x to y in time t given a unit heat source at x. In other words $k_t(x,\cdot) = H_t(\delta_x)$ where δ_x is the Dirac delta function at x: $\delta_x(z) = 0$ for any $z \neq x$, and $\int_M \delta_x(z)dz = 1$. For compact M, the heat kernel has the following eigen-decomposition:

$$k_t(x,y) = \sum_{i=0}^{\infty} e^{-\lambda_i t}\phi_i(x)\phi_i(y), \tag{7.3}$$

where λ_i and ϕ_i are the ith eigenvalue and the ith eigenfunction of the Laplace-Beltrami operator, respectively.

The heat kernel $k_t(x,y)$ has many nice properties. For instance, it is symmetric: $k_t(x,y) = k_t(y,x)$ and satisfies the semigroup identity: $k_{t+s}(x,y) = \int_M k_t(x,z)k_s(y,z)dz$. In fact, one can recover the Riemannian metric from heat kernel as stated in 7.2.1, which means that heat kernel characterizes Riemannian manifolds up to isometry.

Theorem 7.2.1 *Let $T : M \rightarrow N$ be a surjective map between two Riemannian manifolds. T is an isometry if and only if $k_t^N(T(x),T(y)) = k_t^M(x,y)$ for any $x,y \in M$ and any $t > 0$.*

This proposition is a simple consequence of the following equation (see, e.g., [23]). For any x,y on a manifold,

$$\lim_{t\to 0} t \log k_t(x,y) = -\frac{1}{4}d^2(x,y) \tag{7.4}$$

where $d(x,y)$ is the geodesic distance between x and y on M.

Sun et al. [54] observed that for almost all Riemannian manifolds their metric can be recovered from $k_t(x,x)$ which only records the mount of heat remains at a point over the time. Specifically, they showed the following theorem.

Theorem 7.2.2 *If the eigenvalues of the Laplace-Beltrami operators of two compact manifolds M and N are not repeated,*[1] *and T is a homeomorphism from M to N, then T is isometric if and only if $k_t^M(x,x) = k_t^N(T(x),T(x))$ for any $x \in M$ and any $t > 0$.*

[1]Bando and Urakawa [1] show that the simplicity of the eigenvalues of the Laplace-Beltrami operator is a generic property.

7.3 Discrete Heat Kernel

7.3.1 Discrete Laplace–Beltrami Operator

In this work, we focus on discrete surfaces, namely polyhedral surfaces. For example, a triangle mesh is piecewise linearly embedded in $\mathbb{R}^3$.

Definition 7.3.1 (Polyhedral Surface) *A Euclidean polyhedral surface is a triple $(S, T, \mathbf{d})$, where S is a closed surface, T is a triangulation of S and $\mathbf{d}$ is a metric on S, whose restriction to each triangle is isometric to a Euclidean triangle.*

The well-known cotangent edge weight [17, 43] on a Euclidean polyhedral surface is defined as follows:

Definition 7.3.2 (Cotangent Edge Weight) *Suppose $[v_i, v_j]$ is a boundary edge of M, $[v_i, v_j] \in \partial M$, then $[v_i, v_j]$ is incident with a triangle $[v_i, v_j, v_k]$, the angle opposite to $[v_i, v_j]$, at the vertex v_k, is α, then the weight of $[v_i, v_j]$ is given by $w_{ij} = \frac{1}{2}\cot\alpha$. Otherwise, if $[v_i, v_j]$ is an interior edge, the two angles opposite to it are α, β, then the weight is $w_{ij} = \frac{1}{2}(\cot\alpha + \cot\beta)$.*

The discrete Laplace-Beltrami operator is constructed from the cotangent edge weight.

Definition 7.3.3 (Discrete Laplace Matrix) *The discrete Laplace matrix $L = (L_{ij})$ for a Euclidean polyhedral surface is given by*

$$L_{ij} = \begin{cases} -w_{ij}, & i \neq j \\ \sum_k w_{ik}, & i = j \end{cases}.$$

Because L is symmetric, it can be decomposed as

$$L = \Phi \Lambda \Phi^T, \tag{7.5}$$

where $\Lambda = diag(\lambda_0, \lambda_1, \cdots, \lambda_n)$, $0 = \lambda_0 < \lambda_1 \leq \lambda_2 \leq \cdots \leq \lambda_n$, are the eigenvalues of L, and $\Phi = (\phi_0|\phi_1|\phi_2|\cdots|\phi_n)$, $L\phi_i = \lambda_i \phi_i$, are the orthonormal eigenvectors, n is the number of vertices, such that $\phi_i^T \phi_j = \delta_{ij}$.

7.3.2 Discrete Heat Kernel

Definition 7.3.4 (Discrete Heat Kernel) *The discrete heat kernel is defined as follows:*

$$K(t) = \Phi exp(-\Lambda t) \Phi^T. \tag{7.6}$$

7.3.3 Main Theorem

The **Main Theorem**, called *Global Rigidity Theorem*, in this work is as follows:

Theorem 7.3.5 *Suppose two Euclidean polyhedral surfaces $(S, T, \mathbf{d}_1)$ and $(S, T, \mathbf{d}_2)$ are given,*

$$L_1 = L_2,$$

if and only if $\mathbf{d}_1$ and $\mathbf{d}_2$ differ by a scaling.

Corollary 1 *Suppose two Euclidean polyhedral surfaces $(S, T, \mathbf{d}_1)$ and $(S, T, \mathbf{d}_2)$ are given,*

$$K_1(t) = K_2(t), \forall t > 0,$$

if and only if $\mathbf{d}_1$ and $\mathbf{d}_2$ differ by a scaling.

Proof 13 *Note that,*

$$\frac{dK(t)}{dt}|_{t=0} = -L.$$

Therefore, the discrete Laplace matrix and the discrete heat kernel mutually determine each other.

7.3.4 Proof Outline

The main idea for the proof is as follows. We fix the connectivity of the polyhedral surface (S,T). Suppose the edge set of (S,T) is sorted as $E = \{e_1, e_2, \cdots, e_m\}$, where $m = |E|$ is the number of edges and F denotes the face set. A triangle $[v_i, v_j, v_k] \in F$ is also denoted as $\{i,j,k\} \in F$.

By definition, a Euclidean polyhedral metric on (S,T) is given by its edge length function $d : E \to \mathbb{R}^+$. We denote a metric as $\mathbf{d} = (d_1, d_2, \cdots, d_m)$, where $d_i = d(e_i)$ is the length of edge e_i. Let

$$E_{\mathbf{d}}(2) = \{(d_1, d_2, d_3) | d_i + d_j > d_k\}$$

be the space of all Euclidean triangles parameterized by the edge lengths, where $\{i,j,k\}$ is a cyclic permutation of $\{1,2,3\}$. In this work, for convenience, we use $\mathbf{u} = (u_1, u_2, \cdots, u_m)$ to represent the metric, where $u_k = \frac{1}{2}d_k^2$.

Definition 7.3.6 (Admissible Metric Space) *Given a triangulated surface (S,K), the admissible metric space is defined as*

$$\Omega_u = \{(u_1, u_2, u_3 \cdots, u_m) | \sum_{k=1}^{m} u_k = m, (\sqrt{u_i}, \sqrt{u_j}, \sqrt{u_k}) \in E_{\mathbf{d}}(2), \forall \{i,j,k\} \in F\}.$$

We show that Ω_u is a convex domain in $\mathbb{R}^m$.

Definition 7.3.7 (Energy) *An energy $E : \Omega_u \to \mathbb{R}$ is defined as:*

$$E(u_1, u_2 \cdots, u_m) = \int_{(1,1,\cdots,1)}^{(u_1,u_2\cdots,u_m)} \sum_{k=1}^{m} w_k(\mu) d\mu_k, \tag{7.7}$$

where $w_k(\mu)$ is the cotangent weight on the edge e_k determined by the metric μ, d is the exterior differential operator.

Next we show this energy is convex in Lemma 5. According to the following lemma, the gradient of the energy $\nabla E(\mathbf{d}) : \Omega \to \mathbb{R}^m$

$$\nabla E : (u_1, u_2 \cdots, u_m) \to (w_1, w_2, \cdots w_m)$$

is an embedding. Namely the metric is determined by the edge weight uniquely up to a scaling.

Lemma 1 *Suppose $\Omega \subset \mathbb{R}^n$ is an open convex domain in $\mathbb{R}^n$, $h : \Omega \to \mathbb{R}$ is a strictly convex function with positive definite Hessian matrix, then $\nabla h : \Omega \to \mathbb{R}^n$ is a smooth embedding.*

Proof 14 *If $\mathbf{p} \neq \mathbf{q}$ in Ω, let $\gamma(t) = (1-t)\mathbf{p} + t\mathbf{q} \in \Omega$ for all $t \in [0,1]$. Then $f(t) = h(\gamma(t)) : [0,1] \to \mathbb{R}$ is a strictly convex function, so that*

$$\frac{df(t)}{dt} = \nabla h|_{\gamma(t)} \cdot (\mathbf{q} - \mathbf{p}).$$

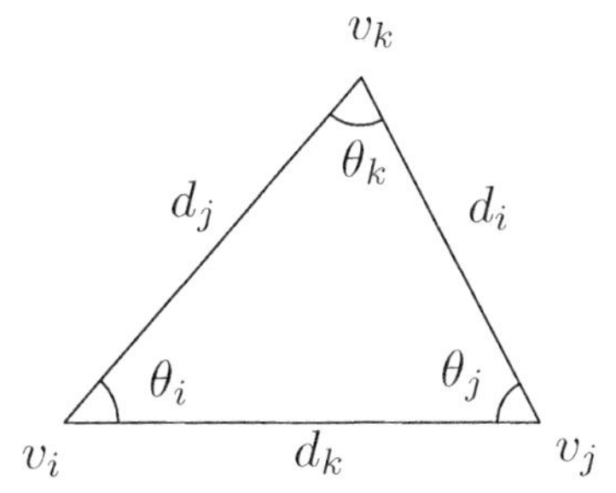

Figure 7.1: A Euclidean triangle.

Because

$$\frac{d^2 f(t)}{dt^2} = (\mathbf{q}-\mathbf{p})^T H|_{\gamma(t)}(\mathbf{q}-\mathbf{p}) > 0,$$

$\frac{df(0)}{dt} \neq \frac{df(1)}{dt}$, *therefore*

$$\nabla h(\mathbf{p}) \cdot (\mathbf{q}-\mathbf{p}) \neq \nabla h(\mathbf{q}) \cdot (\mathbf{q}-\mathbf{p}).$$

This means $\nabla h(\mathbf{p}) \neq \nabla h(\mathbf{q})$, *therefore* ∇h *is injective.*

On the other hand, the Jacobi matrix of ∇h *is the Hessian matrix of* h, *which is positive definite. It follows that* $\nabla h : \Omega \to \mathbb{R}^n$ *is a smooth embedding.*

From the discrete Laplace-Beltrami operator (Eqn. (7.5)) or the heat kernel (Eqn. (7.6)), we can compute all the cotangent edge weights, then because the edge weight determines the metric, we attain the Main Theorem 7.3.5.

7.3.5 Rigidity on One Face

In this section, we show the proof for the simplest case, a Euclidean triangle; in the next section, we generalize the proof to all types of triangle meshes.

Given a triangle $\{i, j, k\}$, three corner angles denoted by $\{\theta_i, \theta_j, \theta_k\}$, and three edge lengths denoted by $\{d_i, d_j, d_k\}$, as shown in Figure 7.1. In this case, the problem is trivial. Given $(w_i, w_j, w_k) = (\cot\theta_i, \cot\theta_j, \cot\theta_k)$, we can compute $(\theta_i, \theta_j, \theta_k)$ by taking the *arccot* function. Then the normalized edge lengths are given by

$$(d_i, d_j, d_k) = \frac{3}{\sin\theta_i + \sin\theta_j + \sin\theta_k}(\sin\theta_i, \sin\theta_j, \sin\theta_k).$$

Although this approach is direct and simple, it can not be generalized to more complicated polyhedral surfaces. In the following, we use a different approach, which can be generalized to all polyhedral surfaces.

Lemma 2 *Suppose a Euclidean triangle is with angles* $\{\theta_i, \theta_j, \theta_k\}$ *and edge lengths* $\{d_i, d_j, d_k\}$, *angles are treated as the functions of the edge lengths* $\theta_i(d_i, d_j, d_k)$, *then*

$$\frac{\partial\theta_i}{\partial d_i} = \frac{d_i}{2A} \tag{7.8}$$

and

$$\frac{\partial\theta_i}{\partial d_j} = -\frac{d_i}{2A}\cos\theta_k, \tag{7.9}$$

where A *is the area of the triangle.*

Proof 15 *According to Euclidean cosine law,*

$$\cos\theta_i = \frac{d_j^2 + d_k^2 - d_i^2}{2d_j d_k}, \tag{7.10}$$

we take derivative on both sides with respective to d_i,

$$-\sin\theta_i \frac{\partial\theta_i}{\partial d_i} = \frac{-2d_i}{2d_j d_k}$$

$$\frac{\partial\theta_i}{\partial d_i} = \frac{d_i}{d_j d_k \sin\theta_i} = \frac{d_i}{2A}, \tag{7.11}$$

where $A = \frac{1}{2}d_j d_k \sin\theta_i$ is the area of the triangle. Similarly,

$$\frac{\partial}{\partial d_j}(d_j^2 + d_k^2 - d_i^2) = \frac{\partial}{\partial d_j}(2d_j d_k \cos\theta_i)$$

$$2d_j = 2d_k \cos\theta_i - 2d_j d_k \sin\theta_i \frac{\partial\theta_i}{\partial d_j}$$

$$2A\frac{\partial\theta_i}{\partial d_j} = d_k\cos\theta_i - d_j = -d_i\cos\theta_k.$$

We get

$$\frac{\partial\theta_i}{\partial d_j} = -\frac{d_i\cos\theta_k}{2A}.$$

Lemma 3 *In a Euclidean triangle, let $u_i = \frac{1}{2}d_i^2$ and $u_j = \frac{1}{2}d_j^2$ then*

$$\frac{\partial\cot\theta_i}{\partial u_j} = \frac{\partial\cot\theta_j}{\partial u_i}. \tag{7.12}$$

Proof 16

$$\begin{aligned}\frac{\partial\cot\theta_i}{\partial u_j} &= \frac{1}{d_j}\frac{\partial\cot\theta_i}{\partial d_j} = -\frac{1}{d_j}\frac{1}{\sin^2\theta_i}\frac{\partial\theta_i}{\partial d_j} \\ &= \frac{1}{d_j}\frac{1}{\sin^2\theta_i}\frac{d_i\cos\theta_k}{2A} = \frac{d_i^2}{\sin^2\theta_i}\frac{\cos\theta_k}{2Ad_id_j} = \frac{4R^2}{2A}\frac{\cos\theta_k}{d_id_j},\end{aligned} \tag{7.13}$$

where R is the radius of the circumcircle of the triangle. The righthand side of Eqn. (7.13) is symmetric with respect to the indices i and j.

In the following, we introduce a differential form. We are going to use them for proving that the integration involved in computing energy is independent of paths. This follows from the fact that the forms which are integrated are closed, and the integration domain is simply connected.

Corollary 2 *The differential form*

$$\omega = \cot\theta_i du_i + \cot\theta_j du_j + \cot\theta_k du_k \tag{7.14}$$

is a closed 1-form.

Proof 17 *By the above Lemma 3 regarding symmetry,*

$$\begin{aligned} d\omega &= (\frac{\partial \cot\theta_j}{\partial u_i} - \frac{\partial \cot\theta_i}{\partial u_j})du_i \wedge du_j + (\frac{\partial \cot\theta_k}{\partial u_j} - \frac{\partial \cot\theta_j}{\partial u_k})du_j \wedge du_k \\ &+ (\frac{\partial \cot\theta_i}{\partial u_k} - \frac{\partial \cot\theta_k}{\partial u_i})du_k \wedge du_i \\ &= 0. \end{aligned}$$

Definition 7.3.8 (Admissible Metric Space) *Let* $u_i = \frac{1}{2}d_i^2$*, the admissible metric space is defined as*

$$\Omega_u := \{(u_i, u_j, u_k) | (\sqrt{u_i}, \sqrt{u_j}, \sqrt{u_k}) \in E_{\mathbf{d}}(2),\ u_i + u_j + u_k = 3\}.$$

Lemma 4 *The admissible metric space* Ω_u *is a convex domain in* $\mathbb{R}^3$.

Proof 18 *Suppose* $(u_i, u_j, u_k) \in \Omega_u$ *and* $(\tilde{u}_i, \tilde{u}_j, \tilde{u}_k) \in \Omega_u$*, then from* $\sqrt{u_i} + \sqrt{u_j} > \sqrt{u_k}$*, we get* $u_i + u_j + 2\sqrt{u_i u_j} > u_k$*. Define* $(u_i^\lambda, u_j^\lambda, u_k^\lambda) = \lambda(u_i, u_j, u_k) + (1-\lambda)(\tilde{u}_i, \tilde{u}_j, \tilde{u}_k)$*, where* $0 < \lambda < 1$*. Then*

$$\begin{aligned} u_i^\lambda u_j^\lambda &= (\lambda u_i + (1-\lambda)\tilde{u}_i)(\lambda u_j + (1-\lambda)\tilde{u}_j) \\ &= \lambda^2 u_i u_j + (1-\lambda)^2 \tilde{u}_i \tilde{u}_j + \lambda(1-\lambda)(u_i \tilde{u}_j + u_j \tilde{u}_i) \\ &\geq \lambda^2 u_i u_j + (1-\lambda)^2 \tilde{u}_i \tilde{u}_j + 2\lambda(1-\lambda)\sqrt{u_i u_j \tilde{u}_i \tilde{u}_j} \\ &= (\lambda\sqrt{u_i u_j} + (1-\lambda)\sqrt{\tilde{u}_i \tilde{u}_j})^2. \end{aligned}$$

It follows

$$\begin{aligned} u_i^\lambda + u_j^\lambda + 2\sqrt{u_i^\lambda u_j^\lambda} &\geq \lambda(u_i + u_j + 2\sqrt{u_i u_j}) + (1-\lambda)(\tilde{u}_i + \tilde{u}_j + 2\sqrt{\tilde{u}_i \tilde{u}_j}) \\ &> \lambda u_k + (1-\lambda)\tilde{u}_k = u_k^\lambda. \end{aligned}$$

This shows $(u_i^\lambda, u_j^\lambda, u_k^\lambda) \in \Omega_u$.

Similarly, we define the edge weight space as follows.

Definition 7.3.9 (Edge Weight Space) *The edge weights of a Euclidean triangle form the edge weight space*

$$\Omega_\theta = \{(\cot\theta_i, \cot\theta_j, \cot\theta_k) | 0 < \theta_i, \theta_j, \theta_k < \pi, \theta_i + \theta_j + \theta_k = \pi\}.$$

Note that,

$$\cot\theta_k = -\cot(\theta_i + \theta_j) = \frac{1 - \cot\theta_i \cot\theta_j}{\cot\theta_i + \cot\theta_j}.$$

Lemma 5 *The energy* $E : \Omega_u \to \mathbb{R}$

$$E(u_i, u_j, u_k) = \int_{(1,1,1)}^{(u_i, u_j, u_k)} \cot\theta_i d\tau_i + \cot\theta_j d\tau_j + \cot\theta_k d\tau_k \tag{7.15}$$

is well defined on the admissible metric space Ω_u *and is convex.*

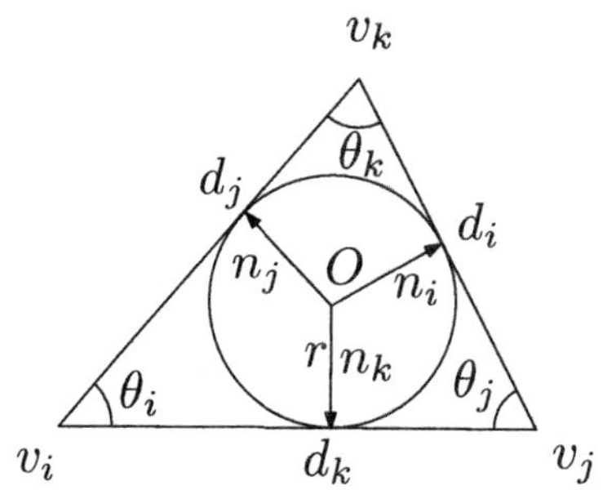

Figure 7.2: The geometric interpretation of the Hessian matrix. The geometric interpretation of the Hessian matrix. The in circle of the triangle is centered at O, with radius r. The perpendiculars, n_i, n_j, and n_k, are from the incenter of the triangle and orthogonal to the edge, e_i, e_j, and e_k, respectively.

Proof 19 *According to Corollary 2, the differential form is closed. Furthermore, the admissible metric space Ω_u is a simply connected domain and the differential form is exact. Therefore, the integration is path independent, and the energy function is well defined.*

Then we compute the Hessian matrix of the energy,

$$H = -\frac{2R^2}{A}\begin{bmatrix} \frac{1}{d_i^2} & -\frac{\cos\theta_k}{d_i d_j} & -\frac{\cos\theta_j}{d_i d_k} \\ -\frac{\cos\theta_k}{d_j d_i} & \frac{1}{d_j^2} & -\frac{\cos\theta_i}{d_j d_k} \\ -\frac{\cos\theta_j}{d_k d_i} & -\frac{\cos\theta_i}{d_k d_j} & \frac{1}{d_k^2} \end{bmatrix} = -\frac{2R^2}{A}\begin{bmatrix} (\eta_i,\eta_i) & (\eta_i,\eta_j) & (\eta_i,\eta_k) \\ (\eta_j,\eta_i) & (\eta_j,\eta_j) & (\eta_j,\eta_k) \\ (\eta_k,\eta_i) & (\eta_k,\eta_j) & (\eta_k,\eta_k) \end{bmatrix}.$$

As shown in Figure 7.2, $d_i\mathbf{n}_i + d_j\mathbf{n}_j + d_k\mathbf{n}_k = 0$,

$$\eta_i = \frac{\mathbf{n}_i}{rd_i}, \eta_j = \frac{\mathbf{n}_j}{rd_j}, \eta_k = \frac{\mathbf{n}_k}{rd_k},$$

where r is the radius of the incircle of the triangle.

Suppose $(x_i, x_j, x_k) \in \mathbb{R}^3$ is a vector in $\mathbb{R}^3$, then

$$[x_i, x_j, x_k]\begin{bmatrix} (\eta_i,\eta_i) & (\eta_i,\eta_j) & (\eta_i,\eta_k) \\ (\eta_j,\eta_i) & (\eta_j,\eta_j) & (\eta_j,\eta_k) \\ (\eta_k,\eta_i) & (\eta_k,\eta_j) & (\eta_k,\eta_k) \end{bmatrix}\begin{bmatrix} x_i \\ x_j \\ x_k \end{bmatrix} = \|x_i\eta_i + x_j\eta_j + x_k\eta_k\|^2 \geq 0.$$

If the result is zero, then $(x_i, x_j, x_k) = \lambda(u_i, u_j, u_k), \lambda \in \mathbb{R}$. That is the null space of the Hessian matrix. In the admissible metric space Ω_u, $u_i + u_j + u_k = C(C = 3)$, then $du_i + du_j + du_k = 0$. If (du_i, du_j, du_k) belongs to the null space, then $(du_i, du_j, du_k) = \lambda(u_i, u_j, u_k)$, therefore, $\lambda(u_i + u_j + u_k) = 0$. Because u_i, u_j, u_k are positive, $\lambda = 0$. This shows the null space of Hessian matrix is orthogonal to the tangent space of Ω_u. Therefore, the Hessian matrix is positive definite on the tangent space. In summary, the energy on Ω_u is convex.

Theorem 7.3.10 *The mapping $\nabla E : \Omega_u \to \Omega_\theta, (u_i, u_j, u_k) \to (\cot\theta_i, \cot\theta_j, \cot\theta_k)$ is a diffeomorphism.*

Proof 20 *The energy $E(u_i, u_j, u_k)$ is a convex function defined on the convex domain Ω_u, according to Lemma 1, $\nabla E : (u_i, u_j, u_k) \to (\cot\theta_i, \cot\theta_j, \cot\theta_k)$ is a diffeomorphism.*

7.3.6 Rigidity for the Whole Mesh

In this section, we consider the whole polyhedral surface.

Closed Surfaces

Given a polyhedral surface $(S, T, \mathbf{d})$, the admissible metric space and the edge weight have been defined in Definitions 7.3.6 and 7.3.2 respectively.

Lemma 6 *The admissible metric space Ω_u is convex.*

Proof 21 *For a triangle $\{i, j, k\} \in F$, define*

$$\Omega_u^{ijk} := \{(u_i, u_j, u_k) | (\sqrt{u_i}, \sqrt{u_j}, \sqrt{u_k}) \in E_{\mathbf{d}}(2)\}.$$

Similar to the proof of Lemma 4, Ω_u^{ijk} is convex. The admissible metric space for the mesh is

$$\Omega_u = \bigcap_{\{i,j,k\} \in F} \Omega_u^{ijk} \bigcap \{(u_1, u_2, \cdots, u_m) | \sum_{k=1}^{m} u_k = m\},$$

the intersection Ω_u is still convex.

Definition 7.3.11 (Differential Form) *The differential form ω defined on Ω_u is the summation of the differential form on each face,*

$$\omega = \sum_{\{i,j,k\} \in F} \omega_{ijk} = \sum_{i=1}^{m} 2 w_i du_i,$$

where ω_{ijk} is given in Eqn. (7.14) in Corollary 2, w_i is the edge weight on e_i, m is the number of edges.

Lemma 7 *The differential form ω is a closed 1-form.*

Proof 22 *According to Corollary 2,*

$$d\omega = \sum_{\{i,j,k\} \in F} d\omega_{ijk} = 0.$$

Lemma 8 *The energy function*

$$E(u_1, u_2, \cdots, u_m) = \sum_{\{i,j,k\} \in F} E_{ijk}(u_1, u_2, \cdots, u_m) = \int_{(1,1,\cdots,1)}^{(u_1, u_2, \cdots, u_m)} \sum_{i=1}^{n} w_i du_i$$

is well defined and convex on Ω_u, where E_{ijk} is the energy on the face, defined in Eqn. (7.15).

Proof 23 *For each face $\{i, j, k\} \in F$, the Hessian matrices of E_{ijk} are semi-positive definite, therefore, the Hessian matrix of the total energy E is semi-positive definite.*

Similar to the proof of Lemma 5, the null space of the Hessian matrix H is

$$kerH = \{\lambda(d_1, d_2, \cdots, d_m), \lambda \in \mathbb{R}\}.$$

The tangent space of Ω_u at $\mathbf{u} = (u_1, u_2, \cdots, u_m)$ is denoted by $T\Omega_u(\mathbf{u})$. Assume $(du_1, du_2, \cdots, du_m) \in T\Omega_u(\mathbf{u})$, then from $\sum_{i=1}^{m} u_i = m$, we get $\sum_{i=1}^{m} du_m = 0$. Therefore,

$$T\Omega_u(\mathbf{u}) \cap KerH = \{0\},$$

hence H is positive definite restricted on $T\Omega_u(\mathbf{u})$. So the total energy E is convex on Ω_u.

Theorem 7.3.12 *The mapping on a closed Euclidean polyhedral surface $\nabla E : \Omega_u \rightarrow \mathbb{R}^m, (u_1, u_2, \cdots, u_m) \rightarrow (w_1, w_2, \cdots, w_m)$ is a smooth embedding.*

Proof 24 *The admissible metric space Ω_u is convex as shown in Lemma 6, the total energy is convex as shown in Lemma 8. According to Lemma 1, ∇E is a smooth embedding.*

Open Surfaces

By the double covering technique [25], we can convert a polyhedral surface with boundaries to a closed surface. First, let $(\bar{S}, \bar{T})$ be a copy of (S, T), then we reverse the orientation of each face in $\bar{M}$, and glue two surfaces S and $\bar{S}$ along their corresponding boundary edges, so that the resulting triangulated surface is a closed one. We get the following corollary

Corollary 3 *The mapping on a Euclidean polyhedral surface with boundaries* $\nabla E : \Omega_u \to \mathbb{R}^m, (u_1, u_2, \cdots, u_m) \to (w_1, w_2, \cdots, w_m)$ *is a smooth embedding.*

Surely, the cotangent edge weights can be uniquely obtained from the discrete heat kernel. By combining Theorem 7.3.12 and Corollary 3, we obtain the major Theorem 7.3.5, *Global Rigidity Theorem*, of this work.

7.4 Heat Kernel Simplification

Recall that heat kernel $k_t(x, y)$ is a function of three variables where x, y are spatial variables and t is a temporal variable. Since heat kernel is the fundamental solution to the heat equation, these variables are not independent. Sun et al. [54] observed that for almost all Riemannian manifolds their metric can be recovered from $k_t(x, x)$ which only records the mount of heat remains at a point over the time. Specifically, they showed the following theorem.

Theorem 7.4.1 *If the eigenvalues of the Laplace-Beltrami operators of two compact manifolds M and N are not repeated,*[2] *and T is a homeomorphism from M to N, then T is isometric if and only if $k_t^M(x, x) = k_t^N(T(x), T(x))$ for any $x \in M$ and any $t > 0$.*

Proof 25 *We prove the theorem in three steps.*

Step 1: *We claim that M and N have the same spectrum and $|\phi_i^M(x)| = |\phi_i^N(T(x))|$ for any eigenfunction ϕ_i and any $x \in M$. We prove the above claim by contradiction. In the following, we sort the eigenvalues in the increasing order. The claim can fail first at the kth eigenvalue for some k, namely $\lambda_k^M \neq \lambda_k^N$ but $\lambda_i^M = \lambda_i^N$ and $|\phi_i^M(x)| = |\phi_i^N(T(x))|$ for any $i < k$ and any $x \in M$, or fail first at the kth eigenfunction for some k, namely there exists a point x such that $|\phi_k^M(x)| \neq |\phi_k^N(T(x))|$ but $\lambda_i^M = \lambda_i^N = \lambda_i$ for any $i \leq k$ and $|\phi_i^M(x)| = |\phi_i^N(T(x))|$ for any $i < k$ and any $x \in M$. In the former case, WLOG, assume $\lambda_k^M < \lambda_k^N$. There must exist a point $x \in M$ such that $\phi_k^M(x)^2 = \epsilon > 0$ for some ϵ. From Eqn. (7.3), we have*

$$\begin{aligned} & k_t^M(x, x) - k_t^N(T(x), T(x)) \\ > \; & e^{-\lambda_k^M t} \phi_k^M(x)^2 - \sum_{i=k}^{\infty} e^{-\lambda_i^N t} \phi_i^N(T(x))^2 \\ = \; & e^{-\lambda_k^M t} \big(\epsilon - \sum_{i=k}^{\infty} e^{-(\lambda_i^N - \lambda_k^M) t} \phi_i^N(T(x))^2\big). \end{aligned} \tag{7.16}$$

By the local Weyl law [28], we have $|\phi_i^N(T(x))| = O((\lambda_i^N)^{(d-1)/4})$ where d is the dimension of N. In addition, the sequence $\{\lambda_i^N\}_{i=0}^{\infty}$ is increasing and hence $\lambda_i^N - \lambda_k^M > 0$ for any $i \geq k$. As the exponential decay can cancel the increasing of any polynomial, we have

[2] Bando and Urakawa[1] show that the simplicity of the eigenvalues of the Laplace-Beltrami operator is a generic property

$$\lim_{t\to\infty}\sum_{i=k}^{\infty} e^{-(\lambda_i^N-\lambda_k^M)t}\phi_i^N(T(x))^2 = 0$$

By choosing a big enough t*, we have* $k_t^M(x,x) - k_t^N(T(x),T(x)) > 0$ *from Eqn. (7.16), which contradicts the hypothesis. In the latter case, WLOG, assume* $\epsilon = \phi_k^M(x)^2 - \phi_k^N(T(x))^2 > 0$*. We have*

$$\begin{aligned}
& k_t^M(x,x) - k_t^N(T(x),T(x)) \\
> \; & e^{-\lambda_k t}(\phi_k^M(x)^2 - \phi_k^N(T(x))^2) - \sum_{i=k+1}^{\infty} e^{-\lambda_i^N t}\phi_i^N(T(x))^2 \\
= \; & e^{-\lambda_k t}(\epsilon - \sum_{i=k+1}^{\infty} e^{-(\lambda_i^N-\lambda_k)t}\phi_i^N(T(x))^2).
\end{aligned} \tag{7.17}$$

Since the sequence $\{\lambda_i^N\}_{i=0}^{\infty}$ *is strictly increasing, similarly for a big enough* t*, we have* $k_t^M(x,x) - k_t^N(T(x),T(x)) > 0$ *from Eqn. (7.17), which contradicts the hypothesis.*

Step 2: *We show that either* $\phi_i^M = \phi_i^N \circ T$ *or* $\phi_i^M = -\phi_i^N \circ T$ *for any* i*. The argument is based on the properties of the nodal domains of the eigenfunction* ϕ*. A nodal domain is a connected component of* $M \setminus \phi^{-1}(0)$*. The sign of* ϕ *is consistent within a nodal domain that is either all positive or all negative. For a fixed eigenfunction, the number of the nodal domains is finite. Since* $|\phi_i^M(x)| = |\phi_i^N(T(x))|$ *and* T *is continuous, the image of a nodal domain under* T *cannot cross two nodal domains, that, is a nodal domain can only be mapped to another nodal domain. A special property of the nodal domains [8] is that a positive nodal domain is only neighbored by negative ones, and vice versa. Pick a fixed point* x_0 *in a nodal domain. If* $\phi_i^M(x_0) = \phi_i^N(T(x_0))$*, we claim that* $\phi_i^M(x) = \phi_i^N(T(x))$ *for any point* x *on the manifold. Certainly the claim holds for the points inside the nodal domain* D *containing* x_0*. Due to the continuity of* T*, the neighboring nodal domains of* D *must be mapped to those next to the one containing* $T(x_0)$*. Because of the alternating property of the signs of neighboring nodal domains, the claims also hold for those neighboring ones. We can continue on expanding nodal domains like this until they are exhausted, which proves the claim. Thus* $\phi_i^M = \phi_i^M \circ T$*. Similarly,* $\phi_i^M(x_0) = -\phi_i^N(T(x_0))$ *leads to* $\phi_i^M = -\phi_i^N \circ T$*.*

Step 3: *We have for any* $x, y \in M$ *and* $t > 0$

$$\begin{aligned}
k_t^M(x,y) &= \sum_{i=0}^{\infty} e^{-\lambda_i t}\phi_i^M(x)\phi_i^M(y) \\
&= \sum_{i=0}^{\infty} e^{-\lambda_i t}\phi_i^N(T(x))\phi_i^N(T(y)) \\
&= k_t^N(T(x),T(y))
\end{aligned} \tag{7.18}$$

which proves the theorem by Eqn. 7.4.

The theorem above assures that the set of functions $HKS_x : \mathbb{R}^+ \to \mathbb{R}^+$ defined by $HKS_x(t) = k_t(x,x)$ for any x on the manifold is almost as informative as the heat kernel $k_t(x,y)$ for any x, y on the manifold and any $t > 0$. In [54], HKS_x is called the Heat Kernel Signature at x. Most notably, the Heat Kernel Signatures at different points are defined over a common temporal domain, which makes them easily commensurable and has been used in many applications in shape analysis [41, 15].

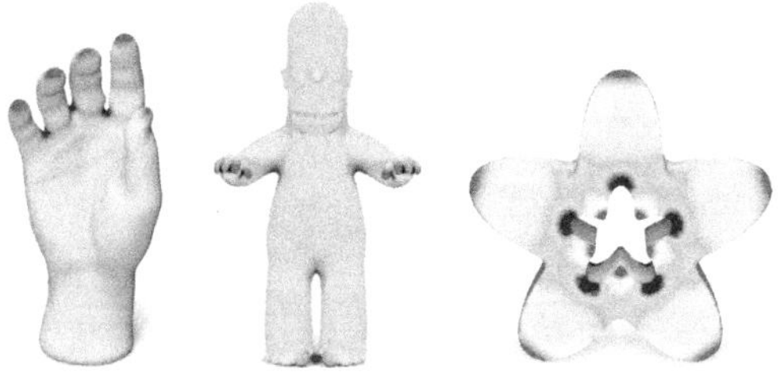

Figure 7.3: (See Color Insert.) Heat kernel function $k_t(x,x)$ for a small fixed t on the hand, Homer, and trim-star models. The function values increase as the color goes from blue to green and to red, with the mapping consistent across the shapes. Note that high and low values of $k_t(x,x)$ correspond to areas with positive and negative Gaussian curvatures, respectively.

In addition to the theorem above, which is rather global in nature, the Heat Kernel Signature for small t at a point x is directly related to the scalar curvature $s(x)$ (twice of Gaussian curvature on a surface) as shown by the following asymptotic expansion which is due to Mckean and Singer [26]

$$k_t(x,x) = (4\pi t)^{-d/2}\sum_{i=0}^{\infty} a_i t^i,$$

where $a_0 = 1$ and $a_1 = \frac{1}{6}s(x)$. This expansion corresponds to the well-known property of the heat diffusion process, which states that heat tends to diffuse slower at points with positive curvature, and faster at points with negative curvature. Figure 7.3 plots the values of $k_t(x,x)$ for a fixed small t on three shapes, where the colors are consistent across the shapes. Note that the values of this function are large in highly curved areas, and small in negatively curved areas. Note that even for the trim-star, which has sharp edges, $k_t(x,x)$ provides a meaningful notion of curvature at all points. For this reason, the function $k_t(x,x)$ can be interpreted as the intrinsic curvature at x at scale t.

Moreover, the Heat Kernel Signature is also closely related to diffusion maps and diffusion distances proposed by Coifman and Lafon [11] for data representation and dimensionality reduction. The diffusion distance between $x, y \in M$ at time scale t is defined as

$$d_t^2(x,y) \quad = \quad k_t(x,x) + k_t(y,y) - 2k_t(x,y).$$

The eccentricity of x in terms of diffusion distance, denoted $ecc_t(x)$, is defined as the average squared diffusion distance over the entire manifold):

$$ecc_t(x) \quad = \quad \frac{1}{A_M}\int_M d_t^2(x,y)dy = k_t(x,x) + H_M(t) - \frac{2}{A_M},$$

where A_M is the surface area of M, and $H_M(t) = \sum_i e^{-\lambda_i t}$ is the heat trace of M. Since both $H_M(t)$ and $\frac{2}{A_M}$ are independent of x, if we consider both $ecc_t(x)$ and $k_t(x,x)$ as functions over M, their level sets, in particular extrema points, coincide. Thus, for small t, we expect the extremal points of $ecc_t(x)$ to be located at the highly curved areas.

7.5 Numerical Experiments

From the above theoretic deduction, we can design the algorithm to compute discrete metric with user prescribed edge weights.

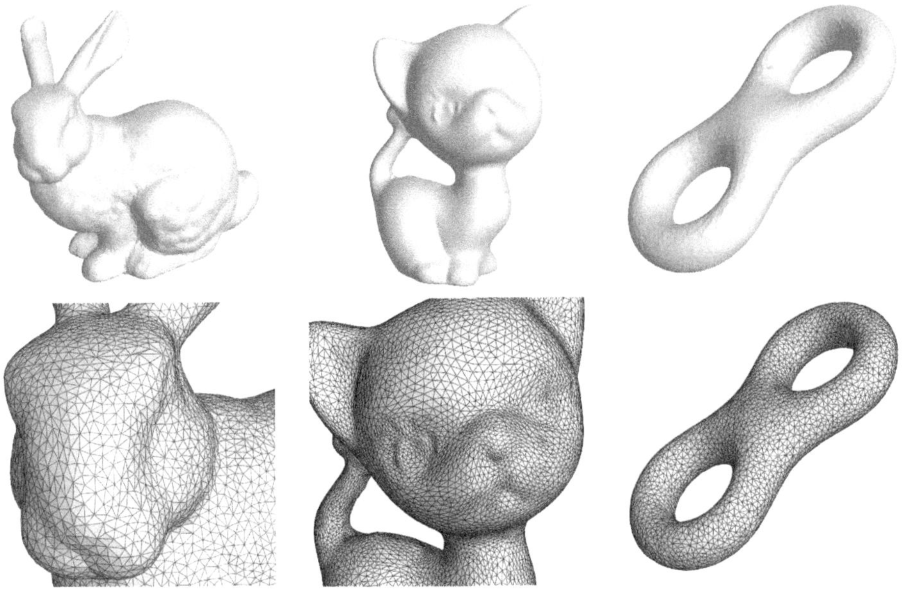

Figure 7.4: Euclidean polyhedral surfaces used in the experiments.

Problem Let (S,T) be a triangulated surface, $\bar{\mathbf{w}}(\bar{w}_1, \bar{w}_2, \cdots, \bar{w}_n)$ are the user prescribed edge weights. The problem is to find a discrete metric $\mathbf{u} = (u_1, u_2, \cdots, u_n)$, such that this metric $\bar{\mathbf{u}}$ induces the desired edge weight $\mathbf{w}$.

The algorithm is based on the following theorem.

Theorem 7.5.1 *Suppose (S,T) is a triangulated surface. If there exists an $\bar{\mathbf{u}} \in \Omega_u$, which induces $\bar{\mathbf{w}}$, then $\mathbf{u}$ is the unique global minimum of the energy*

$$E(\mathbf{u}) = \int_{(1,1,\cdots,1)}^{(u_1,u_2,\cdots,u_n)} \sum_{i=1}^{n} (\bar{w}_i - w_i) d\mu_i. \tag{7.19}$$

Proof 26 *The gradient of the energy $\nabla E(\mathbf{u}) = \bar{\mathbf{w}} - \mathbf{w}$, and since $\nabla E(\bar{\mathbf{u}}) = 0$, therefore $\bar{\mathbf{u}}$ is a critical point. The Hessian matrix of $E(\mathbf{u})$ is positive definite, the domain Ω_u is convex, therefore $\bar{\mathbf{u}}$ is the unique global minimum of the energy.*

In our numerical experiments, as shown in Figure 7.4, we tested surfaces with different topologies, with different genus, with or without boundaries. All discrete polyhedral surfaces are triangle meshes scanned from real objects. Because the meshes are embedded in $\mathbb{R}^3$, they have induced Euclidean metric, which are used as the desired metric $\bar{\mathbf{u}}$. From the induced Euclidean metric, the desired edge weight $\bar{\mathbf{w}}$ can be directly computed. Then we set the initial discrete metric to be the constant metric $(1, 1, \cdots, 1)$. By optimizing the energy in Eqn. (7.19), we can reach the global minimum, and recover the desired metric, which differs from the induced Euclidean metric by a scaling.

7.6 Applications

Laplace–Beltrami operator has been a broad range of applications. It has been applied for mesh parameterization in the graphics field. First order finite element approximations of the

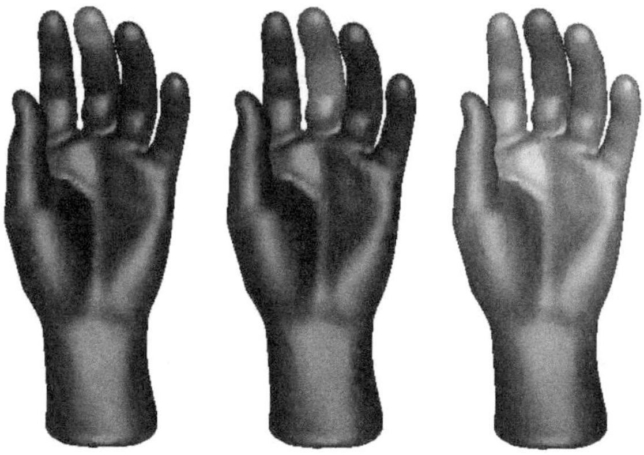

Figure 7.5: (See Color Insert.) From left to right, the function $k_t(x, \cdot)$ with $t = 0.1, 1, 10$ where x is at the tip of the middle figure.

Cauchy-Riemann equations were introduced by Levy et al. [38]. Discrete intrinsic parameterization by minimizing Dirichlet energy was introduced by [14]. Mean value coordinates were introduced in [18] to compute generalized harmonic maps; discrete spherical conformal mappings are used in [9]. Global conformal parameterization based on discrete holomorphic 1-form was introduced in [25]. We refer readers to [19, 36] for thorough surveys.

Laplace-Beltrami operator has been applied for shape analysis. The eigenfunctions of Laplace-Beltrami operator have been applied for global intrinsic symmetry detection in [42]. Heat Kernel Signature was proposed in [54], which is concise and characterizes the shape up to isometry. Spectral methods have been applied for mesh processing and analysis, which rely on the eigenvalues, eigenvectors, or eigenspace projections. We refer readers to [60] for a more detailed survey.

Heat kernel not only determines the metric but also has the advantage of encoding the metric information in a multiscale manner through the parameter t. In particular, consider the heat kernel $k_t(x, y)$. If we fix x, it becomes a function over the manifold. For small values of t, the function $k_t(x, \cdot)$ is mainly determined by small neighborhoods of x, and these neighborhoods grow bigger as t increases; see Figure 7.5. This implies that for small t, the function $k_t(x,)$ only reflects local properties of the shape around x, while for large values of t, $k_t(x,)$ captures the global structure of the manifold from the point of view of x. Therefore heat kernel has the ability to deal with noise, which makes it especially suitable for the applications in shape analysis. To demonstrate this, we will list three of its applications in designing point signature [54], finding correspondences [41], and defining metric in shape space [40].

Shape signature It is desirable to derive shape signatures that are invariant under certain transformations such as isometric transformation to facilitate comparison and differentiation between shapes or parts of a shape. A large amount of work has been done on designing various local point signatures in the context of shape analysis [5, 27, 48]. Recently a point signature based on heat kernel, called heat kernel signature (HKS) [54] has received much attention from the shape analysis community. Specifically, the heat kernel signature at a point x on a shape bounded by a surface M is defined as $HKS(x) = k_t(x, x)$, which basically records the amount of heat remaining at x over time. Sun et al. [54] show that heat kernel signature can recover the metric information for almost all manifolds (see Theorem 7.4.1) and demonstrate many nice properties of heat kernel signature including: encoding geometry in a multiscale way, stable against small perturbation, easy to compare signatures at different points. In addition, Sun et al. show its usage in multiscale matching. For

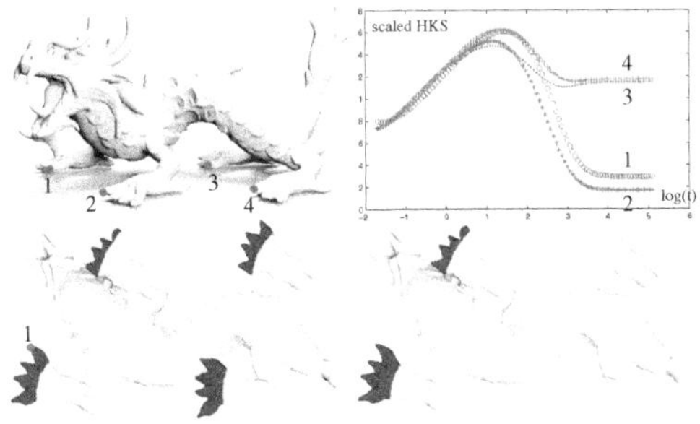

Figure 7.6: Top left: dragon model; top right: scaled HKS at points 1, 2, 3, and 4. Bottom left: the points whose signature is close to the signature of point 1 based on the smaller half of the t's; bottom right: based on the entire range of t's.

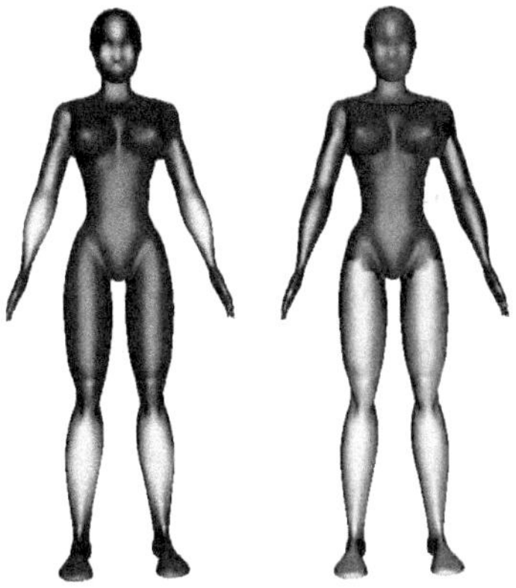

(a) (b)

Figure 7.7: (See Color Insert.) (a) The function of $k_t(x, x)$ for a fixed scale t over a human; (b) The segmentation of the human based on the stable manifold of extreme points of the function shown in (a).

example, in Figure 7.6, the difference between the HKS of the marked point x and the signatures of other points on the model are color plotted. As we can see, at small scales, all four feet of the Dragon are similar to each other. On the other hand, if large values of t, and consequently large neighborhoods are taken into account, the difference function can separate the front feet from the back feet, since the head of the dragon is quite different from its tail. In addition, heat kernel signature can be used for shape segmentation [49, 53, 15] where one considers the heat kernel signature at a relatively large scale, which becomes a function defined over the underlying manifold; see Figure 7.7(a). The segmentation is computed as the stable manifolds of the maximal points of that function. To deal with noise, persistence homology is employed to cancel out those noisy extrema; see Figure 7.7(b).

Shape correspondences Finding a correspondence between the shapes that undergo isometric deformation can find many applications, such as animation reconstruction and human motion estimation. Based on heat kernel, Ovsjanikov et al. [41] show that for a generic Riemannian manifold,[3] any isometry is uniquely determined by the mapping of one generic point. A generic point is a point where the evaluation of any eigenfunction of the Laplace-Beltrami operator is not zero. Specifically, given a point p on a manifold M, the

[3] Its Laplace-Beltrami operator has no repeated eigenvalues.

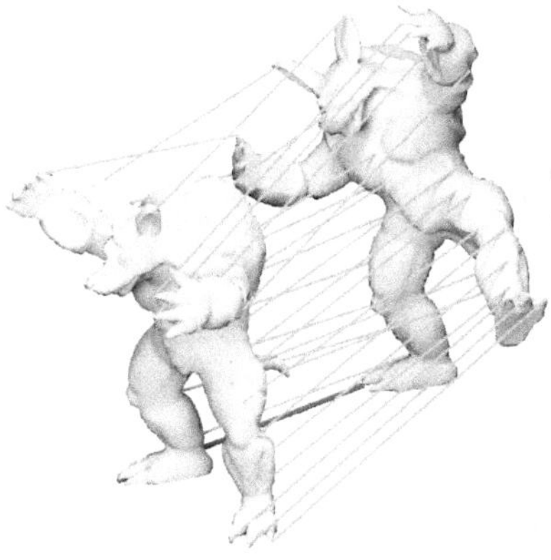

Figure 7.8: Red line: specified corresponding points; green line: corresponding points computed by the algorithm based on heat kernel map.

so-called heat kernel map $\Psi_p^M : M \to C(\mathbb{R}^+)$ is defined by $\psi_p^M(x) = k_t(p, x)$ where $k_t(p, x)$ is considered as a function of t. It is shown that the heat kernel map is injective if both M and p are generic. A generic point is indeed generic as the 0 level set of each eigenfunction is of measure 0 and there are only many countable eigenfunctions. In fact, the above result is constructive: once the user specifies a pair of points on two shapes corresponding to each other, the algorithm can compute a correspondence between them by comparing their heat kernel maps; see Figure 7.8.

Metric in shape space Imposing a good metric over shape space can facilitate many applications such as shape classification and recognition. Memoli[40] used heat kernel to define a spectral version of Gromov-Wasserstein distance. Compared to the standard Gromov-Wasserstein distance, it has the advantages of comparing geometric information according to their scales. Specifically, consider any two Riemannian manifolds (M, g_M) and (N, g_N) equipped with certain measures ν_M and ν_N respectively. A measure ν on $M \times N$ is a coupling of ν_M and ν_N if and only if for any measurable sets $A \subset M$ and $B \subset N$

$$\nu(A \times N) = \nu_M(A) \text{ and } \nu(M \times B) = \nu_N(B).$$

Denote $C(\nu_M, \nu_N)$ the set of all couplings of ν_M and ν_N. The spectral Gromov-Wasserstein distance between M and N is defined as

$$d_{GW}^p(M, N) = \inf_{\nu \in C(\nu_M, \nu_N)} \sup_{t>0} c^2(t) \Big(\int_{M\times N} \int_{M\times N} |k_t(x, x') - k_t(y, y')|^p \nu(dx \times dy) \nu(dx' \times dy') \Big)^{1/p},$$

where $c(t) = e^{-t^{-1}+t}$. $|k_t(x, x') - k_t(y, y')|$ serves as a consistent measure between two pairs x, x' and y, y'. Since the definition of spectral Gromov-Wasserstein distance takes the supremacy over all scales t, it is lower bounded by the distance obtained by only considering a particular scale or a subset of scales. Such scale-wise comparison is useful, especially in the presence of noise, as one can choose proper scales to suppress the effect of noise.

7.7 Summary

We conjecture that the Main Theorem 7.3.5 holds for arbitrary dimensional Euclidean polyhedral manifolds; that means discrete Laplace-Beltrami operator (or equivalently the discrete heat kernel) and the discrete metric for any dimensional Euclidean polyhedral manifold are mutually determined by each other. On the other hand, we will explore the possibility of establishing the same theorem for different types of discrete Laplace-Beltrami operators as in [21]. Also, we will explore further the sufficient and necessary conditions for a given set of edge weights to be admissible.

7.8 Bibliographical and Historical Remarks

Heat kernel is a fundamental geometric object which has been actively studied by mathematicians in the past few decades. Grigor'yan [24] gave a very nice description on the heat equation and the heat kernel of the Laplace-Beltrami operator on Riemannian manifolds. Heat kernel is also closely related to stochastic process on a manifold. A nice description of heat kernel and Brownian motion on manifolds can be found in the book by Hsu [29].

In machine learning, Belkin and Niyogi [2] were the first to employ Laplician eigenfunctions for data representation and dimensionality reduction. Coifman and Lafon [11] later provided a similar framework based on heat diffusion process for finding geometric description of datasets. Jones, Maggioni, and Schul [32] used heat kernels or Laplacian eigenfunctions to construct bi-Lipschitz local coordinates on large classes of Euclidean domains and Riemannian manifolds. In geometry processing, heat kernel was simultaneously introduced by Sun, Ovsjanikov, and Guibas. [54] and Gębal et al. [20] as a robust and multi-scale isometric signature of a shape.

In the discrete setting, heat kernel is approximated as the exponential of discrete Laplace operator. Cotangent scheme is one of the constructions of discrete Laplace operator from meshes, which was proposed by Pinkall and Polthier [44]. The convergence of cotangent scheme was considered by Xu [59] and Wardetzky [57] and proved to be true in a weak sense provided that the elements of the input mesh are well-shaped. Belkin, Sun, and Wang [4] proposed another construction based on heat diffusion process which converges pointwise. On graphs, Laplace operator has been used in many applications including clustering and ranking. There are many good books on graph Laplace operator, including the one by Chung [10].

Bibliography

[1] Shigetoshi Bando and Hajime Urakawa. Generic properties of the eigenvalue of the laplacian for compact Riemannian manifolds. *Tohoku Math. J.*, 35(2):155–172, 1983.

[2] Mikhail Belkin and Partha Niyogi. Laplacian eigenmaps for dimensionality reduction and data representation. *Neural Comput.*, 15(6):1373–1396, 2003.

[3] Mikhail Belkin, Jian Sun, and Yusu Wang. Discrete Laplace operator on meshed surfaces. In *SoCG '08: Proceedings of the Twenty-fourth Annual Symposium on Computational Geometry*, pages 278–287, 2008.

[4] Mikhail Belkin, Jian Sun, and Yusu Wang. Discrete Laplace operator on meshed surfaces. In *Proceedings of SOCG*, pages 278–287, 2008.

[5] Serge Belongie, Jitendra Malik, and Jan Puzicha. Shape context: A new descriptor for shape matching and object recognition. In *In NIPS*, pages 831–837, 2000.

[6] A. I. Bobenko and B. A. Springborn. Variational principles for circle patterns and Koebe's theorem. *Transactions of the American Mathematical Society*, 356:659–689, 2004.

[7] P. L. Bowers and M. K. Hurdal. Planar conformal mapping of piecewise flat surfaces. In *Visualization and Mathematics III (Berlin)*, pages 3–34. Springer, 2003.

[8] Shiu-Yuen Cheng. Eigenfunctions and nodal sets. *Commentarii Mathematici Helvetici*, 51(1):43–55, 1976.

[9] B. Chow and F. Luo. Combinatorial Ricci flows on surfaces. *Journal of Differential Geometry*, 63(1):97–129, 2003.

[10] Fan R. K. Chung. *Spectral Graph Theory (CBMS Regional Conference Series in Mathematics, No. 92)*. American Mathematical Society, 1997.

[11] R. R. Coifman and S. Lafon. Diffusion maps. *Applied and Computational Harmonic Analysis*, 21(1):5 – 30, 2006. Diffusion Maps and Wavelets.

[12] C. Collins and K. Stephenson. A circle packing algorithm. *Computational Geometry: Theory and Applications*, 25:233–256, 2003.

[13] C. de Verdiere Yves. Un principe variationnel pour les empilements de cercles. *Invent. Math*, 104(3):655–669, 1991.

[14] Mathieu Desbrun, Mark Meyer, and Pierre Alliez. Intrinsic parameterizations of surface meshes. *Computer Graphics Forum (Proc. Eurographics 2002)*, 21(3):209–218, 2002.

[15] Tamal K. Dey, K. Li, Chuanjiang Luo, Pawas Ranjan, Issam Safa, and Yusu Wang. Persistent heat signature for pose-oblivious matching of incomplete models. *Comput. Graph. Forum*, 29(5):1545–1554, 2010.

[16] Tamal K. Dey, Pawas Ranjan, and Yusu Wang. Convergence, stability, and discrete approximation of Laplace spectra. In *Proc. ACM/SIAM Symposium on Discrete Algorithms (SODA) 2010*, pages 650–663, 2010.

[17] J. Dodziuk. Finite-difference approach to the Hodge theory of harmonic forms. *American Journal of Mathematics*, 98(1):79–104, 1976.

[18] Michael S. Floater. Mean value coordinates. *Comp. Aided Geomet. Design*, 20(1):19–27, 2003.

[19] Michael S. Floater and Kai Hormann. Surface parameterization: a tutorial and survey. In *Advances in Multiresolution for Geometric Modelling*, pages 157–186. Springer, 2005.

[20] K. Gębal, J. A. Bærentzen, H. Aanæs, and R. Larsen. Shape analysis using the auto diffusion function. In *Proceedings of the Symposium on Geometry Processing*, SGP '09, pages 1405–1413, 2009.

[21] David Glickenstein. A monotonicity property for weighted Delaunay triangulations. *Discrete & Computational Geometry*, 38(4):651–664, 2007.

[22] Craig Gotsman, Xianfeng Gu, and Alla Sheffer. Fundamentals of spherical parameterization for 3D meshes. *ACM Transactions on Graphics*, 22(3):358–363, 2003.

[23] Alexander Grigor'yan. Heat kernels on weighted manifolds and applications. *Cont. Math*, 398:93–191, 2006.

[24] Alexander Grigor'yan. *Heat Kernel and Analysis on Manifolds.* AMS IP Studies in Advanced Mathematics, vol. 47, 2009.

[25] Xianfeng Gu and Shing-Tung Yau. Global conformal parameterization. In *Symposium on Geometry Processing*, pages 127–137, 2003.

[26] H. P. Mckean, Jr., and I. M. Singer. Curvature and the eigenvalues of the Laplacian. *J. Differential Geometry*, 1:43–69, 1967.

[27] Masaki Hilaga, Yoshihisa Shinagawa, Taku Kohmura, and Tosiyasu L. Kunii. Topology matching for fully automatic similarity estimation of 3d shapes. In *Proc. SIGGRAPH*, pages 203–212, 2001.

[28] L. Hörmander. *The Analysis of Linear Partial Differential Operators, IV: Fourier Integral Operators.* Grundlehren. Math. Wiss. 275, Springer, Berlin, 1985.

[29] E. P. Hsu. *Stochastic Analysis on Manifolds.* American Mathematical Society, 2002.

[30] M. Jin, J. Kim, F. Luo, and X. Gu. Discrete surface Ricci flow. *IEEE TVCG*, 14(5):1030–1043, 2008.

[31] M. Jin, F. Luo, and X. Gu. Computing surface hyperbolic structure and real projective structure. In *SPM '06: Proceedings of the 2006 ACM Symposium on Solid and Physical Modeling*, pages 105–116, 2006.

[32] P.W. Jones, M. Maggioni, and R. Schul. Universal local parametrizations via heat kernels and eigenfunctions of the Laplacian. *Ann. Acad. Scient. Fen.*, 35:1–44, 2010.

[33] J. Jorgenson and S. Lang. The ubiquitous heat kernel. *Mathematics unlimited and beyond*, pages 655–83, 2001.

[34] L. Kharevych, B. Springborn, and P. Schröder. Discrete conformal mappings via circle patterns. *ACM Trans. Graph.*, 25(2):412–438, 2006.

[35] Paul Koebe. Kontaktprobleme der konformen abbildung. Ber. Sächs. Akad. Wiss. Leipzig, *Math.-Phys. KI*, 88:141–164, 1936.

[36] Vladislav Kraevoy and Alla Sheffer. Cross-parameterization and compatible remeshing of 3D models. *ACM Transactions on Graphics*, 23(3):861–869, 2004.

[37] Bruno Lévy. Laplace-Beltrami eigenfunctions towards an algorithm that "understands" geometry. In *SMI '06: Proceedings of the IEEE International Conference on Shape Modeling and Applications 2006*, page 13, 2006.

[38] Bruno Lévy, Sylvain Petitjean, Nicolas Ray, and Jérome Maillot. Least squares conformal maps for automatic texture atlas generation. *SIGGRAPH 2002*, pages 362–371, 2002.

[39] Feng Luo. Combinatorial Yamabe flow on surfaces. *Commun. Contemp. Math.*, 6(5):765–780, 2004.

[40] Facundo Memoli. A spectral notion of Gromov-Wasserstein distance and related methods. *Applied and Computational Harmonic Analysis*, 2010.

[41] M. Ovsjanikov, Q. Mérigot, F. Mémoli, and L. Guibas. One point isometric matching with the heat kernel. In *Eurographics Symposium on Geometry Processing (SGP)*, 2010.

[42] Maks Ovsjanikov, Jian Sun, and Leonidas J. Guibas. Global intrinsic symmetries of shapes. *Comput. Graph. Forum*, 27(5):1341–1348, 2008.

[43] Ulrich Pinkall and Konrad Polthier. Computing discrete minimal surfaces and their conjugates. *Experimental Mathematics*, 2(1):15–36, 1993.

[44] Ulrich Pinkall and Konrad Polthier. Computing discrete minimal surfaces and their conjugates. *Experimental Mathematics*, 2(1):15–36, 1993.

[45] Martin Reuter, Franz-Erich Wolter, and Niklas Peinecke. Laplace–Beltrami spectra as 'shape-DNA' of surfaces and solids. *Comput. Aided Des.*, 38(4):342–366, 2006.

[46] B. Rodin and D. Sullivan. The convergence of circle packings to the Riemann mapping. *Journal of Differential Geometry*, 26(2):349–360, 1987.

[47] S. Rosenberg. *Laplacian on a Riemannian manifold.* Cambridge University Press, 1997.

[48] Raif M. Rustamov. Laplace-Beltrami eigenfunctions for deformation invariant shape representation. In *Symposium on Geometry Processing*, pages 225–233, 2007.

[49] P. Skraba, M. Ovsjanikov, F. Chazal, and L. Guibas. Persistence-based segmentation of deformable shapes. In *CVPR Workshop on Non-Rigid Shape Analysis and Deformable Image Alignment*, pages 45–52, June 2010.

[50] Olga Sorkine. Differential representations for mesh processing. *Computer Graphics Forum*, 25(4):789–807, 2006.

[51] Boris Springborn, Peter Schröder, and Ulrich Pinkall. Conformal equivalence of triangle meshes. *ACM Transactions on Graphics*, 27(3):1–11, 2008.

[52] S.Rosenberg. *The Laplacian on a Riemannian manifold.* Number 31 in London Mathematical Society Student Texts. Cambridge University Press, 1998.

[53] Jian Sun, Xiaobai Chen, and Thomas Funkhouser. Fuzzy geodesics and consistent sparse correspondences for deformable shapes. *Computer Graphics Forum (Symposium on Geometry Processing)*, 29(5), July 2010.

[54] Jian Sun, Maks Ovsjanikov, and Leonidas J. Guibas. A concise and provably informative multi-scale signature based on heat diffusion. *Comput. Graph. Forum*, 28(5):1383–1392, 2009.

[55] W. P. Thurston. Geometry and topology of three-manifolds. *Lecture Notes at Princeton university*, 1980.

[56] W. P. Thurston. *The finite Riemann mapping theorem.* 1985. Invited talk.

[57] Max Wardetzky. Convergence of the cotangent formula: An overview. In *Discrete Differential Geometry*, pages 89–112. Birkhäuser Basel, 2005.

[58] Max Wardetzky, Saurabh Mathur, Felix Kälberer, and Eitan Grinspun. Discrete Laplace operators: No free lunch. In *Proceedings of the fifth Eurographics symposium on Geometry processing*, pages 33–37. Eurographics Association, 2007.

[59] Guoliang Xu. Discrete Laplace-Beltrami operators and their convergence. *Comput. Aided Geom. Des.*, 21(8):767–784, 2004.

[60] Hao Zhang, Oliver van Kaick, and Ramsay Dyer. Spectral mesh processing. *Computer Graphics Forum*, 29(6):1865–1894, 2010.

Chapter 8

Discrete Ricci Flow for Surface and 3-Manifold

Xianfeng Gu, Wei Zeng, Feng Luo, and Shing-Tung Yau

8.1 Introduction

Computational conformal geometry is an interdisciplinary field, which has deep roots in pure mathematics fields, such as Riemann surface theory, complex analysis, differential geometry, algebraic topology, partial differential equations, and others. It has been applied to many fields in computer science, such as computer graphics, computer vision, geometric modeling, medical imaging, and computational geometry.

Historically, computational conformal geometry has been broadly applied in many engineering fields [1], such as electromagnetics, vibrating membranes and acoustics, elasticity, heat transfer, and fluid flow. Most of these applications depend on conformal mappings between planar domains. Recently, with the development of 3D scanning technology, increase of computational power, and further advances in mathematical theories, computational conformal geometric theories and algorithms have been greatly generalized from planar domains to surfaces with arbitrary topologies. Besides, the investigation of the topological structures and geometric properties of 3-manifolds is very important. It has great potential for many engineering applications, such as volumetric parameterization, registration, and shape analysis. This work will focus on the methodology of Ricci flow for computing both the conformal structures of metric surfaces with complicated topologies and the hyperbolic geometric structures of 3-manifolds.

Conformal Transformation and Conformal Structure

According to Felix Klein's Erlangen program: *Geometries study those properties of spaces invariant under various transformation groups.* Conformal geometry investigates quantities invariant under the angle preserving transformation group.

Let S_1 and S_2 be two surfaces with Riemannian metrics $\mathbf{g}_1$ and $\mathbf{g}_2$; let $\phi : (S_1, \mathbf{g}_1) \to (S_2, \mathbf{g}_2)$ be a diffeomorphism between them. We say ϕ is *conformal* if it preserves angles. More precisely, as shown in Figure 8.1, let $\gamma_1, \gamma_2 : [0,1] \to S_1$ be two arbitrary curves on S_1, intersecting at an angle θ at the point p. Then under a conformal mapping ϕ, the two curves $\phi \circ \gamma_1(t)$ and $\phi \circ \gamma_2(t)$ still intersect at the angle θ at $\phi(p)$.

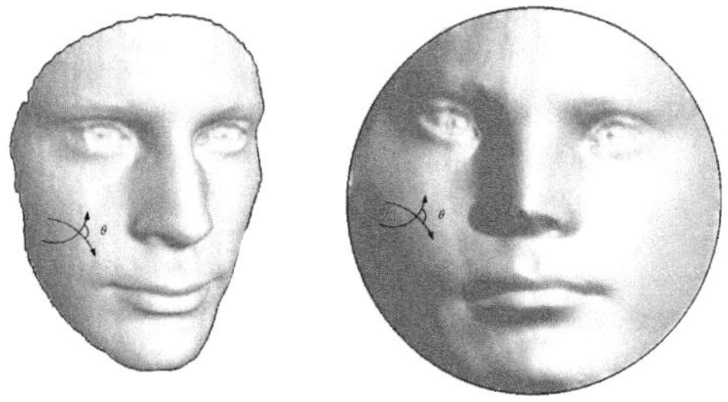

Figure 8.1: Conformal mappings preserve angles.

Infinitesimally, a conformal mapping is a scaling transformation; it preserves local shapes. For example, it maps infinitesimal circles to infinitesimal circles. As shown in Figure 8.2 frame (a), the bunny surface is mapped to the plane via a conformal mapping ϕ. If a circle packing is given on the plane and pulled back by ϕ, it produces a circle packing on the bunny surface. If we put a checkerboard on the plane, then pull back by ϕ, on the bunny surface the checkerboard pattern is such that all the right angles of the squares are preserved. See the same figure frame (b).

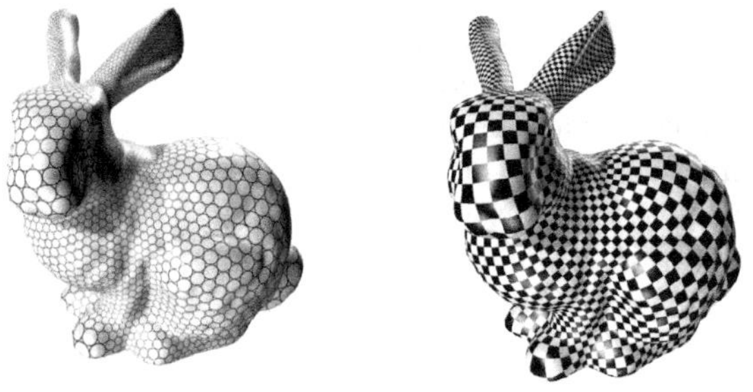

(a) Circle Packing (b) Checkerboard

Figure 8.2: Conformal mappings preserve infinitesimal circle fields.

Two Riemannian metrics g_1 and g_2 on a surface S are *conformal* if they define the same notion of angles in S (i.e., the identity map is a conformal mapping from (S, g_1) to (S, g_2)). The conformal equivalence class of a Riemannian metric on a surface is called a *conformal structure*. A *Riemann surface* is a smooth surface together with a conformal structure. Thus, in a Riemann surface, one can measure angles, but not lengths. Each surface with a Riemannian metric is automatically a Riemann surface.

Two surfaces with Riemannian metrics are *conformally equivalent* if there exists a conformal mapping between them. Conformal mapping is the natural equivalence relation for Riemann surfaces. The goal of conformal geometry is to classify Riemann surfaces up to conformal mappings (or *biholomorphism* in complex geometric terminology). Theoretically, this is called the moduli space problem. Given a smooth surface S, one considers the set of all conformal structures on S modulo conformal mappings. The set is called the moduli space of S. For a closed surface of positive genus, its moduli space is known to be a *finite dimensional space* of positive dimension. Thus, conformal geometry is *between* topology and Riemannian geometry, it is more *rigid* than topology and more *flexible* than Riemannian geometry.

Fundamental Tasks

The following computational problems are some of the most fundamental tasks for computational conformal geometry. These problems are intrinsically inter-dependent:

1. ***Conformal Structure*** Given a surface with a Riemannian metric, compute various representations of its intrinsic conformal structure. One approach is to compute the group of Abelian differentials, the other approach is to compute the canonical Riemannian metrics.

2. ***Conformal Modulus*** The complete conformal invariants are called the conformal modulus of the Riemann surface. As aforementioned, theoretically, it is known that there is a *finite* set of numbers which completely determine a Riemann surface (up to conformal mapping). These are called the *conformal modulus* of the Riemann surface. The difficult task is to explicitly compute this conformal modulus for any given Riemann surface.

3. ***Canonical Riemannian Metric*** All Riemannian metrics on a topological surface can be classified according to conformal equivalence. A fundamental theorem for Riemann surfaces, called the *uniformization theorem*, says that each Riemannian metric is conformal to a (completer) Riemannian metric of constant Gaussian curvature. This metric is unique unless the surface is the 2-sphere or the torus. Computing such metrics is of fundamental importance for computational conformal geometry.

4. ***Conformal Mapping*** Compute the conformal mapping between two given conformal equivalent surfaces. This can be reduced to compute the conformal mapping of each surface to a canonical shape, such as circular domain on the sphere, plane, or hyperbolic space.

5. ***Quasi-Conformal Mapping*** Most diffeomorphisms between Riemann surfaces are not conformal. They send infinitesimal ellipses to infinitesimal circles. If these ellipses have a uniformly bounded ratio of major and minor axes, then the diffeomorphism is called a *quasi-conformal map.* The differential of a quasi-conformal map on a Riemann surface is coded by the so-called Beltrami differential which records the direction of the major axes of the ellipses and the ratio of the major and minor axes. A fundamental theorem says one can recover the quasi-conformal mapping from its Beltrami differential (up to conformal mapping). How to compute the quasi-conformal mapping from its Beltrami differential is another major task which has many applications.

6. ***Conformal Welding*** Glue Riemann surfaces with boundaries to form a new Riemann surface and study the relation between the shape of the sealing curve and the gluing pattern. This is closely related to the quasi-conformal mapping problem.

In this work, we will explain the methods for solving these fundamental problems in detail.

Merits of Conformal Geometry for Engineering Applications

Computational conformal geometry has been proven to be very valuable for a broad range of engineering fields. The following are some of the major intrinsic reasons:

1. ***Canonical Domain*** All metric surfaces can be conformally mapped to canonical domains of either sphere, plane, or hyperbolic disk. This helps to convert 3D geometric

processing problems to 2D ones. The solutions to some partial differential equations have closed forms on the canonical domains, such as the Poisson formula on the unit disk.

2. ***Design Metric by Curvature*** Each conformal structure has a canonical Riemannian metric with constant Gaussian curvature. This metric is valuable for many geometric problems. For example, each homotopy class of a non-trivial loop has a unique closed geodesic representative in a hyperbolic metric. Furthermore, one can design a Riemannian metric with prescribed curvatures, which is useful for geometric modeling purposes.

3. ***General Geometric Structures*** Conformal geometric methods lead to the construction of other general geometric structures, such as affine structure, projective structure, etc. These structures are crucial for geometric modeling applications.

4. ***Construction of Diffeomorphisms*** Conformal mapping and quasi-conformal mapping can be applied to construct diffeomorphisms between surfaces. Surface registration and comparison are the most fundamental tasks in computer vision and medical imaging.

5. ***Isothermal Coordinates*** Conformal structure can be treated as a special atlas, such that all the local coordinates are isothermal coordinates. Under this type of coordinate, the Riemannian metric has the simplest form. Therefore all the differential operators, such as Laplace-Beltrami operator, can be expressed nicely in these coordinates. This helps to simplify the partial differential equations. Isothermal coordinates preserve local shapes. They are preferable for visualization and texture mapping purposes.

In the later discussion, we will demonstrate the powerful conformal geometric methods for various engineering applications.

8.2 Theoretic Background

This section briefly introduces elementary theories of conformal geometry. Conformal geometry is the intersection of many fields in mathematics. We will recall the most fundamental theories in each field, which are closely related to our computational methodologies.

We refer readers to the following classical books in each field for detailed proofs for the theories. A thorough introduction to topological manifolds can be found in Lee's book [79]; differential forms, exterior calculus, Hodge decomposition theorems, and de Rham cohomology can be found in Lang's book [78]; Farkas and Kra's textbook on Riemann surfaces is excellent [80]; do Carmo's textbook [83] on differential geometry of curves and surfaces gives a good introduction to global differential geometry; Ahlfors' classical lecture note [68] on quasi-conformal mapping gives a good introduction; more advanced quasi-conformal Teichmüller theories can be found in Gardiner and Lakic's book [85]; Schoen and Yau's lectures on harmonic maps [74] and differential geometry [75] give thorough explanations on harmonic maps and conformal metric deformation; recent developments on Ricci flow can be found in [84].

Any non-orientable surface has a two-fold cover which is orientable. In the following discussion, by replacing a non-orientable surface by its orientable double cover, we will always assume surfaces are orientable. The symbols used for presentation are listed in Table 8.1.

Table 8.1: Symbol List

S	smooth surface	Σ	triangular mesh
π_1	fundamental group	v_i	*ith* vertex
$\mathbf{g}$	Riemannian metric	$[v_i, v_j]$	edge connecting v_i and v_j
K	Gaussian curvature	g_{ij}	Riemannian metric on $[v_i, v_j]$
$\bar{K}$	target curvature	θ_i	corner angle at v_i
U	neighborhoods on S	l_{ij}	edge length on $[v_i, v_j]$
ϕ	coordinates mapping	$\mathbf{k}$	curvature vector
(U, ϕ)	isothermal coordinate chart	$\mathbf{u}$	discrete metric vector
z	isothermal coordinates	H	Hessian matrix
$\phi_{\alpha\beta}$	holomorphic coordinates transition	C_k	k-dimensional chain space

8.2.1 Conformal Deformation

All oriented compact connected two dimensional topological manifolds can be classified by their genus g and number of boundaries b. Therefore, we use (g, b) to represent the topological type of a surface S.

Suppose q is a base point in S. All the oriented closed curves (loops) in S through q can be classified by homotopy. The set of all such homotopy classes is the so-called *fundamental group* of S, or *the first homotopy group*, denoted as $\pi_1(S, q)$. The group structure of $\pi_1(S, q)$ determines the homotopy type of S.

For a genus g closed surface, one can find *canonical homotopy group generators* $\{a_1, b_1, a_2, b_2, \cdots, a_g, b_g\}$, such that $a_i \cdot a_j = 0$, $b_i \cdot b_j = 0$, $a_i \cdot b_j = \delta_{ij}$, where the operator $r_1 \cdot r_2$ represents the algebraic intersection number between the two loops γ_1 and γ_2, and δ_{ij} is the Kronecker symbol.

Theorem 8.2.1 (Fundamental Group) *[79, Pro. 6.12, p.136 Exp. 6.13, p. 137] For genus g closed surface with a set of canonical basis, the fundamental group is given by*

$$< a_1, b_1, a_2, b_2, \cdots, a_g, b_g | a_1 b_1 a_1^{-1} b_1^{-1} a_2 b_2 a_2^{-1} b_2^{-1} \cdots a_g b_g a_g^{-1} b_g^{-1} = e > .$$

Recall that *covering map* $p : \tilde{S} \to S$ is defined as follows. First, the map p is surjective. Second, each point $q \in S$ has a neighborhood U with its preimage $p^{-1}(U) = \cup_i \tilde{U}_i$ a disjoint union of open sets $\tilde{U}_i$ so that the restriction of p on each $\tilde{U}_i$ is a homeomorphism. We call $(\tilde{S}, p)$ a *covering space* of S. Homeomorphisms of $\tilde{S}$, $\tau : \tilde{S} \to \tilde{S}$, are called *deck transformations*, if they satisfy $p \circ \tau = p$. All the deck transformations form a group, *covering group*, and denoted as $Deck(\tilde{S})$.

Suppose $\tilde{q} \in \tilde{S}$, $p(\tilde{q}) = q$ and surface $\tilde{S}$ is connected. The projection map $p : \tilde{S} \to S$ induces a homomorphism between their fundamental groups, $p_* : \pi_1(\tilde{S}, \tilde{q}) \to \pi_1(S, q)$, if $p_* \pi_1(\tilde{S}, \tilde{q})$ is a normal subgroup of $\pi_1(S, q)$ then the following theorem holds.

Theorem 8.2.2 (Covering Group Structure) *[79, Thm 11.30, Cor. 11.31, p.250] The quotient group of $\pi_1(S)/p_*\pi_1(\tilde{S}, \tilde{q})$ is isomorphic to the deck transformation group of $\tilde{S}$.*

$$\pi_1(S, q)/p_*\pi_1(\tilde{S}, \tilde{q}) \cong Deck(\tilde{S}).$$

If a covering space $\tilde{S}$ is simply connected (i.e. $\pi_1(\tilde{S}) = \{e\}$), then $\tilde{S}$ is called a *universal covering space* of S. For universal covering space

$$\pi_1(\pi) \cong Deck(\tilde{S}).$$

The existence of the universal covering space is given by the following theorem,

Theorem 8.2.3 (Existence of the Universal Covering Space) *[79, Thm. 12.8, p. 262] Every connected and locally simply connected topological space (in particular, every connected manifold) has a universal covering space.*

The concept of universal covering space is essential in Poincaré-Klein-Koebe Uniformization theorem 8.2.10, and the Teichmüller space theory [76]. It plays an important role in computational algorithms as well.

8.2.2 Uniformization Theorem

Classical textbooks for Riemann surface theory are [72] and [73].

Intuitively, a Riemann surface is a topological surface with an extra structure, which can measure angles, but not the lengths.

Definition 8.2.4 (Holomorphic Function) *Suppose a complex function* $f : \mathbb{C} \to \mathbb{C}$, $x + iy \to u(x,y) + iv(x,y)$, *satisfies the* Cauchy-Riemann *equation*

$$\frac{\partial u}{\partial x} = \frac{\partial v}{\partial y}, \quad \frac{\partial u}{\partial y} = -\frac{\partial v}{\partial x},$$

then f *is a* holomorphic function.

Equivalently, define complex differential operators by

$$\frac{\partial}{\partial z} = \frac{1}{2}(\frac{\partial}{\partial x} - i\frac{\partial}{\partial y}), \quad \frac{\partial}{\partial \bar{z}} = \frac{1}{2}(\frac{\partial}{\partial x} + i\frac{\partial}{\partial y}).$$

Then the Cauchy-Riemann equation is

$$\frac{\partial f}{\partial \bar{z}} = 0.$$

A holomorphic function between planar domains preserves angles. For a general manifold covered by local coordinate charts, if all transitions are holomorphic, then angles can still be consistently defined and measured.

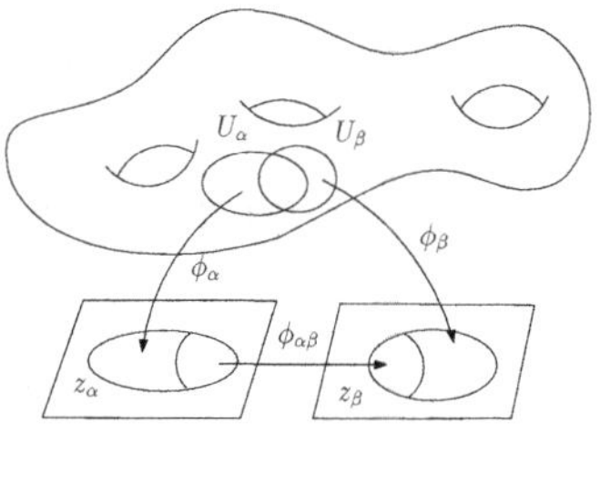

Figure 8.3: Manifold and atlas.

Suppose S is a surface covered by a collection of open sets $\{U_\alpha\}$, $S \subset \bigcup_\alpha U_\alpha$. A chart is (U_α, ϕ_α), where $\phi_\alpha : U_\alpha \to \mathbb{C}$ is a homeomorphism. The chart transition function $\phi_{\alpha\beta}$: $\phi_\alpha(U_\alpha \cap U_\beta) \to \phi_\beta(U_\alpha \cap U_\beta), \phi_{\alpha\beta} = \phi_\beta \circ \phi_\alpha^{-1}$. The collection of the charts $\mathcal{A} = \{(U_\alpha, \phi_\alpha)\}$ is called the atlas of S. The geometric illustration is shown in Figure 8.3.

Definition 8.2.5 (Conformal Atlas) *Suppose* S *is a topological surface with an atlas* $\mathcal{A}$. *If all the coordinate transitions are holomorphic, then* $\mathcal{A}$ *is called a conformal atlas.*

Because all the local coordinate transitions are holomorphic, the measurements of angles are independent of the choice of coordinates. Therefore angles are well defined on the surface. The maximal conformal atlas is a conformal structure,

Definition 8.2.6 (Conformal Structure) *Two conformal atlases are equivalent if their union is still a conformal atlas. Each equivalence class of conformal atlases is called a* conformal structure.

Definition 8.2.7 (Riemann Surface) *A topological surface with a conformal structure is called a Riemann surface.*

The groups of different types of differential forms on the Riemann surface are crucial in designing the computational methodologies.

Definition 8.2.8 (Conformal Mapping) *Suppose S_1 and S_2 are two Riemann surfaces, a mapping $f : S_1 \to S_2$ is called a conformal mapping (holomorphic mapping), if in the local analytic coordinates, it is represented as $w = g(z)$ where g is holomorphic.*

Definition 8.2.9 (Conformal Equivalence) *Suppose S_1 and S_2 are two Riemann surfaces. If a mapping $f : S_1 \to S_2$ is holomorphic, then S_1 and S_2 are conformally equivalent.*

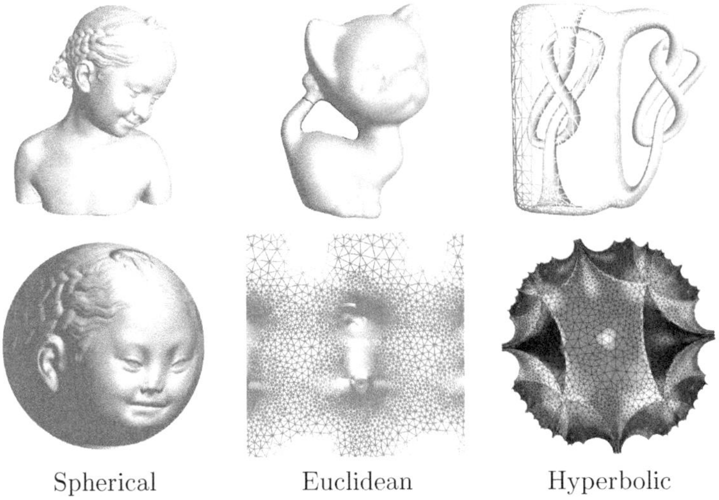

Figure 8.4: Uniformization for closed surfaces.

A *Möbius transformation* is given by

$$\rho(z) = \frac{az+b}{cz+d}, \quad a,b,c,d \in \mathbb{C}, \quad ad - bc = 1. \tag{8.1}$$

All Riemann surfaces can be unified by the following theorem:

Theorem 8.2.10 (Poincaré-Klein-Koebe Uniformization) *[80, p.206] Every connected Riemann surface S is conformally equivalent to D/G with D as one of the three canonical spaces:*

1. *Extended complex plane $\mathbb{C} \cup \{\infty\}$;*

2. *Complex plane* $\mathbb{C}$;

3. *Unit disk* $D = \{z \in \mathbb{C} || z| < 1\}$

where G *is a subgroup of Möbius transformations that acts freely discontinuous on* D. *Furthermore,* $G \cong \pi_1(S)$.

Definition 8.2.11 (Circle Domain) *A circle domain in a Riemann surface is a domain so that the components of the complement of the domain are closed geodesic disks and points. Here a geodesic disk in a Riemann surface is a topological disk whose lifts in the universal cover are a round desk in* $\mathbf{E^2}$ *or* $\mathbf{S^2}$ *or* $\mathbf{H^2}$.

Theorem 8.2.12 (He and Schramm) *[81, Thm. 0.1] Let* S *be an open Riemann surface with finite genus and at most countably many ends. Then there is a closed Riemann surface* $\tilde{S}$*, such that* S *is conformally homeomorphic to a circle domain* Ω *in* $\tilde{S}$. *Moreover, the pair* $(\tilde{S}, \Omega)$ *is unique up to conformal homeomorphisms.*

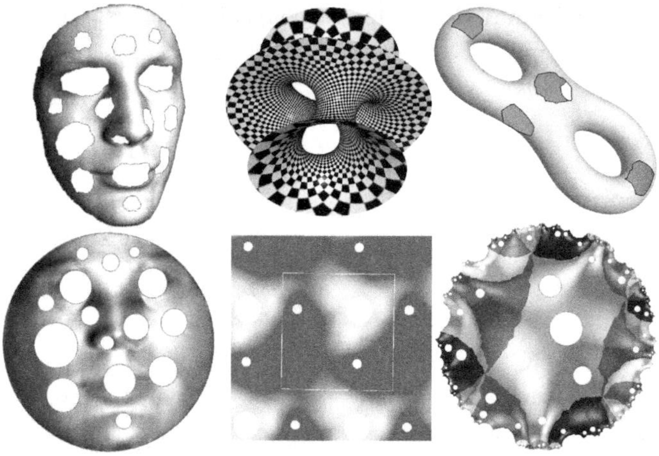

Figure 8.5: Uniformization for surfaces with boundaries.

The uniformization theorem states that the universal covering space of closed metric surfaces can be conformally mapped to one of three canonical spaces, the sphere $\mathbb{S}^2$, the plane $\mathbb{E}^2$, or the hyperbolic space $\mathbb{H}^2$, as shown in Figure 8.4. Similarly, uniformization theorem holds for surfaces with boundaries as shown in Figure 8.5, the covering space can be conformally mapped to a circle domain in $\mathbb{S}^2$, $\mathbb{E}^2$ or $\mathbb{H}^2$.

8.2.3 Yamabe Equation

The details for the following discussion can be found in Schoen and Yau's lectures on differential geometry [75].

Definition 8.2.13 (Isothermal Coordinates) *Let* S *be a smooth surface with a Riemannian metric* $\mathbf{g}$. *Isothermal coordinates* (u, v) *for* $\mathbf{g}$ *satisfy*

$$\mathbf{g} = e^{2\lambda(u,v)}(du^2 + dv^2).$$

Locally, isothermal coordinates always exist [82]. An atlas with all local coordinates being isothermal is a conformal atlas. Therefore a Riemannian metric uniquely determines a conformal structure, namely

Theorem 8.2.14 *All oriented metric surfaces are Riemann surfaces.*

The Gaussian curvature of the surface is given by

$$K(u,v) = -\Delta_{\mathbf{g}}\lambda, \tag{8.2}$$

where $\Delta_{\mathbf{g}} = e^{-2\lambda(u,v)}(\frac{\partial^2}{\partial u^2} + \frac{\partial^2}{\partial v^2})$ is the Laplace–Beltrami operator induced by $\mathbf{g}$. Although the Gaussian curvature is intrinsic to the Riemannian metric, the total Gaussian curvature is a topological invariant:

Theorem 8.2.15 (Gauss–Bonnet) *[83, p. 274] The total Gaussian curvature of a closed metric surface is*

$$\int_S K dA = 2\pi\chi(S),$$

where $\chi(S)$ is the Euler number of the surface.

Suppose $\mathbf{g}_1$ and $\mathbf{g}_2$ are two Riemannian metrics on the smooth surface S. If there is a differential function $\lambda : S \to \mathbb{R}$, such that

$$\mathbf{g}_2 = e^{2\lambda}\mathbf{g}_1,$$

then the two metrics are *conformal equivalents.* Let the Gaussian curvatures of $\mathbf{g}_1$ and $\mathbf{g}_2$ be K_1 and K_2, respectively. Then they satisfy the following *Yamabe equation*

$$K_2 = \frac{1}{e^{2\lambda}}(K_1 - \Delta_{\mathbf{g}_1}\lambda).$$

Detailed treatment of the Yamabe equation can be found in Schoen and Yau's [75] Chapter V: *conformal deformation of scalar curvatures.*

Consider all possible Riemannian metrics on S. Each conformal equivalence class defines a *conformal structure.* Suppose a mapping $f : (S_1, \mathbf{g}_1) \to (S_2, \mathbf{g}_2)$ is differentiable. If the pull back metric is conformal to the original metric $\mathbf{g}_1$

$$f^*\mathbf{g}_2 = e^{2\lambda}\mathbf{g}_1,$$

then f is a *conformal mapping.*

8.2.4 Ricci Flow

Suppose the metric $\mathbf{g} = (g_{ij})$ in a local coordinate. Hamilton introduced the Ricci flow as

$$\frac{dg_{ij}}{dt} = -Kg_{ij}.$$

For surface case, Ricci flow is equivalent to Yamabe flow. During the flow, the Gaussian curvature will evolve according to a heat diffusion process.

Theorem 8.2.16 (Hamilton and Chow) *[84, Thm. B.1, p. 504] Suppose S is a closed surface with a Riemannian metric. If the total area is preserved, the surface Ricci flow will converge to a Riemannian metric of constant Gaussian curvature.*

This gives another approach to prove the Poincaré uniformization theorem 8.2.10. As shown in Figure 8.4, all closed surfaces can be conformally deformed to one of the three canonical spaces: the unit sphere $\mathbb{S}^2$, the plane $\mathbb{E}^2$, or the hyperbolic space $\mathbb{H}^2$.

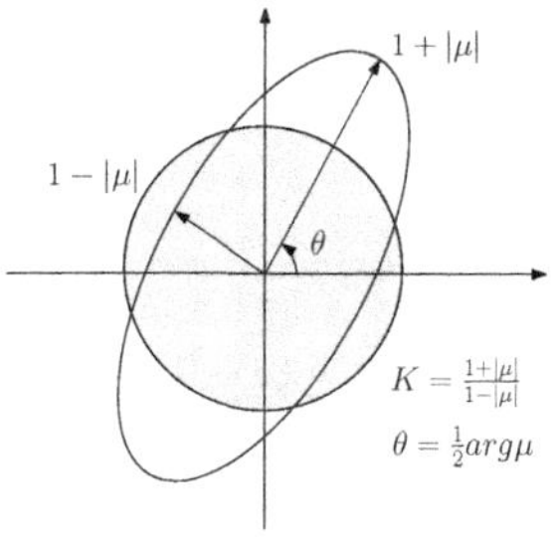

Figure 8.6: Illustration of how Beltrami coefficient μ measures the distortion by a quasi-conformal map that is an ellipse with dilation K.

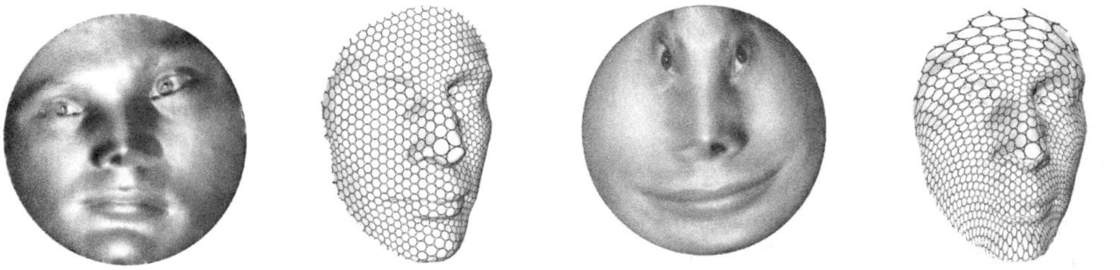

(a) Conformal mapping | (b) Circle Packing induced by (a) | (c) Quasi-conformal mapping | (d) Circle packing induced by (c)

Figure 8.7: Conformal and quasi-conformal mappings for a topological disk.

8.2.5 Quasi-Conformal Maps

A generalization of a conformal map is called the *quasi-conformal* map, which is an orientation-preserving homeomorphism between Riemann surfaces with bounded conformality distortion. The latter means that the first order approximation of the quasi-conformal homeomorphism takes small circles to small ellipses of bounded eccentricity. In particular, a conformal homeomorphism is quasi-conformal.

Mathematically, ϕ is quasi-conformal provided that it satisfies Beltrami's equation 8.3. In a local conformal chart, the Beltrami equation takes the form

$$\frac{\partial \phi}{\partial \bar{z}} = \mu(z)\frac{\partial \phi}{\partial z}, \tag{8.3}$$

where μ, called the *Beltrami coefficient*, is a complex valued Lebesgue measurable function satisfying $|\mu|_\infty < 1$. The Beltrami coefficient measures the deviation of ϕ from conformality. In particular, the map ϕ is conformal at a point p if and only if $\mu(p) = 0$. In general, ϕ maps an infinitesimal circle to a infinitesimal ellipse. Geometrically, the Beltrami coefficient $\mu(p)$ codes the direction of the major axis and the ratio of the major and minor axes of the infinitesimal ellipses. Specifically, the angle of major axis with respect to the x-axis is $arg\mu(p)/2$ and the ratio of the major and minor axes is $1 + |\mu(p)|$. The angle between the minor axis and the x-axis is $(arg\mu(p) - \pi)/2$. The distortion or dilation is given by:

$$K = \frac{1 + |\mu(p)|}{1 - |\mu(p)|}. \tag{8.4}$$

Thus, the Beltrami coefficient μ gives us all the information about the conformality of the map (see Figure 8.6).

If equation 8.3 is defined on the extended complex plane (the complex plane plus the point at infinity), Ahlfors proved the following theorem.

Theorem 8.2.17 (The Measurable Riemann Mapping) *[85, Thm. 1, p. 10] The equation 8.3 gives a one-to-one correspondence between the set of quasi-conformal homeomorphisms of $\mathbb{C} \cup \{\infty\}$ that fix the points $0, 1$, and ∞ and the set of measurable complex-valued functions μ for which $|\mu|_\infty < 1$ on $\mathbf{C}$.*

Suppose $\phi : \Omega \to \mathbf{C}$ is a quasi conformal map with Beltrami coefficient μ defined on some domain $\Omega \subset \mathbf{C}$. Then the pullback of the canonical Euclidean metric $\mathbf{g}_0$ under ϕ is the metric given by:

$$\phi^*(\mathbf{g}_0) = |\frac{\partial \phi}{\partial z}|^2 |dz + \mu(z) d\bar{z}|^2. \tag{8.5}$$

Figure 8.7 shows the conformal and quasi-conformal mappings for a topological disk, where the Beltarmi coefficient is set to be $\mu = z$. From the texture mappings in frames (c) and (d), we can see the conformality is maintained well around the nose tip (circles to circles), while it is changed a lot along the boundary area (circles to quasi-ellipses).

8.3 Surface Ricci Flow

Surface Ricci flow is a powerful tool to construct conformal Riemannian metrics with prescribed Gaussian curvatures. Discrete surface Ricci flow generalizes the curvature flow method from smooth surface to discrete triangular meshes. The key insight to discrete Ricci flow is based on the following observation: conformal mappings transform infinitesimal circle fields to infinitesimal circle fields, see Figure 8.2. Discrete Ricci flow replaces infinitesimal circles by circles with finite radii, and modifies the circle radii to deform the discrete metric, to achieve the desired curvature. Readers can refer to [43] and [45] for more details.

Background Geometry In engineering fields, surfaces are approximated by their triangulations, the triangle meshes.

Definition 8.3.1 (Triangle Mesh) *A triangle mesh Σ is a 2 dimensional simplical complex, which is homeomorphic to a surface.*

It is generally assumed that a mesh Σ is embedded in the three dimensional Euclidean space $\mathbb{R}^3$, and therefore each face is Euclidean. In this case, we say the mesh is with Euclidean background geometry. Similarly, we can assume that a mesh is embedded in the three dimensional sphere $\mathbb{S}^3$ or hyperbolic space $\mathbb{H}^3$, then each face is a spherical or a hyperbolic triangle. We say the mesh is with spherical or hyperbolic background geometry.

8.3.1 Derivative Cosine Law

Discrete Riemannian Metric A discrete Riemannian metric on a mesh Σ is a piecewise constant curvature metric with cone singularities at the vertices. The edge lengths are sufficient to define a discrete Riemannian metric,

$$l : E \to \mathbb{R}^+, \tag{8.6}$$

as long as, for each face $[v_i, v_j, v_k]$, the edge lengths satisfy the triangle inequality: $l_{ij} + l_{jk} > l_{ki}$ for all the three background geometries, and another inequality: $l_{ij} + l_{jk} + l_{ki} < 2\pi$ for spherical geometry.

In the smooth case, the curvatures are determined by the Riemannian metrics as in Equation 8.2. In the discrete case, the angles of each triangle are determined by the edge lengths. According to different background geometries, there are different cosine laws. For simplicity, we use e_i to denote the edge across from the vertex v_i, namely $e_i = [v_j, v_k]$, and l_i the edge length of e_i. The cosine laws are given as:

$$\begin{array}{lcll} l_k^2 & = & l_i^2 + l_j^2 - 2l_il_j \cos\theta_k & \mathbb{E}^2 \\ \cosh l_k & = & \cosh l_i \cosh l_j - \sinh l_i \sinh l_j \cos\theta_k & \mathbb{H}^2 \\ \cos l_k & = & \cos l_i \cos l_j - \sin l_i \sin l_j \cos\theta_k & \mathbb{S}^2 \end{array}$$

Discrete Gaussian Curvature The discrete Gaussian curvature K_i at a vertex $v_i \in \Sigma$ can be computed as the angle deficit,

$$K_i = \begin{cases} 2\pi - \sum_{[v_i,v_j,v_k]\in\Sigma} \theta_i^{jk}, & v_i \notin \partial\Sigma \\ \pi - \sum_{[v_i,v_j,v_k]\in\Sigma} \theta_i^{jk}, & v_i \in \partial\Sigma \end{cases} \tag{8.7}$$

where θ_i^{jk} represents the corner angle attached to vertex v_i in the face $[v_i, v_j, v_k]$, and $\partial\Sigma$ represents the boundary of the mesh.

Discrete Gauss-Bonnet Theorem The Gauss-Bonnet theorem 8.2.15 states that the total curvature is a topological invariant. It still holds on meshes as follows.

$$\sum_{v_i\in V} K_i + \lambda \sum_{f_i\in F} A_i = 2\pi\chi(M), \tag{8.8}$$

where the second term is the integral of the ambient constant Gaussian curvature on the faces; A_i denotes the area of face f_i, and λ represents the constant curvature for the background geometry; $+1$ for the spherical geometry, 0 for the Euclidean geometry, and -1 for the hyperbolic geometry.

8.3.2 Circle Pattern Metric

In the smooth case, *Conformal deformation of a Riemannian metric* is defined as

$$\mathbf{g} \to e^{2\lambda}\mathbf{g}, \quad \lambda : S \to \mathbb{R}.$$

In the discrete case, there are many ways to define conformal metric deformation. Figure 8.8 illustrates some of them. Generally, we associate each vertex v_i with a circle (v_i, γ_i) centered at v_i with radius γ_i. On an edge $[v_i, v_j]$, two circles intersect at an angle Θ_{ij}. During the conformal deformation, the radii of circles can be modified, but the intersection angles are preserved. Geometrically, the discrete conformal deformation can be interpreted as follows [67]: see Figure 8.9: there exists a unique circle, the so called *radial circle*, that is orthogonal to three vertex circles. The radial circle center is denoted as o. We connect the radial circle center to three vertices, to get three rays $\overrightarrow{ov_i}, \overrightarrow{ov_j}$, and $\overrightarrow{ov_k}$. We deform the triangle by infinitesimally moving the vertex v_i along $\overrightarrow{ov_i}$ to ov_i', and construct a new circle (v_i', γ_i'), such that the intersection angles among the circles are preserved, $\Theta_{ij}' = \Theta_{ij}$, $\Theta_{ki}' = \Theta_{ki}$.

The discrete conformal metric deformation can be generalized to all other configurations, with different circle intersection angles (including zero or virtual angles), and different circle radii (including zero radii). In Figure 8.8, the radial circle is well defined for all cases, as are the rays from the radial circle center to the vertices. Therefore, discrete conformal metric deformations are well defined as well. The precise analytical formulae for discrete conformal

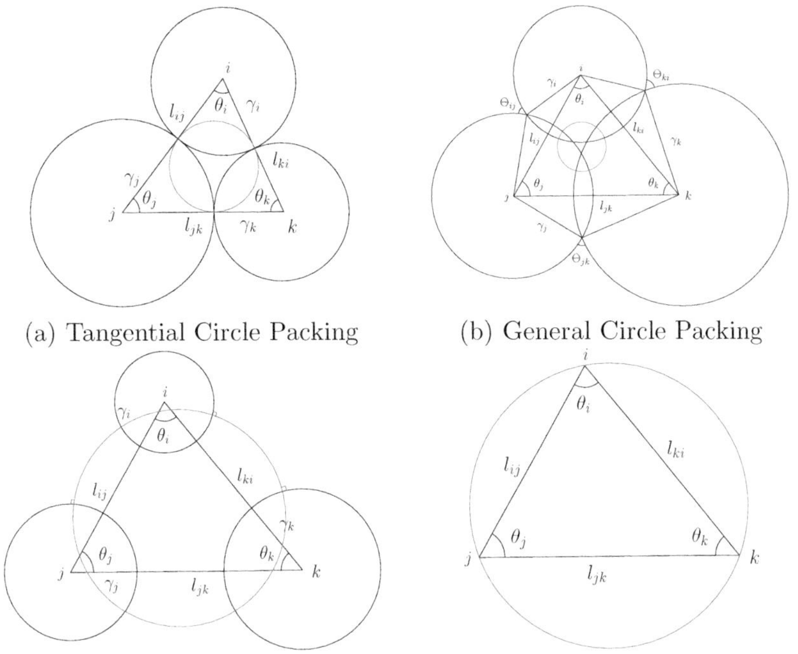

(a) Tangential Circle Packing (b) General Circle Packing

(c) Inversive Distance Circle Packing (d) Combinatorial Yamabe flow

Figure 8.8: Different configurations for discrete conformal metric deformation.

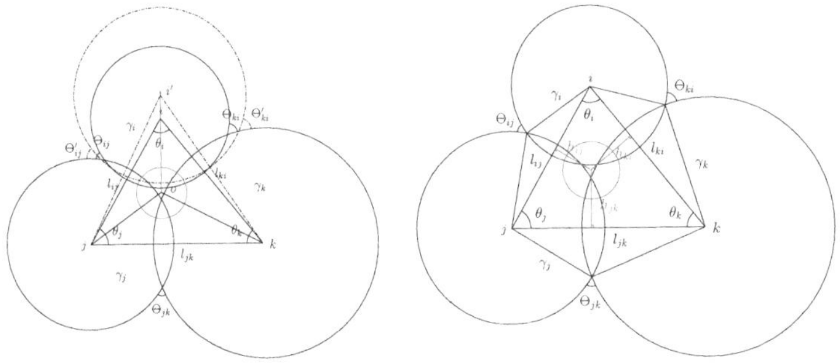

Figure 8.9: Geometric interpretation of discrete conformal metric deformation.

metric deformation are explained as the follows: let $u : V \to \mathbb{R}$ be the *discrete conformal factor*, which measures the local area distortion. If the vertex circles are with finite radii, then u_i can be formulated as

$$u_i = \begin{cases} \log \gamma_i & \mathbb{E}^2 \\ \log \tanh \frac{\gamma_i}{2} & \mathbb{H}^2 \\ \log \tan \frac{\gamma_i}{2} & \mathbb{S}^2 \end{cases} \tag{8.9}$$

1. ***Tangential Circle Packing*** In Figure 8.8 (a), the intersection angles are 0's. Therefore, the edge length is given by

$$l_{ij} = \gamma_i + \gamma_j,$$

for both the Euclidean case and the hyperbolic case, e.g., [30].

2. ***General Circle Packing*** In Figure 8.8 (b), the intersection angles are acute, $\Theta_{ij} \in (0, \frac{\pi}{2})$. The edge length is

$$l_{ij} = \sqrt{\gamma_i^2 + \gamma_j^2 + 2\gamma_i\gamma_j \cos \Theta_{ij}}$$

for the Euclidean case, and

$$l_{ij} = \cosh^{-1}(\cosh \gamma_i \cosh \gamma_j + \sinh \gamma_i \sinh \gamma_j \cos \Theta_{ij})$$

in hyperbolic geometry, e.g. [31] and [32].

3. ***Inversive Distance Circle Packing*** In Figure 8.8 (c), all the circles intersect at "virtual" angles. The $\cos \Theta_{ij}$ is replaced by the so-called *inversive distance* I_{ij}, and during the deformation, I_{ij}'s are never changed. The edge lengths are given by

$$l_{ij} = \sqrt{\gamma_i^2 + \gamma_j^2 + 2\gamma_i\gamma_j I_{ij}}$$

for Euclidean case, and

$$l_{ij} = \cosh^{-1}(\cosh \gamma_i \cosh \gamma_j + \sinh \gamma_i \sinh \gamma_j I_{ij})$$

in hyperbolic geometry, e.g. [50] and [51].

4. ***Combinatorial Yamabe Flow*** Figure 8.8 (d), all the circles are degenerated to points, $\gamma_i = 0$. The discrete conformal factor is still sensible. The edge length is given by

$$l_{ij} = e^{u_i} e^{u_j} l_{ij}^0,$$

in Euclidean background geometry, e.g. [41] and [42], and

$$\sinh \frac{l_{ij}}{2} = e^{u_i} \sinh \frac{l_{ij}^0}{2} e^{u_j},$$

in hyperbolic background geometry, e.g., [18] and [44], where l_{ij}^0 is the initial edge length of $[v_i, v_j]$, as an example.

8.3.3 Discrete Metric Surface

In the following, we want to clarify the spaces of all possible metrics and all possible curvatures of a discrete surface.

Let the vertex set be $V = \{v_1, v_2, \cdots, v_n\}$; we represent a discrete metric on Σ by a vector $\mathbf{u} = (u_1, u_2, \cdots, u_n)^T$. Similarly, we represent the Gaussian curvatures at mesh vertices by the curvature vector $\mathbf{k} = (K_1, K_2, \cdots, K_n)^T$. All the possible $\mathbf{u}$'s form the *admissible metric space*, and all the possible $\mathbf{k}$'s form the *admissible curvature space*.

According to the Gauss-Bonnet theory (see Eqn. 8.8), the total curvature must be $2\pi\chi(\Sigma)$, and therefore the curvature space is $n-1$ dimensional. We add one linear constraint to the metric vector $\mathbf{u}$, $\sum u_i = 0$, for the normalized metric. As a result, the metric space is also $n-1$ dimensional. For the circle packing metric, if all the intersection angles are acute including zero, then the edge lengths induced by a circle packing satisfy the triangle inequality. There is no further constraint on $\mathbf{u}$. Therefore, the admissible metric space is simply $\mathbb{R}^{n-1}$.

A curvature vector $\mathbf{k}$ is *admissible* if there exists a metric vector $\mathbf{u}$, which induces $\mathbf{k}$. The admissible curvature space is a convex polytope. The detailed proof can be found in [31]. The admissible curvature space for weighted meshes with hyperbolic or spherical background geometries is more complicated. We refer the readers to [31] for a detailed discussion.

Unfortunately, admissible metric space for inversive distance circle packing with both Euclidean and hyperbolic background geometries are non-convex. The admissible metric space for the combinatorial Yamabe flow with both Euclidean and Hyperbolic background geometries are non-convex.

For tangential and general circle packing cases with both $\mathbb{E}^2$ and $\mathbb{H}^2$ background geometries, see Figure 8.8 (a) and (b); the correspondence between the curvature $\mathbf{k}$ and metric $\mathbf{u}$ is globally one-to-one. This is called the *global rigidity* property. For inversive distance circle packing and combinatorial Yamabe flow cases with both $\mathbb{E}^2$ and $\mathbb{H}^2$ background geometries, see Figure 8.8 (c) and (d); only local rigidity holds. This is caused by the non-convexity of their metric spaces. In practice, non-global rigidity causes many difficulties.

8.3.4 Discrete Ricci Flow

In all configurations, the discrete Ricci flow is defined as follows:

$$\frac{du_i(t)}{dt} = (\bar{K}_i - K_i), \tag{8.10}$$

where $\bar{K}_i$ is the user defined target curvature and K_i is the curvature induced by the current metric. The discrete Ricci flow has exactly the same form as the smooth Ricci flow, which conformally deforms the discrete metric according to the Gaussian curvature.

8.3.5 Discrete Ricci Energy

The discrete Ricci flow can be formulated in the variational setting, namely, it is a negative gradient flow of a special energy form, the so-called *entropy energy*. The energy is given by

$$f(\mathbf{u}) = \int_{\mathbf{u}_0}^{\mathbf{u}} \sum_{i=1}^{n} (\bar{K}_i - K_i) du_i, \tag{8.11}$$

where $\mathbf{u}_0$ is an arbitrary initial metric.

Computing the desired metric with user-defined curvature $\{\bar{K}_i\}$ is equivalent to minimizing the discrete entropy energy. In the case of the tangential circle packing metric with both

Euclidean and hyperbolic background geometries, the discrete Ricci energy (see Equation 8.11) was first proved to be strictly convex in the seminal work of Colin de Verdiere [29]. It was generalized to the general circle packing metric in [31]. The global minimum uniquely exists, corresponding to the desired metric, which induces the prescribed curvature. The discrete Ricci flow converges to this global minimum. Although the spherical Ricci energy is not strictly convex, the desired metric $\bar{\mathbf{u}}$ is still a critical point of the energy.

The Hessian matrices for discrete entropy are positive definite for both the Euclidean case (with one normalization constraint $\sum_i u_i = 0$) and the hyperbolic case. The energy can be optimized using Newton's method. The Hessian matrix can be computed using the following formula. For all configurations with Euclidean metric, suppose the distance from the radial circle center to edge $[v_i, v_j]$ is d_{ij} as shown in Figure 8.9 (b), then

$$\frac{\partial \theta_i}{\partial u_j} = \frac{d_{ij}}{l_{ij}},$$

furthermore

$$\frac{\partial \theta_j}{\partial u_i} = \frac{\partial \theta_i}{\partial u_j}, \quad \frac{\partial \theta_i}{\partial u_i} = -\frac{\partial \theta_i}{\partial u_j} - \frac{\partial \theta_i}{\partial u_k}.$$

We define the edge weight w_{ij} for edge $[v_i, v_j]$, which is adjacent to $[v_i, v_j, v_k]$ and $[v_j, v_i, v_l]$ as

$$w_{ij} = \frac{d_{ij}^k + d_{ij}^l}{l_{ij}}.$$

The Hessian matrix $H = (h_{ij})$ is given by the discrete Laplace form

$$h_{ij} = \begin{cases} 0, & [v_i, v_j] \notin E \\ -w_{ij}, & i \neq j \\ \sum_k w_{ik}, & i = j \end{cases}$$

With hyperbolic background geometry, the computation of Hessian matrix is much more complicated. In the following, we give the formula for one face directly, for both circle packing cases:

$$\begin{pmatrix} d\theta_i \\ d\theta_j \\ d\theta_k \end{pmatrix} = \frac{-1}{A} \begin{pmatrix} 1-a^2 & ab-c & ca-b \\ ab-c & 1-b^2 & bc-a \\ ca-b & bc-a & 1-c^2 \end{pmatrix} \begin{pmatrix} \frac{1}{a^2-1} & 0 & 0 \\ 0 & \frac{1}{b^2-1} & 0 \\ 0 & 0 & \frac{1}{c^2-1} \end{pmatrix}$$

$$\begin{pmatrix} 0 & ay-z & az-y \\ bx-z & 0 & bz-x \\ cx-y & cy-x & 0 \end{pmatrix} \begin{pmatrix} du_i \\ du_j \\ du_k \end{pmatrix}$$

where $(a, b, c) = (\cosh l_i, \cosh l_j, \cosh l_k)$ and $(x, y, z) = (\cosh \gamma_i, \cosh \gamma_j, \cosh \gamma_k)$, A is double the area of the triangle $A = \sinh l_i \sinh l_j \sin \theta_k$.

For hyperbolic Yamabe flow case,

$$\frac{\partial \theta_i}{\partial u_j} = \frac{\partial \theta_j}{\partial u_i} = \frac{-1}{A} \frac{1+c-a-b}{1+c}$$

and

$$\frac{\partial \theta_i}{\partial u_i} = \frac{-1}{A} \frac{2abc - b^2 - c^2 + ab + ac - b - c}{(1+b)(1+c)}.$$

For tangential and general circle packing cases, with both $\mathbb{R}^2$ and $\mathbb{H}^2$ background geometries, the Newton's method leads to the solution efficiently. For inversive distance circle packing case and the combinatorial Yamabe flow case, with both $\mathbb{R}^2$ and $\mathbb{H}^2$ background geometries, because of the non-convexity of the metric space, Newton's method may get stuck at the boundary of the metric space; this raises intrinsic difficulty in practical computation.

Algorithmic details for general combinatorial Ricci flow can be found in [32], inversive distance circle packing metric in [51], combinatorial Yamabe flow in [44].

8.3.6 Quasi-Conformal Mapping by Solving Beltrami Equations

This section generalizes the methods for conformal mappings to general diffeomorphisms. If two surfaces are with different conformal moduli, there is no conformal mappings between them. All the diffeomorphisms between them are *quasi-conformal mappings*. This section focuses on the computational algorithms for quasi-conformal mappings.

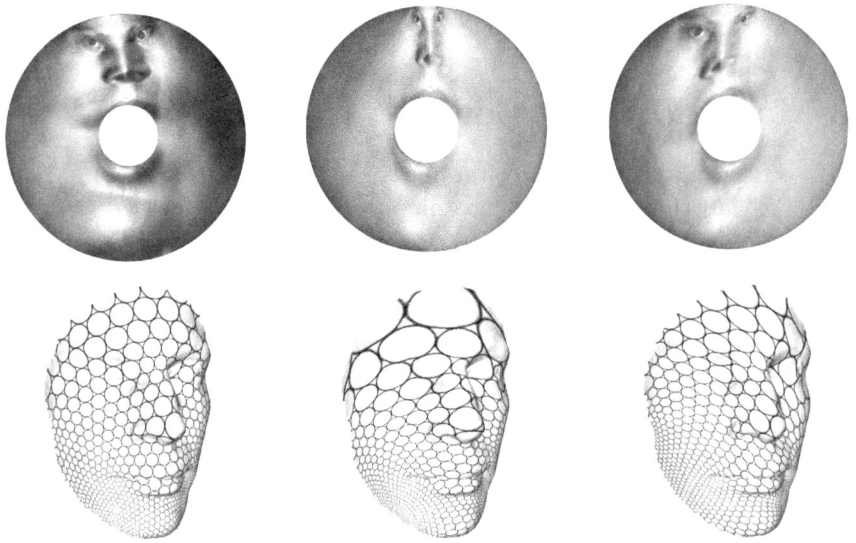

(a) $\mu = 0$ (b) $\mu = 0.25 + 0.0i$ (c) $\mu = 0.0 + 0.25i$

Figure 8.10: Quasi-conformal mapping for doubly connected domain.

Given a surface S, with a Riemannian metric $\mathbf{g}$, also given a measurable complex value function defined on the surface $\mu : S \to \mathbb{C}$, we want to find a quasi-conformal map $\phi : S \to \mathbb{C}$, such that ϕ satisfies the Beltrami equation:

$$\frac{\partial \phi}{\partial \bar{z}} = \mu \frac{\partial \phi}{\partial z}.$$

First we construct a conformal mapping $\phi_1 : (S, \mathbf{g}) \to (\mathbb{D}_1, \mathbf{g}_0)$, where $\mathbb{D}_1$ is a planar domain on $\mathbb{C}$ with the canonical Euclidean metric

$$\mathbf{g}_0 = dzd\bar{z}.$$

Then we construct a new metric, called *auxiliary metric*, on $(\mathbb{D}_1, \mathbf{g}_0)$, such that

$$\mathbf{g}_1 = |dz + \mu d\bar{z}|^2.$$

Then we construct another conformal map $\phi_2 : (\mathbb{D}_1, \mathbf{g}_1) \to (\mathbb{D}_2, \mathbf{g}_0)$. The composition

$$\phi = \phi_2 \circ \phi_1 : (S, \mathbf{g}) \to (\mathbb{D}_2, \mathbf{g}_0)$$

is the desired quasi-conformal mapping.

On discrete surface Σ, we first conformally map it to the planar domain $\mathbb{D}_1$, and the Beltrami coefficient is a piecewise linear complex value function $\mu : V \to \mathbb{C}$. We denote $\mu(v_i)$ as μ_i. Then Riemannian metrics are represented as edge lengths. For an edge $[v_i, v_j]$, its length under the canonical Euclidean metric $\mathbf{g}_0$ is $|z_j - z_i|$, where z_i, z_j are the complex planar coordinates of v_i and v_j respectively. The length under the metric $\mathbf{g}_1$ is

$$|(z_j - z_i) + \frac{1}{2}(\mu_i + \mu_j)(\bar{z}_j - \bar{z}_i)|.$$

Figure 8.10 illustrates quasi-conformal mappings for a doubly connected domain with different Beltrami coefficients.

We use the auxiliary metric for Ricci flow, and the resulting mapping is the desired quasi-conformal mapping. This method is called *quasi-conformal curvature flow.* Algorithmic details for solving the Beltrami-equation can be learned from [49] and [85].

8.4 3-Manifold Ricci Flow

All surfaces admit constant Gauss curvature metrics. This fact also holds for 3-manifolds. According to Poincaré conjecture and Thurston's geometrization conjecture, all 3-manifolds can be canonically decomposed to prime 3-manifolds. All prime 3-manifolds can be further decomposed by tori into pieces so that each piece has one of eight canonical geometries.

Studying the topological and geometric structures of three dimensional manifolds has fundamental importance in science and engineering. Computational algorithms for 3-manifolds can help topologists and geometers to investigate the complicated structures of 3-manifolds, and they also have great potential for wide applications in the engineering world. The most direct applications include volumetric parameterizations, volumetric shape analysis, volumetric deformation, solid modeling, etc. Figure 8.11 shows a simple example for volumetric parameterization for the volumetric Max Planck model, which is a topological ball.

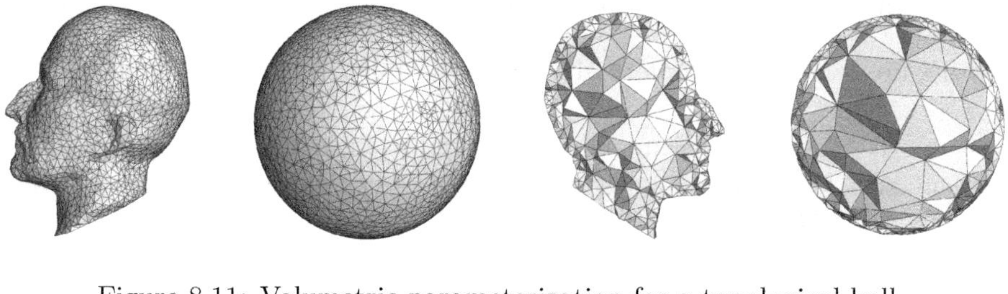

Figure 8.11: Volumetric parameterization for a topological ball.

8.4.1 Surface and 3-Manifold Curvature Flow

Similar to the surface case, most 3-manifolds have hyperbolic metric, which induces constant sectional curvature. A hyperbolic 3-manifold with boundaries is shown in Figure 8.12, where the 3-manifold is the 3-ball with a knotted pipe removed, which is called *Thurston's knotted Y-shape.* The hyperbolic 3-manifold with complete geodesic boundaries have the following topological properties:

1. The genus of boundary surfaces are greater than one.

2. For any closed curve on the boundary surface, if it cannot shrink to a point on the boundary, then it cannot shrink to a point inside the volume.

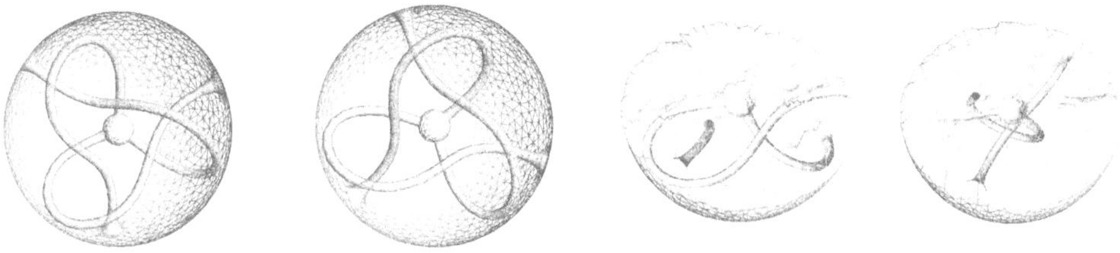

(a)Boundary surface (b) Boundary surface (c) Cut view (d) Cut view

Figure 8.12: Thurston's knotted Y-shape.

Similarities between Surface and 3-Manifold Curvature Flow

Discrete surface curvature flow can be naturally generalized to the 3-manifold case. In the following, we directly generalize discrete hyperbolic surface Ricci flow to discrete curvature flow for hyperbolic 3-manifolds with geodesic boundaries. The 3-manifold is triangulated to tetrahedra with hyperbolic background geometry, and the edge lengths determine the metric. The edge lengths are deformed according to the curvature. At the steady state, the metric induces the constant sectional curvature.

For the purpose of comparison, first we illustrate the discrete hyperbolic Ricci flow for surface case using Figure 8.13. A surface with negative Euler number is parameterized and conformally embedded in the hyperbolic space $\mathbb{H}^2$. The three boundaries are mapped to geodesics. Given two arbitrary boundaries, there exists a unique geodesic orthogonal to both boundaries. Three such geodesics partition the whole surface into two right-angled hexagons as shown in (c). The universal covering space of the surface is embedded in $\mathbb{H}^2$, frame (c) shows one fundamental polygon, frame (d) shows the finite portion of the whole universal covering space.

The hyperbolic 3-manifold with boundaries is quite similar. Given a hyperbolic 3-manifold with geodesic boundaries, such as the Thurston's knotted Y-shape in Figure 8.12, discrete curvature flow can lead to the hyperbolic metric. The boundary surface become hyperbolic planes (geodesic submanifolds). Hyperbolic planes orthogonal to the boundary surfaces segment the 3-manifold to several hyperbolic truncated tetrahedra 8.14. The universal covering space of the 3-manifold with the hyperbolic metric can be embedded in $\mathbb{H}^3$ as shown in Figure 8.25.

There are many intrinsic similarities between surface curvature flow and volumetric curvature flow. We summarize the corresponding concepts for surfaces and 3-manifolds respectively in Table 8.2: the building blocks for surfaces are right-angled hyperbolic hexagons as shown in figure Figure 8.13 frame (c); for 3-manifolds they are truncated hyperbolic tetrahedra as shown in Figure 8.14. Both cases require performing curvature flows. The curvature used in the surface case is the vertex curvature in Figure 8.15, which in the 3-manifold case is the edge curvature in Figure 8.16. The parameter domain for the surface case is the hyperbolic space $\mathbb{H}^2$ using the upper half plane model; the domain for 3-manifold case is the hyperbolic space $\mathbb{H}^3$ using the upper half space model.

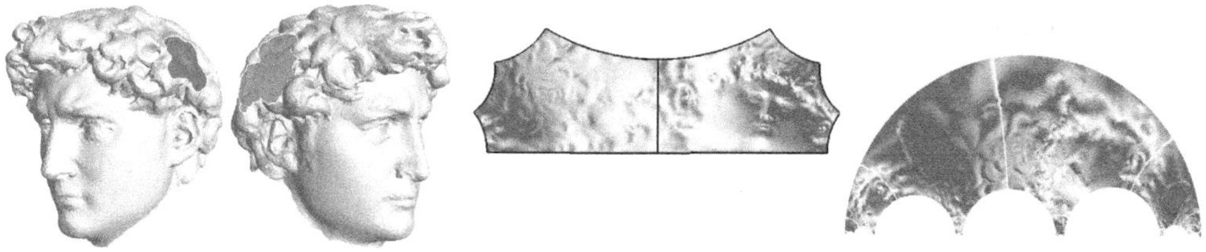

(a)Left view (b) Right view (c) Fundamental domain (d) Periodic embedding

Figure 8.13: Surface with boundaries with negative Euler number can be conformally periodically mapped to the hyperbolic space $\mathbb{H}^2$.

Differences between Surface and 3-Manifold Curvature Flow

There are fundamental differences between surfaces and 3-manifolds. The *Mostow rigidity* is the most prominent one [90]. Mostow rigidity states that the geometry of a finite volume hyperbolic manifold (for dimension greater than two) is determined by the fundamental group. Namely, suppose M and N are complete finite volume hyperbolic n-manifolds with $n > 2$. If there exists an isomorphism $f : \pi_1(M) \to \pi_1(N)$ then it is induced by a unique isometry from M to N. For surface case, the geometry of the surface is not determined by the fundamental group. Suppose M and N are two surfaces with hyperbolic metrics. If M and N share the same topology, then there exist isomorphisms $f : \pi_1(M) \to \pi_1(N)$. But there may not exist an isometry from M to N. If we fix the fundamental group of the surface M, then there are infinite many pairwise non-isometric hyperbolic metrics on M, each of them corresponding to a conformal structure of M.

Namely, surfaces have conformal geometry, and 3-manifolds don't have conformal geometry. All the Riemannian metrics on the topological surface S can be classified by the conformal equivalence relation, and each equivalence class is a *conformal structure*. If the surface is with a negative Euler number, then there exists a unique hyperbolic metric in each conformal structure.

Conformality is an important criteria for surface parameterization. Conformal surface parameterization is equivalent to finding a metric with constant Gaussian curvature conformal to the induced Euclidean metric. For 3-manifold parameterizations, conformality cannot be achieved in general. Surface parameterizations need the original induced Eu-

Table 8.2: Correspondence between surface and 3-manifold parameterizations.

	Surface	3-Manifold
Manifold	with negative Euler number with boundaries Figure 8.13	Hyperbolic 3-manifold with geodesic boundaries Figure 8.12
Building Block	hyperbolic right-angled hexagons Figure 8.13	Truncated hyperbolic tetrahedra Figure 8.14
Curvature	Gaussian curvature Fig 8.15	Sectional curvature Figure 8.15, Figure 8.16
Algorithm	Discrete Ricci flow	Discrete curvature flow
Parameter domain	Upper half plane $\mathbb{H}^2$ Figure 8.13	Upper half space $\mathbb{H}^3$ Figure 8.25

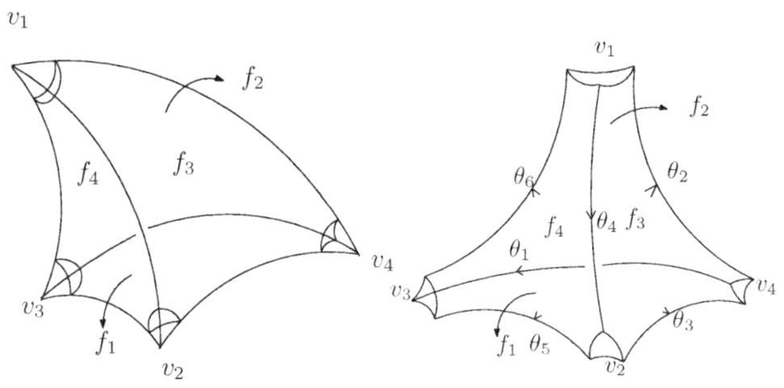

Figure 8.14: Hyperbolic tetrahedron and truncated tetrahedron.

clidean metric, namely, the vertex positions or the edge lengths are essential parts of the input. In contrast, for 3-manifolds, only topological information is required. The tessellation of a surface will affect the conformality of the parameterization result. The tessellation doesn't affect the computational results of 3-manifolds. In order to reduce the computational complexity, we can use the simplest triangulation for a 3-manifold. For example, the 3-manifold of Thurston's Knotted Y-Shape in Figure 8.12 can be either represented as a high resolution tetrahedral mesh or a mesh with only 2 truncated tetrahedra; the resulting canonical metrics are identical. Meshes with very few tetrahedra are highly desired for the sake of efficiency.

In practice, on discrete surfaces, there are only vertex curvatures, which measure the angle deficient at each vertex. On discrete 3-manifolds, like a tetrahedral mesh, there are both vertex curvatures and edge curvatures. The vertex curvature equals 4π minus all the surrounding solid angles; the edge curvature equals 2π minus all the surrounding dihedral angles. The vertex curvatures are determined by the edge curvatures. In our computational algorithm, we mainly use the edge curvature.

8.4.2 Hyperbolic 3-Manifold with Complete Geodesic Boundaries

2-manifolds (surfaces) are approximated by triangular meshes with different background geometries. Similarly, 3-manifolds are approximated by tetrahedron meshes with different background geometry. 3-manifolds with boundaries can also be approximated by truncated tetrahedron meshes; where the face hexagons are glued together, the vertex triangles form the boundary surface.

Hyperbolic Tetrahedron and Truncated Hyperbolic Tetrahedron

A closed 3-manifold can be triangulated to tetrahedra. The left frame in Figure 8.14 shows a hyperbolic tetrahedron $[v_1v_2v_3v_4]$. Each face f_i of a hyperbolic tetrahedron is a hyperbolic plane, each edge e_{ij} is a hyperbolic line segment. The right frame in Figure 8.14 shows a truncated hyperbolic tetrahedron, where the four vertices are truncated by hyperbolic planes. The cutting plane at vertex v_i is perpendicular to the edges e_{ij}, e_{ik}, e_{il}. Therefore, each face of a truncated hyperbolic tetrahedron is a right-angled hyperbolic hexagon, and each cutting section is a hyperbolic triangle.

As shown in Figure 8.14, the dihedral angles are $\{\theta_1, \theta_2, \cdots, \theta_6\}$. The geometry of the truncated tetrahedron is determined by these angles. For example, the hyperbolic triangle at v_2 has inner angles $\theta_3, \theta_4, \theta_5$, and its edge lengths can be determined using the formula in section 8.3.2. For face f_4, the edge lengths e_{12}, e_{23}, e_{31} are determined by the hyperbolic triangles at v_1, v_2, v_3 using the right-angled hyperbolic hexagon cosine law in section 8.3.1.

On the other hand, the geometry of a truncated tetrahedron is determined by the length of edges $e_{12}, e_{13}, e_{14}, e_{23}, e_{34}, e_{42}$. Due to the fact that each face is a right angled hexagon, the above six edge lengths will determine the edge lengths of each vertex triangle, and therefore determine its three inner angles, which are equal to the corresponding dihedral angles.

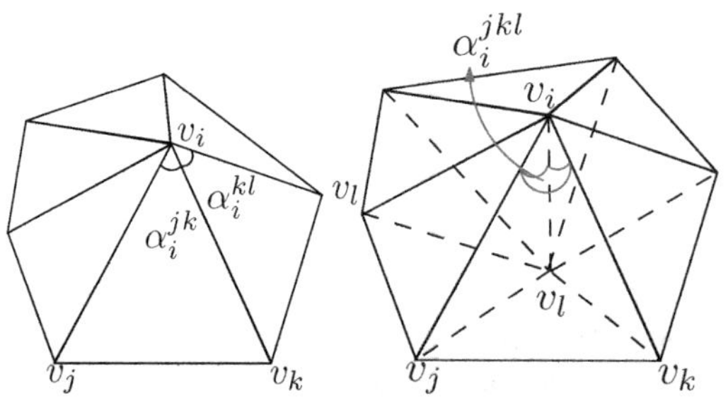

Figure 8.15: Discrete vertex curvature for 2-manifold and 3-manifold.

Discrete Curvature

In a 3-manifold case, as shown in Figure 8.15, each tetrahedron $[v_i, v_j, v_k, v_l]$ has four solid angles at their vertices, denoted as $\{\alpha_i^{jkl}, \alpha_j^{kli}, \alpha_k^{lij}, \alpha_l^{ijk}\}$; for an interior vertex, the vertex curvature is 4π minus the surrounding solid angles,

$$K(v_i) = 4\pi - \sum_{jkl} \alpha_i^{jkl}.$$

For a boundary vertex, the vertex curvature is 2π minus the surrounding solid angles.

In a 3-manifold case, there is another type of curvature, *edge curvature*. Suppose $[v_i, v_j, v_k, v_l]$ is a tetrahedron; the dihedral angle on edge e_{ij} is denoted as β_{ij}^{kl}. If edge e_{ij} is an interior edge (i.e. e_{ij} is not on the boundary surface), its curvature is defined as

$$K(e_{ij}) = 2\pi - \sum_{kl} \beta_{ij}^{kl}.$$

If e_{ij} is on the boundary surface, its curvature is defined as

$$K(e_{ij}) = \pi - \sum_{kl} \beta_{ij}^{kl}.$$

For 3-manifolds, edge curvature is more essential than vertex curvature. The latter is determined by the former.

Theorem 8.4.1 *Suppose M is a tetrahedron mesh, and v_i is an interior vertex of M. Then*

$$\sum_j K(e_{ij}) = K(v_i).$$

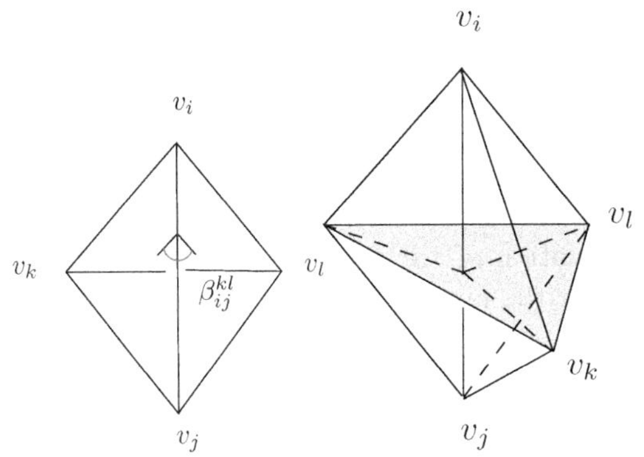

Figure 8.16: Discrete edge curvature for a 3-manifold.

Discrete Curvature Flow

Given a hyperbolic tetrahedron in $\mathbb{H}^3$ with edge lengths x_{ij} and dihedral angles θ_{ij}, the *volume* of the tetrahedron V is a function of the dihedral angles $V = V(\theta_{12}, \theta_{13}, \theta_{14}, \theta_{23}, \theta_{24}, \theta_{34})$, and the Schlaefli formula can be expressed as

$$\frac{\partial V}{\partial \theta_{ij}} = \frac{-x_{ij}}{2}, \tag{8.12}$$

namely, the differential 1-form dV is $\frac{-1}{2}\sum_{ij} x_{ij} d\theta_{ij}$. It can be further proved that the volume of a hyperbolic truncated tetrahedron is a strictly concave function of the dihedral angles.

Given an ideal triangulated 3-manifold (M, T), let E be the set of edges in the triangulation. An assignment $x : E \to \mathbb{R}^+$ is called a *hyperbolic cone metric associated with the triangulation* T if for each tetrahedron t in T with edges $e_1, e_2, \cdots, e_6$, and the $x(e_i)$ are the edge lengths of a hyperbolic truncated tetrahedron in $\mathbb{H}^3$. The set of all hyperbolic cone metrics associated with T is denoted as $L(M, T)$, which is an open set. The discrete curvature of a cone metric is a map $K(x) : L \to R$, mapping each edge e to its discrete curvature. The discrete curvature flow is then defined by

$$\frac{dx_{ij}}{dt} = K_{ij}, \tag{8.13}$$

where x_{ij} is the edge length of e_{ij}, and K_{ij} is the edge curvature of e_{ij}. The curvature flow is the gradient flow of the hyperbolic volume of the M,

$$V(\mathbf{x}) = \int_{\mathbf{x}_0}^{\mathbf{x}} \sum_{e_{ij}} K_{ij} dx_{ij}, \tag{8.14}$$

where $\mathbf{x}_0 = (1, 1, \cdots, 1)$ or another other initial metric.

Theorem 8.4.2 *For any ideal triangulated 3-manifold (M, T), the equilibrium points of the discrete curvature flow Eqn.8.13 are the complete hyperbolic metric with totally geodesic boundary. Each equilibrium is a local attractor of the flow.*

Furthermore, a hyperbolic cone metric associated with an ideal triangulation is locally determined by its cone angles. For any ideal triangulated 3-manifold, under the discrete curvature flow, the discrete curvature $K_{ij}(t)$ evolves according to the discrete heat equation. Furthermore, the total curvature $\sum_{ij} K_{ij}^2$ is strictly decreasing until all edge curvatures (also the vertex curvatures) are zeros. The theoretic proofs can be found in [89].

8.4.3 Discrete Hyperbolic 3-Manifold Ricci Flow

The input to the algorithm is the boundary surface of a 3-manifold, represented as a triangular mesh. The output is a realization of (fundamental domain of) the 3-manifold in the hyperbolic space $\mathbb{H}^3$. The algorithm pipeline is as the following:

1. Compute the triangulation of the 3-manifold as a tetrahedral mesh. Simplify the triangulation such that the number of the tetrahedra is minimal.

2. Run discrete curvature flow on the tetrahedral mesh to obtain the hyperbolic metric.

3. Realize the mesh with the hyperbolic metric in the hyperbolic space $\mathbb{H}^3$.

Triangulation and Simplification

In geometric processing, surfaces are approximated by triangular meshes. 3-manifolds are approximated by tetrahedral meshes. In general, given the boundary surfaces of a 3-manifold, there are existing methods to tessellate the interior and construct the tetrahedral mesh. In this work, we use tetrahedral tessellation based on volumetric Delaunay triangulation.

In order to simplify the triangulation, we use the following algorithm.

1. Denote the boundary of a 3-manifold M as $\partial M = \{S_1, S_2, \cdots, S_n\}$. For each boundary surface S_i, create a cone vertex v_i, and connect each face $f_j \in S_i$ to form a tetrahedron T_j^i. Therefore, M is augmented to $\tilde{M}$.

2. Use *edge collapse* as shown in Figure 8.18 to simplify the triangulation, such that all vertices are removed except for those cone vertices $\{v_1, v_2, \cdots, v_n\}$ generated in the previous step. Denote the simplified tetrahedral mesh still as $\tilde{M}$.

3. For each tetrahedron $\tilde{T}_i \in \tilde{M}$, cut $\tilde{T}$ by the boundary surfaces to form a truncated tetrahedron (hyper ideal tetrahedron), denoted as T_i.

The simplified triangulation is represented as a collection of truncated tetrahedra and their gluing pattern. As shown in Figure 8.17, the simplified tetrahedral mesh has only two truncated tetrahedra T_1, T_2. Let A_i, B_i, C_i, D_i represent the four faces of the tetrahedron T_i; let a_i, b_i, c_i, d_i represent the truncated vertices of T_i. The gluing pattern is given as

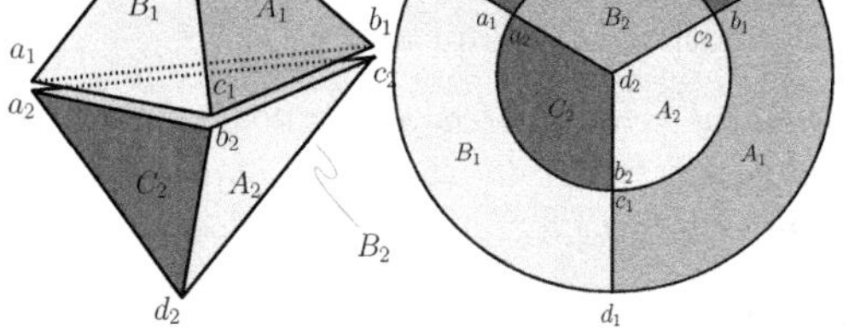

Figure 8.17: Simplified triangulation and gluing pattern of Thurston's knotted-Y. The two faces with the same color are glued together.

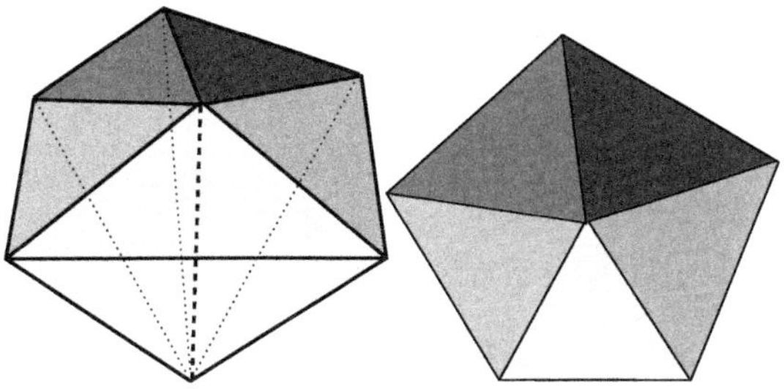

Figure 8.18: (See Color Insert.) Edge collapse in tetrahedron mesh.

follows:

$$\begin{array}{ll} A_1 \to B_2 & \{b_1 \to c_2, d_1 \to a_2, c_1 \to d_2\} \\ B_1 \to A_2 & \{c_1 \to b_2, d_1 \to c_2, a_1 \to d_2\} \\ C_1 \to C_2 & \{a_1 \to a_2, d_1 \to b_2, b_1 \to d_2\} \\ D_1 \to D_2 & \{a_1 \to a_2, b_1 \to c_2, c_1 \to b_2\} \end{array}$$

The first row means that face $A_1 \in T_1$ is glued with $B_2 \in T_2$, such that the truncated vertex b_1 is glued with c_2, d_1 with a_2, and c_1 with d_2. Other rows can be interpreted in the same way.

Hyperbolic Embedding of 3-Manifolds

Once the edge lengths of the tetrahedron mesh have been obtained, we can realize it in the hyperbolic space $\mathbb{H}^3$. First, we introduce how to construct a single truncated tetrahedron; then we explain how to glue multiple truncated tetrahedra by hyperbolic rigid motion.

Construction of a Truncated Hyperbolic Tetrahedron The geometry of a truncated hyperbolic tetrahedron is determined by its dihedral angles. This section explains the algorithm to construct a truncated tetrahedron in the upper half space model of $\mathbb{H}^3$. The algorithm consists of two steps. First, construct a circle packing on the plane; second, compute a CSG (Constructive Solid Geometry) surface. The resulting surface is the boundary of the truncated tetrahedron.

Step 1: Construct a Circle Packing. Suppose the dihedral angles of a truncated tetrahedron are given. The tetrahedron can be realized in $\mathbb{H}^3$ uniquely, up to rigid motion. The tetrahedron is the intersection of half spaces, the boundaries of these half spaces are the hyper planes on faces f_1, f_2, f_3, f_4, and the cutting planes at the vertices v_1, v_2, v_3, v_4. Each plane intersects the infinity plane at a hyperbolic line, which is a Euclidean circle on the xy-plane. By abusing the symbols, we use f_i to represent the intersection circle between the hyperbolic plane through the face f_i and the infinity plane. Similarly, we use v_j to represent the intersection circle between the cutting plane at v_j and the infinity plane. The goal of this step is to find planar circles (or lines) f_i's and v_j's, such that

1. f_i and circle f_j intersect at the given corresponding angle β_{ij}^{kl}.

2. circle v_i is orthogonal to circles f_j, f_k, f_l.

As shown in Figure 8.19, all the circles can be computed explicitly with two extra constraints, f_1 and f_2 are lines with two intersection points 0 and ∞, and the radius of f_3 equals one. The dihedral angle on edges $\{e_{34}, e_{14}, e_{24}, e_{12}, e_{23}, e_{13}\}$ are $\{\theta_1, \theta_2, \theta_3, \theta_4, \theta_5, \theta_6\}$

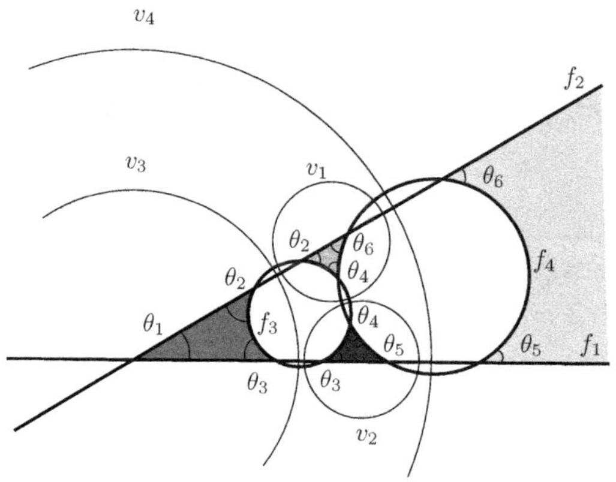

Figure 8.19: (See Color Insert.) Circle packing for the truncated tetrahedron.

as shown in Figure 8.14.

After finding v_1, v_2, v_3, v_4, we transform them back using ϕ. Let w_1, w_2, w_3 be points on the circle v_1, the $\phi(w_1), \phi(w_2), \phi(w_3)$ are the points on the circle $\phi(v_1)$.

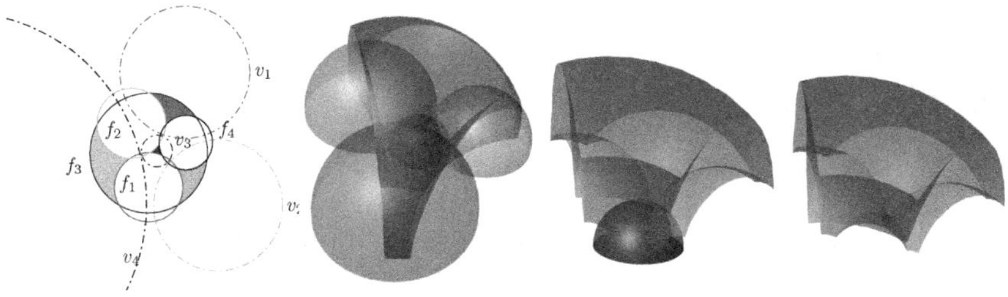

Figure 8.20: (See Color Insert.) Constructing an ideal hyperbolic tetrahedron from circle packing using CSG operators.

Step 2: CSG Modeling. After we obtain the circle packing, we can construct hemispheres whose equators are those circles. If the circle is a line, then we construct a half plane orthogonal to the xy-plane through the line. Computing CSG among these hemispheres and half-planes, we can get the truncated tetrahedron as shown in Figure 8.20.

Each hemisphere is a hyperbolic plane, and separates $\mathbb{H}^3$ to two half-spaces. For each hyperbolic plane, we select one half-space; the intersection of all such half-spaces is the desired truncated tetrahedron embedded in $\mathbb{H}^3$. We need to determine which half-space of the two is to be used. We use f_i to represent both the face circle and the hemisphere whose equator is the face circle f_i. Similarly, we use v_k to represent both the vertex circle and the hemisphere whose equator is the vertex circle. As shown in Figure 8.19, three face circles f_i, f_j, f_k bound a curved triangle Δ_{ijk}, which is color coded; one of them is infinite. If Δ_{ijk} is inside the circle f_i, then we choose the half space inside the hemisphere f_i; otherwise we choose the half-space outside the hemisphere f_i. Suppose vertex circle v_k is orthogonal to the face circles f_i, f_j, f_k; if Δ_{ijk} is inside the circle v_k, then we choose the half-space inside the hemisphere v_k; otherwise we choose the half-space outside the hemisphere v_k.

Figure 8.21 demonstrates a realization of a truncated hyperbolic tetrahedron in the upper half space model of $\mathbb{H}^3$, based on the circle packing in Figure 8.19.

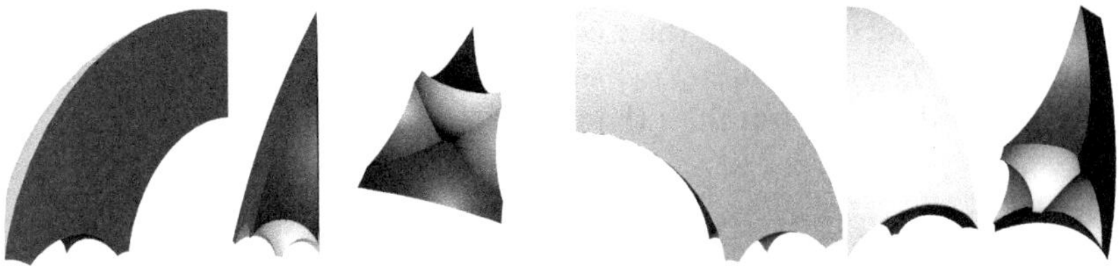

Figure 8.21: (See Color Insert.) Realization of a truncated hyperbolic tetrahedron in the upper half space model of $\mathbb{H}^3$, based on the circle packing in Figure 8.19.

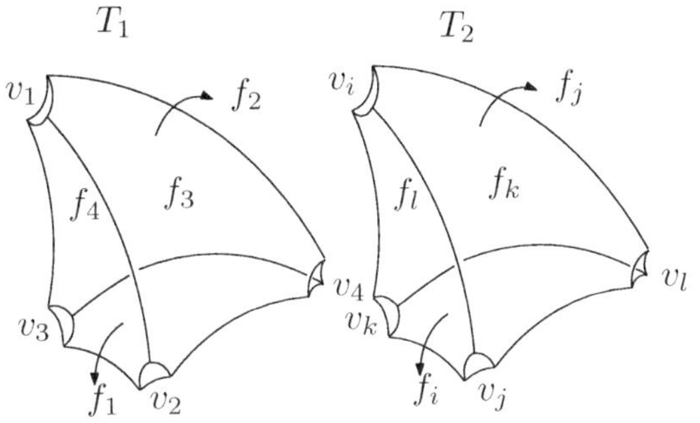

Figure 8.22: Glue T_1 and T_2 along $f_4 \in T_1$ and $f_l \in T_2$, such that $\{v_1, v_2, v_3\} \subset T_1$ are attached to $\{v_i, v_j, v_k\} \subset T_2$.

Gluing Two Truncated Hyperbolic Tetrahedra Suppose we want to glue two truncated hyperbolic tetrahedra, T_1 and T_2, along their faces. We need to specify the correspondence between the vertices and faces between T_1 and T_2. As shown in Figure 8.22, suppose we want to glue $f_4 \in T_1$ to $f_l \in T_2$, such that $\{v_1, v_2, v_3\} \subset T_1$ are attached to $\{v_i, v_j, v_k\} \subset T_2$. Such a gluing pattern can be denoted as a permutation $\{1, 2, 3, 4\} \to \{i, j, k, l\}$. The right-angled hyperbolic hexagon of f_4 is congruent to the hexagon of f_l.

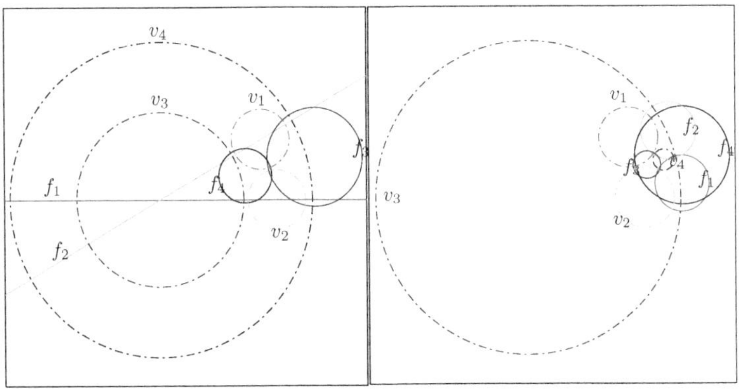

Figure 8.23: (See Color Insert.) Glue two tetrahedra by using a Möbius transformation to glue their circle packings, such that $f_3 \to f_4$, $v_1 \to v_1, v_2 \to v_2, v_4 \to v_3$.

As shown in Figure 8.23, the gluing can be realized by a rigid motion in $\mathbb{H}^3$, which induces a Möbius transformation on the xy-plane. The Möbius transformation aligns the corresponding circles, $f_3 \to f_4$, $\{v_1, v_2, v_4\} \to \{v_1, v_2, v_3\}$. The Möbius transformation can be explicitly computed, and determines the rigid motion in $\mathbb{H}^3$.

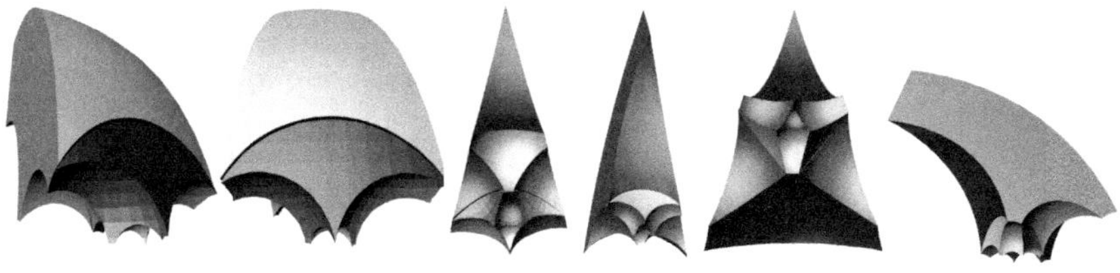

Figure 8.24: (See Color Insert.) Glue T_1 and T_2. Frames (a)(b)(c) show different views of the gluing $f_3 \to f_4$, $\{v_1, v_2, v_4\} \to \{v_1, v_2, v_3\}$. Frames (d) (e) (f) show different views of the gluing $f_4 \to f_3$,$\{v_1, v_2, v_3\} \to \{v_2, v_1, v_4\}$.

Figure 8.25: (See Color Insert.) Embed the 3-manifold periodically in the hyperbolic space $\mathbb{H}^3$.

Figure 8.24 shows the gluing between two truncated hyperbolic tetrahedra. By repeating the gluing process, we can embed the universal covering space of the hyperbolic 3-manifold in $\mathbb{H}^3$. Figure 8.25 shows different views of the embedding of the (finite portion) universal covering space of Thurston's knotted Y-Shape in $\mathbb{H}^3$ with the hyperbolic metric. More computation details can be found in [91].

8.5 Applications

Computational conformal geometry has been broadly applied in many engineering fields. In the following, we briefly introduce some of our recent projects, which are the most direct applications of computational conformal geometry in the computer science field.

Graphics

Conformal geometric methods have broad applications in computer graphics. Isothermal coordinates are natural for global surface parameterization purposes [11]. Because conformal mapping doesn't distort the local shapes, it is desirable for texture mapping. Figure 8.26 shows one example of using holomorphic 1-forms for texture mapping.

Special flat metrics are valuable for designing vector fields on surfaces, which plays an important role for non-photorealistic rendering and special art form design. Figure 8.27

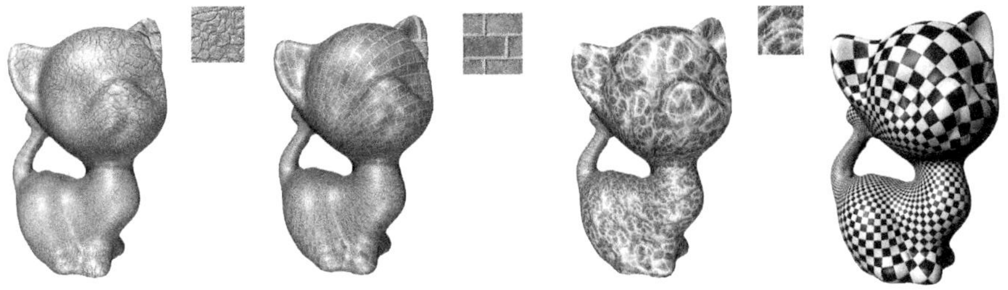

Figure 8.26: Global conformal surface parameterization using holomorphic 1-forms.

shows the examples for vector fields design on surfaces using the curvature flow method [92].

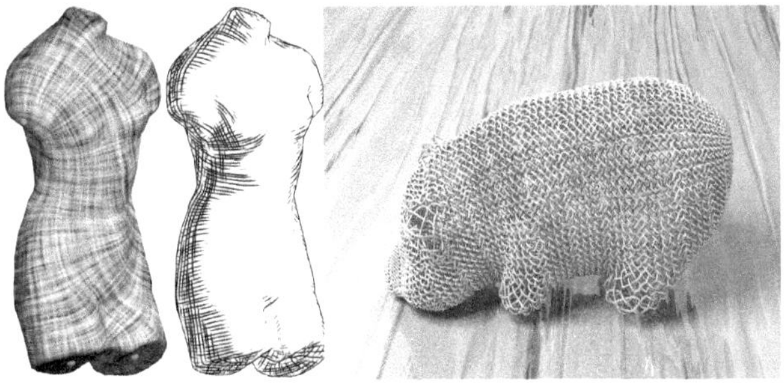

Figure 8.27: Vector field design using special flat metrics.

Geometric Modeling

One of the most fundamental problems in geometric modeling is to systematically generalize conventional spline schemes from Euclidean domains to manifold domains. This relates to the general geometric structures on the surface.

Definition 8.5.1 ((G, X) structure) *Suppose X is a topological space, G is a transformation group of X. Let M be a manifold with an atlas $\mathbb{A}$; if all the coordinate charts (U_α, ϕ_α) are defined on the space X, $\phi_\alpha : U_\alpha \to X$, and all chart transition functions $\phi_{\alpha\beta}$ are in group G, then the atlas is a (G, X) atlas. The maximal (G, X) atlas is a (G, X) structure.*

For example, suppose the manifold is a surface; if X is the affine plane A and G is the affine transformation group $Aff(A)$, then the (G, X) structure is the affine structure. Similarly, if X is the hyperbolic plane $\mathbb{H}^2$, and G is the hyperbolic isometric transformation (Möbius transformation), then (G, X) is a hyperbolic structure; if X is the real projective plane $\mathbb{RP}^2$, and G is the real projective transformation group $PGL(2, R)$, then the (G, X) structure is a real projective structure of the surface. Real projective structure can be constructed from the hyperbolic structure.

Conventional spline schemes are constructed based on affine invariance. If the manifold has an affine structure, then affine geometry can be defined on the manifold and conventional splines can be directly defined on the manifold. Due to the topological obstruction, general manifolds don't have affine structures, but by removing several singularities, general surfaces can admit affine structures. Details can be found in [22].

Affine structures can be explicitly constructed using conformal geometric methods. For example, we can concentrate all the curvatures at the prescribed singularity positions, and set the target curvatures to be zeros everywhere else. Then we use curvature flow to compute a flat metric with cone singularities from the prescribed curvature. The flat metric induces an atlas on the punctured surface (with singularities removed), such that all the transition functions are rigid motions on the plane. Another approach is to use holomorphic 1-forms; a holomorphic 1-form induces a flat metric with cone singularities at the zeros, where the curvatures are $-2k\pi$. Figure 8.28 shows the manifold splines constructed using the curvature flow method.

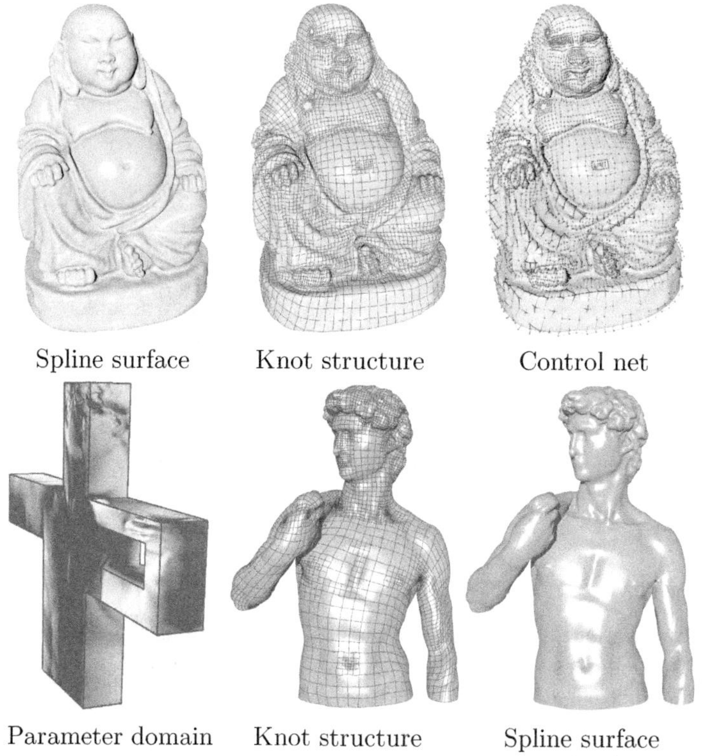

Figure 8.28: Manifold splines with extraordinary points.

Compared to other methods for constructing domains with prescribed singularity positions, such as the one based on trivial connection [88], the major advantage of this one is that it gives global conformal parameterizations of the spline surface, namely, the isothermal coordinates. Differential operators, such as gradient and Laplace–Beltrami operators, have the simplest form under isothermal coordinates, which greatly simplifies the downstream physical simulation tasks based on the splines.

Medical Imaging

Conformal geometry has been applied for many fields in medical imaging. For example, in the field of brain imaging, it is crucial to register different brain cortex surfaces. Because brain surfaces are highly convoluted, and different people have different anatomic structures, it is quite challenging to find a good matching between cortex surfaces. Figure 8.29

illustrates one solution [10] by mapping brains to the unit sphere in a canonical way. Then by finding an automorphism of the sphere, the registration between surfaces can be easily established.

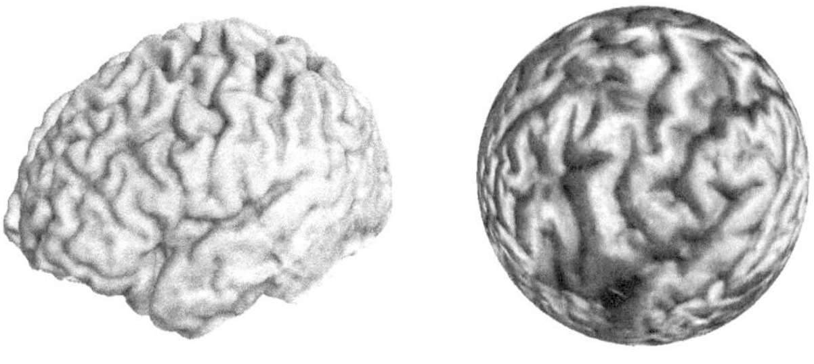

Figure 8.29: Brain spherical conformal mapping.

In virtual colonoscopy [19], the colon surface is reconstructed from CT images. By using conformal geometric methods, one can flatten the whole colon surface onto a planar rectangle. Then polyps and other abnormalities can be found efficiently on the planar image. Figure 8.30 shows an example for virtual colon flattening based on conformal mapping.

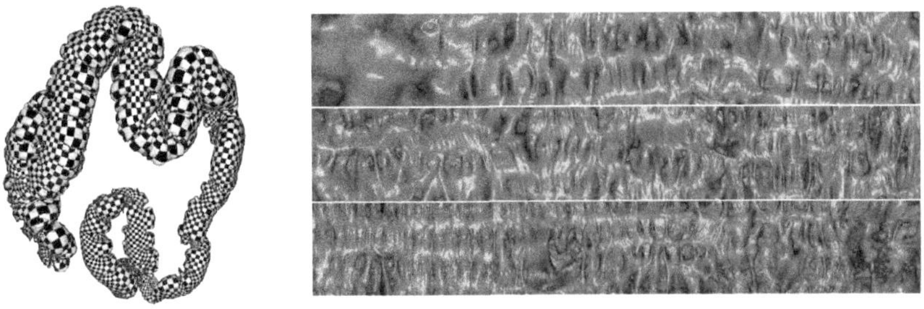

Figure 8.30: Colon conformal flattening.

Vision

Surface matching is a fundamental problem in computer vision. The main framework of surface matching can be formulated in the commutative diagram in Figure 8.31.

S_1, S_2 are two given surfaces, $f : S_1 \to S_2$ is the desired matching. We compute $\phi_i : S_i \to D_i$ which maps S_i conformally onto the canonical domain D_i. We construct a diffeomorphism map $\bar{f} : D_1 \to D_2$, which incorporates the feature constraints. The final map ϕ is induced by $f = \phi_2 \circ \bar{f} \circ \phi_1^{-1}$. Figure 8.32 shows one example of surface matching among views of a human face with different expressions. The first row shows the surfaces in $\mathbb{R}^3$. The second row illustrates the matching results using consistent texture mapping. The intermediate conformal slit mappings are shown in the third row. For details, we refer readers to [21],[20]. Conformal geometric invariants can also be applied for shape analysis and recognition; details can be found in [93].

Teichmüller theory can be applied for surface classification in [46, 47]. By using Ricci curvature flow, we can compute the hyperbolic uniformization metric. Then we compute the pants decomposition using geodesics and compute the Fenchel-Nielsen coordinates. In Figure 8.33, a set of canonical fundamental group basis is computed (a). Then a fundamental domain is isometrically mapped to the Poincaré disk with the uniformization metric (b).

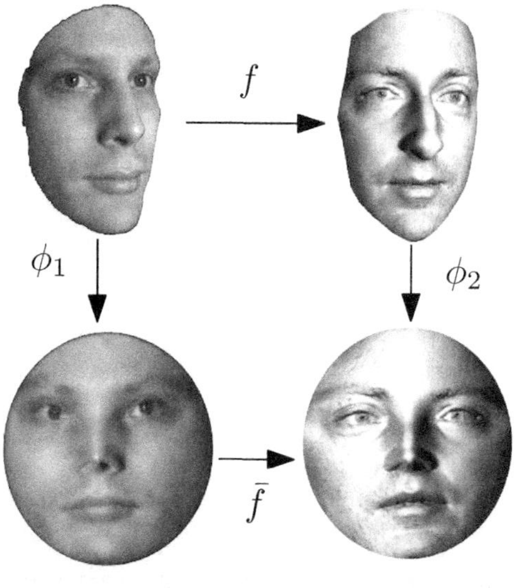

Figure 8.31: Surface matching framework.

By using Fuchsian transformation, the fundamental domain is transferred (c) and a finite portion of the universal covering space is constructed in (d). Figure 8.34 shows the pipeline for computing the Teichmüller coordinates. The geodesics on the hyperbolic disk are found in (a), and the surface is decomposed by these geodesics (b). The shortest geodesics between two boundaries of each pair of hyperbolic pants are computed in (c),(d), and (e). The twisting angle is computed in (f). Details can be found in [47].

Computational Geometry

In computational geometry, homotopy detection is an important problem: given a loop on a high genus surface, compute its representation in the fundamental group, or to verify whether two loops are homotopic to each other.

We use Ricci flow to compute the hyperbolic uniformization metric in [44]. According to the Gauss-Bonnet theorem, each homotopy class has a unique closed geodesic. Given a loop γ, we compute the Möbius transformation τ corresponding to the homotopy class of γ. The axis of τ is a closed geodesic $\tilde{\gamma}$ on the surface under the hyperbolic metric. We use $\tilde{\gamma}$ as the canonical representation as the homotopy class of $[\gamma]$. As shown in Figure 8.35, if two loops γ_1 and γ_2 are homotopic to each other, then their canonical representations $\tilde{\gamma}_1$ and $\tilde{\gamma}_2$ are equal.

Wireless Sensor Network

In the wireless sensor network field, it is important to design a Riemannian metric to ensure the delivery of packets and balance the computational load among all the sensors. Because each sensor can only collect the information in its local neighbors, it is desirable to use greedy routing. Basically, each node has virtual coordinates. The sensor sends the packet to its direct neighbor, which is the closest one to the destination. If the network has concave holes, as shown in Figure 8.36, the routing may get stuck at the nodes along the inner boundaries. We use Ricci flow to compute the virtual coordinates, such that all inner holes become circles or hyperbolic geodesics, and then greedy routing delivery is guaranteed. The delivery path is guided by geodesics under the special Riemannian metric.

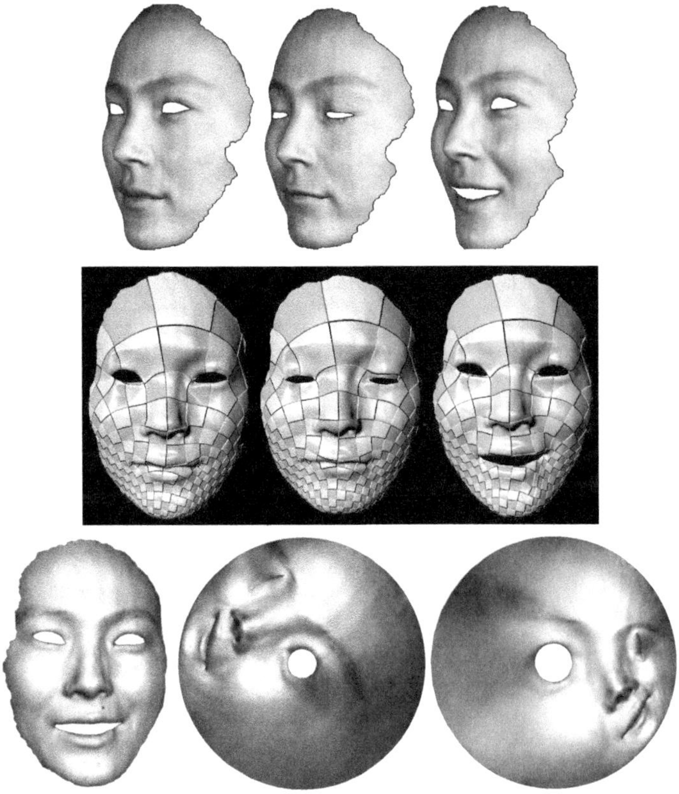

Figure 8.32: Matching among faces with different expressions.

The covering spaces with Euclidean and hyperbolic geometry pave a new way to handle load balancing and data storage problems. Using the virtual coordinates, many shortest paths will pass through the the nodes on the inner boundaries. Therefore, the nodes on the inner boundaries will be overloaded. Then, we can reflect the network about the inner circular boundaries or hyperbolic geodesics. All such reflections form the so-called Schottky group in a Euclidean case (b), the so-called Fuchsian group in a hyperbolic case (a), and then perform the routing on the covering space. This method ensures delivery and improves load balancing using greedy routing. Implementation details can be found in [94], [95], and [96].

8.6 Summary

Computational conformal geometry is an interdisciplinary field between mathematics and computer science. This work explains the fundamental concepts and theories for the subject. Major tasks in computational conformal geometry and their solutions are explained. Both the holomorphic differential method and the Ricci flow method are elaborated in detail. Some engineering applications are briefly introduced.

There are many fundamental open problems in computational conformal geometry, which will require deeper insights and more sophisticated and accurate computational methodologies. The following problems are just a few samples which have important implications for both theory and application.

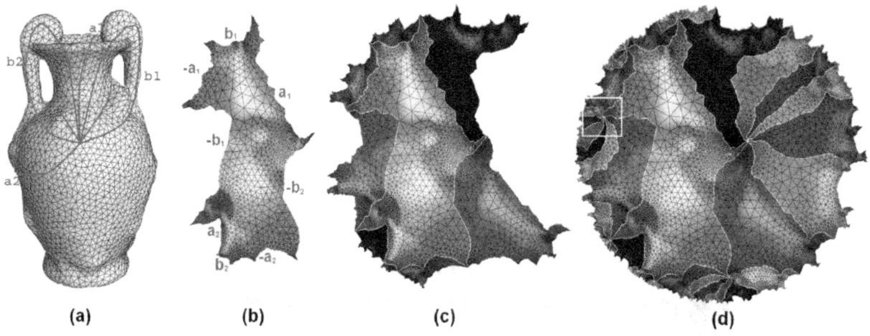

Figure 8.33: Computing finite portion of the universal covering space on the hyperbolic space.

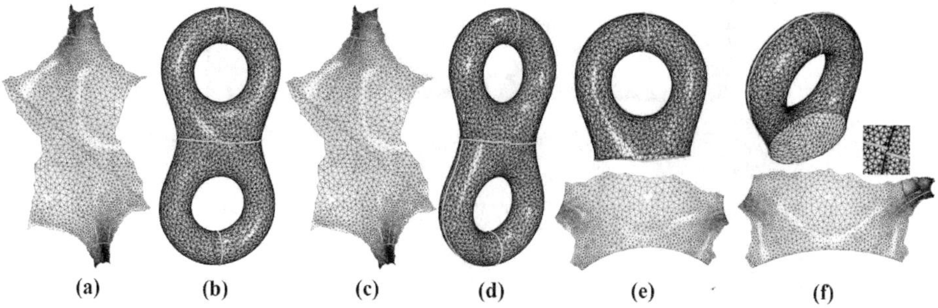

Figure 8.34: Computing the Fenchel–Nielsen coordinates in the Teichmüller space for a genus two surface.

1. ***Teichmüller Map*** Given two metric surfaces and the homotopy class of the mapping between them, compute the unique one with minimum angle distortion, the so-called *Teichmüller map.*

2. ***Abel Differential*** Compute the group of various types of Abel differentials, especially the holomorphic quadratic differentials.

3. ***Relation between Combinatorial Structure and Conformal Structure*** Given a topological surface, each triangulation has a natural conformal structure determined by tangential circle packing. Study the relation between the two structures.

4. ***Approximation Theories*** Although the algorithms for computing conformal invariants have been developed, the approximation theoretic results have not been fully developed. For conformal mappings between planar domains, the convergence of different discrete methods has been established; for those between surfaces, the convergence analysis is still open.

5. ***Accuracy and Stability*** The hyperbolic geometric computation is very sensitive to numerical error. It is challenging to improve the computational accuracy. Exact arithmetic methods in computational geometry show the promise to conquer this problem.

   In the inversive distance circle packing method and the combinatorial Yamabe flow method, the non-convexity of the admissible curvature space causes instability of the algorithm. Therefore, they require higher mesh triangulation quality. In practice, it is

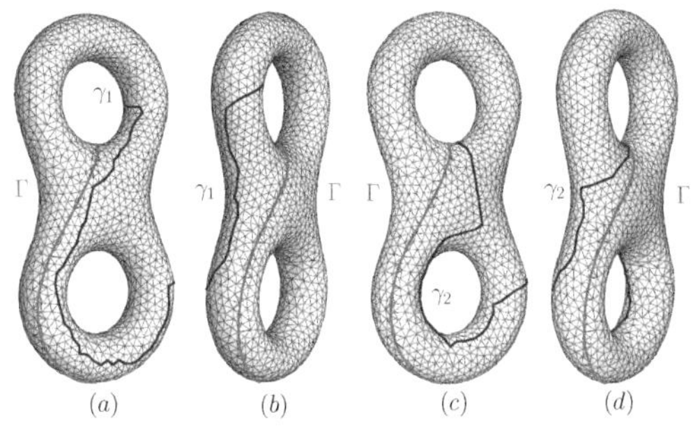

Figure 8.35: Homotopy detection using hyperbolic metric.

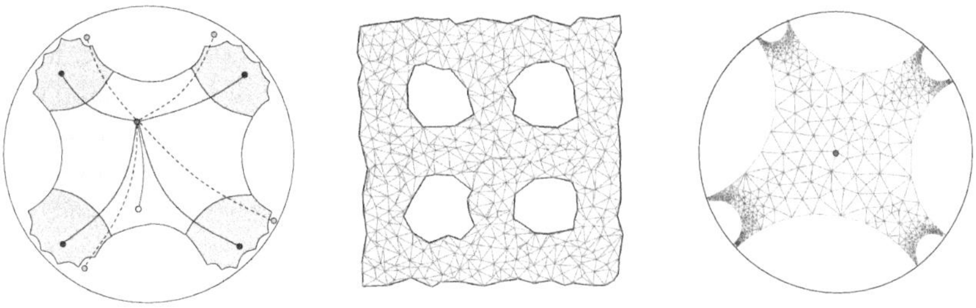

(a) Hyperbolic universal covering space

(b) Euclidean covering space

Figure 8.36: (See Color Insert.) Ricci flow for greedy routing and load balancing in wireless sensor network.

important to improve the triangulation quality for these methods. The circle packing method with acute intersection angles is more stable, the holomorphic differential form method is the most stable.

Furthermore, designing discrete curvature flow algorithms for general 3-manifolds is a challenging problem. The rigorous algorithms lead to a discrete version of a constructive proof of Poincaré's conjecture and Thurston's geometrization conjecture. One approach is to study the property of the map from the edge length to the edge curvature. If the map is globally invertible, then one can design metrics by curvatures. If the map is locally invertible, then by carefully choosing a special path in the curvature space, one can design metrics by special curvatures. One of the major difficulties is to verify whether the prescribed curvature is admissible by the mesh. The degenerated tetrahedra may emerge in the process of the curvature flow. The understanding of the formation of the degeneracies will be the key to design the discrete 3-manifold curvature flow.

We expect to see greater theoretic breakthroughs and broader applications in computational conformal geometry in the near future.

8.7 Bibliographical and Historical Remarks

Computational conformal geometry is an interdisciplinary field, one which has deep roots in pure mathematics fields, such as Riemann surface theory, complex analysis, differential geometry, algebraic topology, partial differential equations, and others.

The Ricci flow was first proposed by Hamilton [23] as a tool to conformally deform the metric according to the curvature. In [31] Chow and Luo developed the theories of the combinatorial surface Ricci flow, which was later implemented and applied for surface parameterization [32], shape classification [97], and surface matching [93].

In [41] Luo studied the discrete Yamabe flow on surfaces. He introduced a notion of discrete conformal change of polyhedral metric, which plays a key role in developing the discrete Yamabe flow and the associated variational principle in the field. Based on the discrete conformal class and geometric consideration, Luo gave the discrete Yamabe energy as an integration of a differential 1-form and proved that this energy is a locally convex function. He also deduced from it that the curvature evolution of the Yamabe flow is a heat equation. In a very nice recent work of Springborn et al. [42] they were able to identify the Yamabe energy introduced by Luo with the Milnor-Lobachevsky function and the heat equation for the curvature evolution with the cotangent Laplace equation. They constructed an algorithm based on their explicit formula.

Theories of Yamabe flow on a discrete hyperbolic surface can be found in [18]. This is the first work to develop the computational algorithm for *hyperbolic Yamabe flow*, which is used to compute the uniform hyperbolic metric for surfaces with negative Euler number.

Historically, computational conformal geometry has been broadly applied in many engineering fields [1], such as electromagnetics, vibrating membranes and acoustics, elasticity, heat transfer, and fluid flow. In recent years, it has been applied to a broad range of fields in computer science, such as computer graphics, computer vision, geometric modeling, medical imaging, and computational geometry.

Another intrinsic curvature flow is called Yamabe flow. It has the same physical intuition with the Ricci flow, except for the fact that it is driven by the scalar curvature instead of Ricci curvature. For two manifolds, the Yamabe flow is essentially equivalent to the Ricci flow. But for higher dimensional manifolds, Yamabe flow is much more flexible than the Ricci flow to reach constant-scalar-curvature metrics. In the discrete case, there is a subtle difference caused by a different notion of discrete conformal classes.

Acknowledgements We want to thank our collaborators: Arie Kaufman, Hong Qin, Dimitris Samaras, Jie Gao, Paul Thompson, Tony Chan, Yalin Wang, Lok Ming Lui, and many other colleagues. We want to thank all the students, especially Miao Jin, Ying He, Xin Li, and Xiaotian Yin. The research has been supported by NSF CCF-0448399, NSF DMS-0528363, NSF DMS-0626223, NSF IIS-0713145, NSF CCF-0830550, NSF CCF-0841514, ONR N000140910228, NSF III 0916286, NSF CCF-1081424,NSF Nets 1016829, NIH R01EB7530 and NSFC 60628202.

Bibliography

[1] R. Schinzinger and P. A. Laura, *Conformal Mapping: Methods and Applications*, Mineola, NY: Dover Publications, 2003.

[2] P. Henrici, *Applied and Computational Complex Analysis, Power Series Integration Conformal Mapping Location of Zero*, vol. 1, Wiley-Interscience, 1988.

[3] M. S. Floater and K. Hormann, *Surface parameterization: a tutorial and survey*, Advances in Multiresolution for Geometric Modelling, pp. 157–186, Springer, 2005.

[4] V. Kraevoy and A. Sheffer, *Cross-parameterization and compatible remeshing of 3D models*, ACM Transactions on Graphics, vol. 23, no. 3, pp. 861–869, 2004.

[5] U. Pinkall and K. Polthier, *Computing discrete minimal surfaces and their conjugates*, Experimental Mathematics, vol. 2, no. 1, pp. 15–36, 1993.

[6] B. Lévy, S. Petitjean, N. Ray, and J. Maillot, *Least squares conformal maps for automatic texture atlas generation*, SIGGRAPH 2002, pp. 362–371, 2002.

[7] M. Desbrun, M. Meyer, and P. Alliez, *Intrinsic parameterizations of surface meshes*, Computer Graphics Forum (Proc. Eurographics 2002), vol. 21, no. 3, pp. 209–218, 2002.

[8] M. S. Floater, *Mean value coordinates*, Computer Aided Geometric Design, vol. 20, no. 1, pp. 19–27, 2003.

[9] C. Gotsman, X. Gu, and A. Sheffer, *Fundamentals of spherical parameterization for 3D meshes*, ACM Transactions on Graphics, vol. 22, no. 3, pp. 358–363, 2003.

[10] X. Gu, Y. Wang, T. F. Chan, P. M. Thompson, and S.-T. Yau, *Genus zero surface conformal mapping and its application to brain surface mapping*, IEEE Trans. Med. Imaging, vol. 23, no. 8, pp. 949–958, 2004.

[11] X. Gu and S.-T. Yau, *Global conformal parameterization*, Symposium on Geometry Processing, pp. 127–137, 2003.

[12] C. Mercat, *Discrete Riemann surfaces and the Ising model*, Communications in Mathematical Physics, vol. 218, no. 1, pp. 177–216, 2004.

[13] A. N. Hirani, *Discrete exterior calculus*. PhD thesis, California Institute of Technology, 2003.

[14] M. Jin, Y.Wang, S.-T. Yau, and X. Gu, *Optimal global conformal surface parameterization*, IEEE Visualization 2004, pp. 267–274, 2004.

[15] S. J. Gortler, C. Gotsman, and D. Thurston, *Discrete one-forms on meshes and applications to 3D mesh parameterization*, Computer Aided Geometric Design, vol. 23, no. 2, pp. 83–112, 2005.

[16] G. Tewari, C. Gotsman, and S. J. Gortler, *Meshing genus-1 point clouds using discrete one-forms*, Comput. Graph., vol. 30, no. 6, pp. 917–926, 2006.

[17] Y. Tong, P. Alliez, D. Cohen-Steiner, and M. Desbrun, *Designing quadrangulations with discrete harmonic forms*, Symposium on Geometry Processing, pp. 201–210, 2006.

[18] A. Bobenko, B. Springborn, and U. Pinkall, *Discrete conformal equivalence and ideal hyperbolic polyhedra*, In preparation.

[19] W. Hong, X. Gu, F. Qiu, M. Jin, and A. E. Kaufman, *Conformal virtual colon flattening*, Symposium on Solid and Physical Modeling, pp. 85–93, 2006.

[20] S. Wang, Y. Wang, M. Jin, X. D. Gu, and D. Samaras, *Conformal geometry and its applications on 3D shape matching, recognition, and stitching*, IEEE Trans. Pattern Anal. Mach. Intell., vol. 29, no. 7, pp. 1209–1220, 2007.

[21] W. Zeng, Y. Zeng, Y. Wang, X. Yin, X. Gu, and D. Samaras, *3D non-rigid surface matching and registration based on holomorphic differentials*, The 10th European Conference on Computer Vision (ECCV) 2008, pp. 1–14, 2008.

[22] X. Gu, Y. He, and H. Qin, *Manifold splines*, Graphical Models, vol. 68, no. 3, pp. 237–254, 2006.

[23] R. S. Hamilton, *Three manifolds with positive Ricci curvature*, Journal of Differential Geometry, vol. 17, pp. 255–306, 1982.

[24] R. S. Hamilton, *The Ricci flow on surfaces*, Mathematics and general relativity (Santa Cruz, CA, 1986), Contemp. Math. Amer. Math. Soc., Providence, RI, vol. 71, 1988.

[25] W. P. Thurston, *Geometry and Topology of Three-Manifolds*, lecture notes at Princeton University, 1980.

[26] P. Koebe, *Kontaktprobleme der Konformen Abbildung*, Ber. Sächs. Akad. Wiss. Leipzig, Math.-Phys. Kl., vol. 88, pp. 141–164, 1936.

[27] W. P. Thurston, *The finite Riemann mapping theorem*, 1985.

[28] B. Rodin and D. Sullivan, *The convergence of circle packings to the Riemann mapping*, Journal of Differential Geometry, vol. 26, no. 2, pp. 349–360, 1987.

[29] Y. Colin de verdiére, *Un principe variationnel pour les empilements de cercles*, Invent. Math., vol. 104, no. 3, pp. 655–669, 1991.

[30] C. Collins and K. Stephenson, *A circle packing algorithm*, Computational Geometry: Theory and Applications, vol. 25, pp. 233–256, 2003.

[31] B. Chow and F. Luo, *Combinatorial Ricci flows on surfaces*, Journal Differential Geometry, vol. 63, no. 1, pp. 97–129, 2003.

[32] M. Jin, J. Kim, F. Luo, and X. Gu, *Discrete surface Ricci flow*, IEEE Transactions on Visualization and Computer Graphics, 2008.

[33] P. L. Bowers and M. K. Hurdal, *Planar conformal mapping of piecewise flat surfaces*, Visualization and Mathematics III (Berlin), pp. 3–34, Springer-Verlag, 2003.

[34] A. I. Bobenko and B. A. Springborn, *Variational principles for circle patterns and Koebe's theorem*, Transactions of the American Mathematical Society, vol. 356, pp. 659–689, 2004.

[35] L. Kharevych, B. Springborn, and P. Schröder, *Discrete conformal mappings via circle patterns*, ACM Trans. Graph., vol. 25, no. 2, pp. 412–438, 2006.

[36] H. Yamabe, *The Yamabe problem*, Osaka Math. J., vol. 12, no. 1, pp. 21–37, 1960.

[37] N. S. Trudinger, *Remarks concerning the conformal deformation of Riemannian structures on compact manifolds*, Ann. Scuola Norm. Sup. Pisa, vol. 22, no. 2, pp. 265-274, 1968.

[38] T. Aubin, *Équations diffréntielles non linéaires et problème de Yamabe concernant la courbure scalaire*, J. Math. Pures Appl., vol. 55, no. 3, pp. 269–296, 1976.

[39] R. Schoen, *Conformal deformation of a Riemannian metric to constant scalar curvature*, J. Differential Geom., vol. 20, no. 2, pp. 479–495, 1984.

[40] J. M. Lee and T. H. Parker, *The Yamabe problem*, Bulletin of the American Mathematical Society, vol. 17, no. 1, pp. 37–91, 1987.

[41] F. Luo, *Combinatorial Yamabe flow on surfaces*, Commun. Contemp. Math., vol. 6, no. 5, pp. 765–780, 2004.

[42] B. Springborn, P. Schröder, and U. Pinkall, *Conformal equivalence of triangle meshes*, ACM Transactions on Graphics, vol. 27, no. 3, pp. 1–11, 2008.

[43] X. Gu and S.-T. Yau, *Computational Conformal Geometry*, Advanced Lectures in Mathematics, vol. 3, Boston: International Press and Higher Education Press, 2007.

[44] W. Zeng, M. Jin, F. Luo, and X. Gu, *Computing canonical homotopy class representative using hyperbolic structure*, IEEE International Conference on Shape Modeling and Applications (SMI09), 2009.

[45] F. Luo, X. Gu, and J. Dai, *Variational Principles for Discrete Surfaces*, Advanced Lectures in Mathematics, Boston: Higher Education Press and International Press, 2007.

[46] W. Zeng, L. M. Lui, X. Gu, and S.-T. Yau, *Shape analysis by conformal modulus*, Methods and Applications of Analysis, 2009.

[47] M. Jin, W. Zeng, D. Ning, and X. Gu, *Computing Fenchel-Nielsen coordinates in teichmuller shape space*, IEEE International Conference on Shape Modeling and Applications (SMI09), 2009.

[48] W. Zeng, X. Yin, M. Zhang, F. Luo, and X. Gu, *Generalized Koebe's method for conformal mapping multiply connected domains*, SIAM/ACM Joint Conference on Geometric and Physical Modeling (SPM), pp. 89–100, 2009.

[49] W. Zeng, L. M. Lui, F. Luo, T. Chang, S.-T. Yau and X. Gu, *Computing Quasiconformal Maps Using an Auxiliary Metric with Discrete Curvature Flow* Numeriche Mathematica, 2011.

[50] R. Guo, *Local Rigidity of Inversive Distance Circle Packing*, Tech. Rep. arXiv.org, Mar 8 2009.

[51] Y.-L. Yang, R. Guo, F. Luo, S.-M. Hu. and X. Gu, *Generalized Discrete Ricci Flow*, Comput. Graph. Forum., vol. 28, no. 7, pp. 2005–2014, 2009.

[52] J. Dai, W. Luo, M. Jin, W. Zeng, Y. He, S.-T. Yau and X. Gu, *Geometric accuracy analysis for discrete surface approximation*, Computer Aided Geometric Design, vol. 24, issue 6, pp. 323–338, 2006.

[53] W. Luo, *Error estimates for discrete harmonic 1-forms over Riemann surfaces*, Comm. Anal. Geom., vol. 14, pp. 1027–1035, 2006.

[54] T. A. Driscoll and L. N. Trefethen, *Schwartz-Christoffel Mapping*, Cambridge, UK: Cambridge Press, 2002.

[55] T. K. DeLillo. *The accuracy of numerical conformal mapping methods: a survey of examples and results*, SIAM J. Numer. Anal., 31(3):788–812, 1994.

[56] V. I. Ivanov and M. K. Trubetskov, *Handbook of conformal mapping with computer-aided visualization*, CRC Press, Boca Raton, FL, 1995.

[57] I. Binder, M. Braverman, and M. Yampolsky, *On the computational complexity of the Riemann mapping*, Ark. Mat., 45(2):221–39, 2007.

[58] L. N. Trefethen, editor, *Numerical conformal mapping*. North-Holland Publishing Co., Amsterdam, 1986. Reprint of J. Comput. Appl. Math. 14 (1986), no. 1–2.

[59] R. Wegmann. *Methods for numerical conformal mapping*. In Handbook of complex analysis: geometric function theory, vol. 2, pp. 351–77, Elsevier, Amsterdam, 2005.

[60] P. Henrici, *Applied and Computational Complex Analysis, Discrete Fourier Analysis, Cauchy Integrals, Construction of Conformal Maps, Univalent Functions*, vol 3., Wiley-Interscience, 1993

[61] D. E. Marshall and S. Rohde, *Convergence of a variant of the zipper algorithm for conformal mapping*, SIAM J. Numer., vol. 45, no. 6, pp. 2577–2609, 2007.

[62] T. A. Driscoll and S. A. Vavasis, *Numerical conformal mapping using cross-ratios and Delaunay triangulation*, SIAM Sci. Comp. 19, pp. 1783–803, 1998.

[63] L. Banjai and L. N. Trefethen, *A Multipole Method for Schwarz-Christoffel Mapping of Polygon with Thousands of Sides*, SIAM J. Comput., vol. 25, no. 3, pp. 1042–1065, 2003.

[64] C.J. Bishop, *Conformal Mapping in Linear Time*, Preprint.

[65] T.K. Delillo, A. R. Elcrat, J.A. Pfaltzgraff *Schwarz-Christoffel mapping of multiply connected domains*, Journal d'Analyse Mathématique, vol. 94, no. 1, pp 17–47, 2004.

[66] D. Crowdy, *The Schwartz-Christoffel mapping to bounded multiply connected polygonal domains*, Proc. R. Soc. A(2005) 461, pp. 2653–2678, 2005.

[67] D. Glickenstein, *Discrete conformal variations and scalar curvature on piecewise flat two and three dimensional manifolds*, preprint at arXiv:0906.1560

[68] Lars V. Ahlfors, *Lectures on Quasiconformal Mappings*, of University Lecture Series, vol. 38, American Mathematical Society, 1966.

[69] C. Kosniowski, *A First Course in Algebraic Topology*, Cambridge, U.K.: Cambridge University Press, 1980.

[70] A. Hatcher, *Algebraic Topology*, Cambridge, U.K.: Cambridge University Press, 2002.

[71] S.-S. Chern, W.-H. Chern, and K.S. Lam, *Lectures on Differential Geometry*, World Scientific Publishing Co. Pte. Ltd., 1999.

[72] O. Forster, *Lectures on Riemann Surfaces*, Graduate texts in mathematics, New York: Springer, vol. 81, 1991.

[73] J. Jost, *Compact Riemann Surfaces: An Introduction to Contemporary Mathematics*, Springer Berlin Heidelberg, 2000.

[74] R. Schoen and S.-T. Yau, *Lectures on Harmonic Maps*, Boston: International Press, 1994.

[75] R. Schoen and S.-T. Yau, *Lectures on Differential Geometry*, Boston: International Press, 1994.

[76] A. Fletcher and V. Markovic, *Quasiconformal Maps and Teichmuller Theory*, Cary, N.C.: Oxford University Press, 2007.

[77] Y. Imayoshi and M. Taniguchi, *An Introduction to Teichmüller Spaces*, Springer-Verlag, Berlin/New York, 1992.

[78] S. Lang, *Differential and Riemannian Manifolds*, Graduate Texts in Mathematics 160, Springer-Verlag New York, 1995

[79] J. M. Lee, *Introduction to Topological Manifolds*, Graduate Texts in Mathematics 202, New York: Springer-Verlag, 2000.

[80] H. M. Farkas and I. Kra, *Riemann Surfaces*, Graduate Texts in Mathematics 71, New York: Springer-Verlag, 1991.

[81] Z.-X. He and O. Schramm, *Fixed Points, Koebe Uniformization and Circle Packings*, Annals of Mathematics, vol. 137, no. 2, pp. 369–406, 1993.

[82] S.-S. Chern, *An elementary proof of the existence of isothermal parameters on a surface*, Proc. Amer. Math. Soc. 6, pp. 771–782, 1955.

[83] M. P. do Carmo, *Differential Geometry of Curves and Surfaces*, Upper Saddle River, N.J., Prentice Hall, 1976.

[84] B. Chow, P. Lu, and L. Ni, *Hamilton's Ricci Flow*, Providence R.I.: American Mathematical Society, 2006.

[85] F. P. Gardiner and N. Lakic, *Quasiconformal Teichmüller Theory*, Mathematical Surveys and monographs, vol. 76, Providence, R.I.: American Mathematical Society 1999.

[86] X. Gu, S. Zhang, P. Huang, L. Zhang, S.-T. Yau, and R. Martin, *Holoimages*, Proc. ACM Solid and Physical Modeling, pp. 129–138, 2006.

[87] C. Costa, *Example of a complete minimal immersion in* $\mathbb{R}^3$ *of genus one and three embedded ends*, Bol. Soc. Bras. Mat. 15, pp. 47-54, 1984.

[88] K. Crane, M. Desbrun, and P. Schröder, *Trivial Connections on Discrete Surfaces*, Comput. Graph. Forum, vol. 29, no. 5, pp. 1525–1533, 2010.

[89] Feng Luo, *A combinatorial curvature flow for compact 3-manifolds with boundary*, Electron. Res. Announc. Amer. Math. Soc., vol. 11, pp. 12–20, 2005.

[90] G. D. Mostow, *Quasi-conformal mappings in n-space and the rigidity of the hyperbolic space forms*, Publ. Math. IHES, vol. 34, pp. 53–104, 1968.

[91] X. Yin, M. Jin, F. Luo, and X. Gu, *Discrete Curvature Flow for Hyperbolic 3-Manifolds with Complete Geodesic Boundaries*, Proc. of the International Symposium on Visual Computing (ISVC2008), December 2008.

[92] Y. Lai, M. Jin, X. Xie, Y. He, J. Palacios, E. Zhang, S. Hu, and X. Gu, *Metric-Driven RoSy Fields Design*, IEEE Transaction on Visualization and Computer Graphics (TVCG), vol. 15, no. 3, pp. 95–108, 2010.

[93] W. Zeng, D. Samaras and X. Gu, *Ricci Flow for 3D Shape Analysis*, IEEE Transaction of Pattern Analysis and Machine Intelligence (PAMI), vol. 32, no. 4, pp. 662–677, 2010.

[94] R. Sarkar, X. Yin, J.Gao, and X. Gu, *Greedy Routing with Guaranteed Delivery Using Ricci Flows*, Proc. of the 8th International Symposium on Information Processing in Sensor Networks (IPSN'09), pp. 121–132, April, 2009.

[95] W. Zeng, R. Sarkar, F. Luo, X. Gu, and J. Gao, *Resilient Routing for Sensor Networks Using Hyperbolic Embedding of Universal Covering Space*, Proc. of the 29th IEEE Conference on Computer Communications (INFOCOM'10), Mar. 15–19, 2010.

[96] R. Sarkar, W. Zeng, J.Gao, and X. Gu, *Covering Space for In-Network Sensor Data Storage*, Proc. of the 9th International Symposium on Information Processing in Sensor Networks (IPSN'10), pp. 232–243, April 2010.

[97] M. Jin, W. Zeng, F. Luo, and X. Gu. *Computing Tëichmuller Shape Space*, IEEE Transactions on Visualization and Computer Graphics 2008, 99(2): 1030–1043.

Chapter 9

2D and 3D Objects Morphing Using Manifold Techniques

Chafik Samir, Pierre-Antoine Absil, and Paul Van Dooren

In this chapter we present a framework for morphing 2D and 3D objects. In particular we focus on the problem of smooth interpolation on a shape manifold. The proposed method takes advantage of two recent works on 2D and 3D shape analysis to compute elastic geodesics between any two arbitrary shapes and interpolations on a Riemannian manifold. Given a finite set of frames of the same 2D or 3D object from a video sequence, or different expressions of a 3D face, our goal in this chapter is to interpolate between the given data in a manner that is smooth. Algorithms, examples, and illustrations demonstrate how this framework may be applied in different applications to fit smooth interpolation.

9.1 Introduction

There has been an increasing interest in recent years in analyzing shapes of 3D objects. Advances in shape estimation algorithms, 3D scanning technology, hardware-accelerated 3D graphics, and related tools are enabling access to high-quality 3D data. As such technologies continue to improve, the need for automated methods for analyzing shapes of 3D objects will also grow. In terms of characterizing 3D objects, for detection, classification, morphing, and recognition, their shape is naturally an important feature. It already plays an important role in medical diagnostics, object designs, database search, and some forms of 3D face animation. Focusing on the last topic, our goal in this chapter is to develop a new method for morphing 2D curves and 3D faces in a manner that is smooth and more "natural," i. e. interpolate the given shapes smoothly, and capture the optimal and elastic non-linear deformations when transforming one face to another.

9.1.1 Fitting Curves on Manifolds

Construction of smooth interpolations in non-linear spaces is an interesting theoretical problem [18] which finds many applications. For example, interpolation in 3-dimensional rotation group $SO(3)$ has immediate applications in robotics for smooth motion of 3D rigid objects in space. For more details we refer the reader to the work of Leite et al. ([5], [8]) and references therein.

The De Casteljau algorithm, which was used to generate polynomial curves in Euclidean spaces, has become popular due to its construction based on a successive linear interpolation. A new version of the De Casteljau algorithm, introduced by Popeil et al. [16] generalizes Bézier curves on a connected Riemannian manifold where line-segments in the classical algorithm are replaced by geodesic segments on the manifold. The proposed algorithm was implemented and tested on a data set in a two-dimensional hyperbolic space. Numerous examples in the literature reveal that the key idea of the extension of the De Casteljau algorithm is the existence of minimizing geodesics between the points to be interpolated [1].

Recently Kume et al. [12] proposed to combine the unrolling and unwarping procedure on a landmark-shape manifold. The technique consists of rolling the manifold on its affine tangent space. The interpolation problem is then solved on the tangent space and rolled back to the manifold to insure the smoothness of the interpolation. However, due to the outlined embedding, the method could not be generalized to a more general shape manifold.

9.1.2 Morphing Techniques

Most techniques for morphing 2D and 3D shapes are based on a sparse set of user selected feature points used to establish the correspondences for interpolation [3]. Although many efficient algorithms exist for 2D metamorphosis, 3D morphs, on the other hand, change the geometry of the object and are then harder to compute and control. A good summary of works on the 3D morphing problem as that of Lazarus et al. [14] note that there are unlimited ways to interpolate between different 3D objects. Most of the proposed methods are only extensions of 2D morphing algorithms ([2],[22]) to 3D. For example, working in the Fourier domain provides a novel way to control the morph by treating frequency bands with different functions of time [7].

Existing methods for 3D morphing can be categorized into two broad classes: volume-based and surface-based approaches. Most applications in graphics use surface-based representations, making surface-based modeling more applicable. However, to the best of our knowledge, none of these methods uses more than two interpolated objects. It mainly aligns both the source and the target object as a first step and then estimates the evolution path between the two models. Thus, the interpolation problem could be solved as an optimization problem where the the smooth interpolation is a solution of an evolution equation ([21], [23]).

9.1.3 Morphing Using Interpolation

Given a finite set of points on a shape manifold M, we want to fit the given data with a smooth and continuous curve. One efficient way to reach this goal is to apply the De Casteljau and the Aitken–Neville algorithms [8] to interpolate between the given data. Introduced few decades ago, these interpolations have been defined and applied to the Euclidean spaces. But recently, new versions of the algorithm served as a tool to generalize them on any Riemannian manifold, if there is a way to compute geodesics on this manifold ([16], [15]).

Based on recent work on 2D and 3D shape analysis, we will first introduce an efficient method to compute geodesics on a shape manifold between any two arbitrary closed curves in $\mathbb{R}^n$. We will then generalize it to surfaces of genus zero. To this end, we will choose a representation for curves and surfaces in order to facilitate their analysis as elements of infinite-dimensional non-linear manifolds. Other methods to compute geodesics on a shape manifold could be applied for the same purpose. But we will show that our choice is based on some advantages of this method: the smoothness of the resulting curve, the non-rigidity of the observed object, and the non-linearity of transformations going from one control point to another.

The rest of this chapter is organized as follows. A detailed specific example of $\mathbb{R}^m$ is given in Section 1.2 and and a generalization on any Riemannian manifold M is given in Section 1.3. An interpolation on a classical Riemannian manifold as $SO(3)$ is detailed in Section 1.4. Furthermore, Section 1.5 gives a nice illustration about the motion of a rigid object in space. A Riemannian analysis of closed curves in $\mathbb{R}^3$ is presented in Section 1.6, with its extension to Riemannian analysis of facial surfaces. The notion of smoothing, or morphing of 2D and 3D objects on a shape manifold is applied to curves and facial surfaces in Section 1.7, and the chapter finishes with a brief conclusion in Section 1.8.

9.2 Interpolation on Euclidean Spaces

Since Lagrange and Bézier interpolations are classical techniques, we will give examples, in order to make the reader familiar with the De Casteljau and the Aitken–Neville algorithms [13]. In this section we will give some examples of interpolations on Euclidean spaces to help in understanding the extension of this simple case to Riemannian manifolds. Consider the problem of fitting a finite set of 2D points; the goal is to interpolate between them using the De Casteljau and the Aitken–Neville algorithms.

The classical definitions of Lagrange and Bézier curves give explicit expressions of polynomials. To implement them numerically, one needs to compute polynomial coefficients. It is clear that the computational cost increases substantially with the number of points used to estimate coefficients. In this section we will consider alternative solutions based on successive linear interpolations: the Aitken–Neville algorithm for constructing Lagrange curves, the classical De Casteljau algorithm to generate Bézier curves, and a revisited De Casteljau algorithm for constructing a C^1 smooth cubic spline [16].

9.2.1 Aitken–Neville Algorithm on $\mathbb{R}^m$

The Aitken–Neville algorithm is a geometric algorithm, and is one of the best known algorithms used to generate Lagrange curves numerically in general Euclidean spaces. Its importance also follows from the simple geometric construction based on successive linear interpolations. The classical Aitken–Neville algorithm is used to construct parameterized Lagrange curves joining points in $\mathbb{R}^m$. A sequence of $(n+1)$ points P_i, $i = 0..n$ is used to implement the algorithm and for that reason they are called control points.

Consider the following polynomial function:

$$L_{i,j}(t) = \frac{(t-t_i)L_{i+1,j}(t) - (t-t_j)L_{i,j-1}(t)}{t_j - t_i}, \; 0 \leq i < j \leq n, \; t \in [t_i, t_j]$$

Where $L_{i+1,j}(t)$ and $L_{i,j-1}(t)$ are polynomial functions of degree $j-i-1$ and passing through $P_{i+1}, \ldots, P_j$ at $t_{i+1}, \ldots, t_j$ and through $P_i, \ldots, P_{j-1}$ at $t_i, \ldots, t_{j-1}$, respectively, then the polynomial function $L_{i,j}$ is of degree $j-i$ passing through $P_i, \ldots, P_j$ at $t_i, \ldots, t_j$. We write:

$$\begin{aligned}
L_{i,j}(t_i) &= L_{i,j-1}(t_i) = P_i \\
L_{i,j}(t_j) &= L_{i+1,j}(t_j) = P_j \\
L_{i,j}(t_k) &= \frac{(t_k-t_i)L_{i+1,j}(t_k) - (t_k-t_j)L_{i,j-1}(t_k)}{t_j-t_i} = \frac{(t_k-t_i)P_k - (t_k-t_j)P_k}{t_j-t_i} = P_k, \; i < k < j
\end{aligned}$$

which could be summarized in the following recursive scheme:

$$\begin{array}{llllll}
P_0 & & & & & \\
 & \searrow & & & & \\
P_1 \rightarrow & L_{0,1} & & & & \\
 & \searrow & \searrow & & & \\
P_2 \rightarrow & L_{1,2} \rightarrow & L_{0,2} & & & \\
 & \searrow & \searrow & \searrow & & \\
P_3 \rightarrow & L_{2,3} \rightarrow & L_{1,3} \rightarrow & L_{0,3} & & \\
\vdots & \vdots & \vdots & \vdots & \ddots & \\
 & \searrow & \searrow & \searrow & \cdots \searrow & \\
P_n \rightarrow & L_{n-1,n} \rightarrow & L_{n-2,n} \rightarrow & L_{n-3,n} & \cdots \rightarrow & L_{0,n}
\end{array}$$

The complete algorithm is given in Algorithm 2:

Algorithm 2 Aitken–Neville algorithm on $\mathbb{R}^2$

Require: $P_0, \ldots, P_n \in \mathbb{R}^2$, $t_0 < \ldots < t_n \in \mathbb{R}$, $\mathcal{D}$ a discretization of $[t_0, t_n]$

  for i=0 to $n-1$ **do**

    for $u \in \mathcal{D}$ **do**

      $L_{i,i+1}(u) = \frac{(u-t_i)P_{i+1}-(u-t_{i+1})P_i}{t_{i+1}-t_i}$

    end for

  end for

  for r=1 to n **do**

    for i=0 to $n-r$ **do**

      for $u_1 \in \mathcal{D}$ **do**

        for $u_2 \in \mathcal{D}$ **do**

          $\mathbf{L}^{\mathbf{geo}}_{\mathbf{i,i+r}}(u_1, u_2) = \frac{(u_2-t_i)L_{i+1,i+r}(u_1)-(u_2-t_{i+r})L_{i,i+r-1}(u_1)}{t_{i+r}-t_i}$

        end for

        $L_{i,i+r}(u_1) = \mathbf{L}^{\mathbf{geo}}_{\mathbf{i,i+r}}(u_1, u_1)$

      end for

    end for

  end for

  return $L(u) = L_{0,n}(u)$

9.2.2 De Casteljau Algorithm on $\mathbb{R}^m$

The De Casteljau algorithm is also a geometric algorithm based on recursive linear interpolations. Similar to the Aitken–Neville algorithm, it is given by the following scheme:

$$B_{i,j}(t) = t\,B_{i+1,j}(t) + (1-t)B_{i,j-1}(t), \ \forall 0 \leq i < j \leq n, \ t \in [0,1] \tag{9.1}$$

where $B_{i+1,j}(t)$ and $B_{i,j-1}(t)$ are Bézier curves of degree $j-i-1$ corresponding to control points $P_{i+1}, \ldots, P_j$ and $P_i, \ldots, P_{j-1}$, respectively.

The De Casteljau algorithm is given by the algorithm 3:

Algorithm 3 De Casteljau algorithm on $\mathbb{R}^2$

Require: $P_0, \ldots, P_n \in \mathbb{R}^2$, $\mathcal{D}$ a discretization of $[0,1]$

```
for i=0 to n do
  for t ∈ D do
    B_{i,i}(u) = P_i
  end for
end for
for r = 1 to n do
  for i = 0 to n − r do
    for u1 ∈ D do
      for u2 ∈ D do
        B^geo_{i,i+r}(u1, u2) = u2 B_{i+1,i+r}(u1) + (1 − u2) B_{i,i+r−1}(u1)
      end for
      B_{i,i+r}(u) = B^geo_{i,i+r}(u1, u1)
    end for
  end for
end for
return B(u) = B_{0,n}(u)
```

9.2.3 Example of Interpolations on $\mathbb{R}^2$

A summary of resulting interpolations, using Bézier and Lagrange curves, is given in Figure 9.1(a). Just as a reminder, the Lagrange interpolation passes through the control points, while the Bézier curve starts at the first control point and ends at the last one without passing through the intermediate control points. The velocity and acceleration at the first and last control points are readily related to the position of the control points, which makes it possible to achieve C^2 interpolation by piecing together Bézier curves obtained from adequately-chosen control points.

9.3 Generalization of Interpolation Algorithms on a Manifold M

As shown in the previous section, the construction of Lagrange and Bézier curves in $\mathbb{R}^2$ and generally in $\mathbb{R}^m$ is based on recursive affine combinations. Intermediate points during an iteration are selected on the segment connecting two constructed points obtained in a previous iteration. Moreover, in $\mathbb{R}^m$ the segment connecting two points is the geodesic between these two points. It is then possible to generalize interpolation on $\mathbb{R}^m$ to more general Riemannian manifold M by replacing the straight lines by geodesics in algorithms 2 and 3.

Consider a set of points $P_i \in M$, $i = 0, \ldots, n$; we can apply the recursive affine combinations:

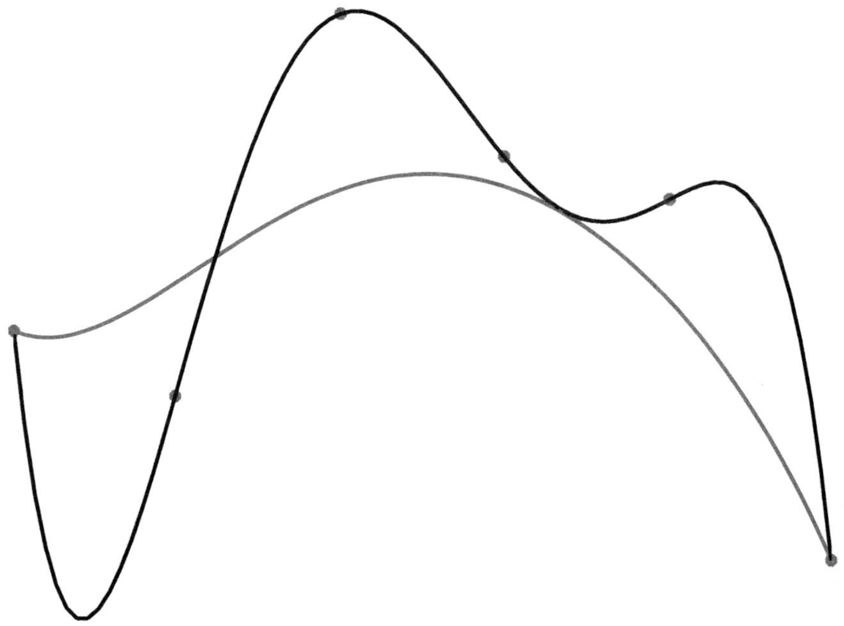

Figure 9.1: Example of Bézier and Lagrange curves on Euclidean plane.

$$
\begin{array}{llllll}
P_0 & & & & & \\
\searrow & & & & & \\
P_1 \to & \alpha_{0,1} & & & & \\
\searrow & \searrow & & & & \\
P_2 \to & \alpha_{1,2} \to & \alpha_{0,2} & & & \\
\searrow & \searrow & \searrow & & & \\
P_3 \to & \alpha_{2,3} \to & \alpha_{1,3} \to & \alpha_{0,3} & & \\
\vdots & \vdots & \vdots & \vdots & \ddots & \\
\searrow & \searrow & \searrow & \cdots \searrow & & \\
P_n \to & \alpha_{n-1,n} \to & \alpha_{n-2,n} \to & \alpha_{n-3,n} & \cdots \to & \alpha_{0,n}
\end{array}
$$

where each element $\alpha_{i,i+r}$ is a curve on M constructed from geodesics between $\alpha_{i,i+r-1}$ and $\alpha_{i+1,i+r}$, $1 < r \leq n$, $0 \leq i \leq n-r$. This recursive scheme could be used to generalize Aitken–Neville and De Casteljau algorithms. The difference comes from the way we construct $\alpha_{i,i+r}$ from $\alpha_{i,i+r-1}$ and $\alpha_{i+1,i+r}$, and the intervals on which they are defined.

Recall that in the algorithms 2 and 3, we defined the maps $L^{geo}_{i,j}$ and $B^{geo}_{i,i+r}$ to show that each point lies on the segment connecting the two points obtained from the previous iteration. In what follows, we will redefine these maps on M by replacing line segments by geodesics in order to obtain a generalization of Aitken–Neville and Casteljau algorithms on M.

9.3.1 Aitken–Neville on M

Given a set of points $P_0, \ldots, P_n$ on a manifold M at times $t_0 < \ldots < t_n$, for $0 \leq i \leq n-r$, $1 \leq r \leq n$, we define the maps $\mathbf{L}^{\mathbf{geo}}_{\mathbf{i,i+r}} : [t_0, t_n] \times [t_0, t_n] \to M$ as follows: for u_1 in $[t_0, t_n]$,

$\mathbf{L}^{\mathbf{geo}}_{\mathbf{i,i+r}}(u_1, u_2)$ is a geodesic on $[t_0, t_n]$ such that:

$$\begin{aligned} \mathbf{L}^{\mathbf{geo}}_{\mathbf{i,i+r}}(u_1, t_i) &= L_{i,i+(r-1)}(u_1) \\ \mathbf{L}^{\mathbf{geo}}_{\mathbf{i,i+r}}(u_1, t_{i+r}) &= L_{i+1,i+r}(u_1) \end{aligned}$$

where $L_{i,i+r}(u) = \mathbf{L}^{\mathbf{geo}}_{\mathbf{i,i+r}}(u, u)$ and $L_{i,i}(u) = P_i, u \in [t_0, t_n]$.

The generalized version of the Aitken–Neville algorithm is given in Algorithm 4:

Algorithm 4 Aitken–Neville algorithm on M

Require: $P_0, \ldots, P_n \in M,\ t_0 < \ldots < t_n \in \mathbb{R},\ \mathcal{D}$ a discretization of $[t_0, t_n]$

```
  for i=0 to n − 1 do
    L_{i,i+1}(u) ≡ geodesic between P_i at u = t_i and P_{i+1} at u = t_{i+1}
  end for
  for r=1 to n do
    for i=0 to n − r do
      for u_1 ∈ D do
1.        L^geo_{i,i+r}(u_1, u_2) ≡ geodesic between L_{i,i+(r−1)}(u_1) at u_2 = t_i and L_{i+1,i+r}(u_1)
          at u_2 = t_{i+r}.
2.        L_{i,i+r}(u_1) = L^geo_{i,i+r}(u_1, u_1)
      end for
    end for
  end for
  return L(u) = L_{0,n}(u)
```

The curve $([0, 1], L)$ obtained using Algorithm 4 is called the **Lagrange geodesic curve**.

9.3.2 De Casteljau Algorithm on M

Given a set of points $P_0, \ldots, P_n$ on a manifold M, for $0 \leq i \leq n - r$, $1 \leq r \leq n$, we define the maps $\mathbf{B}^{\mathbf{geo}}_{\mathbf{i,i+r}} : [0, 1] \times [0, 1] \to M$ as follows: for u_1 in $[0, 1]$, $\mathbf{B}^{\mathbf{geo}}_{\mathbf{i,i+r}}(u_1, u_2)$ is a geodesic on $[0, 1]$ such that:

$$\begin{aligned} \mathbf{B}^{\mathbf{geo}}_{\mathbf{i,i+r}}(u_1, 0) &= B_{i,i+(r-1)}(u_1) \\ \mathbf{B}^{\mathbf{geo}}_{\mathbf{i,i+r}}(u_1, 1) &= B_{i+1,i+r}(u_1) \end{aligned}$$

where $B_{i,i+r}(u) = \mathbf{B}^{\mathbf{geo}}_{\mathbf{i,i+r}}(u, u)$ and $B_{i,i}(u) = P_i, u \in [0, 1]$.

For $0 \leq i \leq n - r$, $1 \leq r \leq n$, each curve $B_{i,i+r}(u)$, passes through P_i at $u = 0$ and P_{i+r} at $u = 1$. Indeed, $B_{i,i+r}(u) = \mathbf{B}^{\mathbf{geo}}_{\mathbf{i,i+r}}(u, u)$ and $\mathbf{B}^{\mathbf{geo}}_{\mathbf{i,i+r}}(u_1, u_2)$ is a geodesic between $B_{i,i+(r-1)}(u_1)$ at $u_2 = 0$ and $B_{i+1,i+r}(u_1)$ at $u_2 = 1$. Thus:

$$\begin{aligned} B_{i,i+r}(0) &= B_{i,i+(r-1)}(0),\ 0 \leq i \leq n-r,\ 1 \leq r \leq n \\ B_{i,i+r}(1) &= B_{i+1,i+r}(1),\ 0 \leq i \leq n-r,\ 1 \leq r \leq n \end{aligned}$$

which gives us the following recursive and explicit expressions:

$$\begin{aligned} B_{i,i+r}(0) &= B_{i,i}(0) = P_i,\ 0 \leq i \leq n-r,\ 1 \leq r \leq n \\ B_{i,i+r}(1) &= B_{i+r,i+r}(1) = P_{i+r},\ 0 \leq i \leq n-r,\ 1 \leq r \leq n \end{aligned}$$

Algorithm 5 De Casteljau algorithm on M

Require: $P_0, \ldots, P_n \in M$, $\mathcal{D}$ a discretization of $[0,1]$

for i=0 to n **do**
  $B_{i,i}(u) = P_i$
end for

for r=1 to n **do**
  for i=0 to $n-r$ **do**
    for $u_1 \in \mathcal{D}$ **do**
1. $\mathbf{B}^{\mathbf{geo}}_{\mathbf{i,i+r}}(u_1, u_2) \equiv$ geodesic between $B_{i,i+(r-1)}(u_1)$ at $u_2 = 0$ and $B_{i+1,i+r}(u_1)$ at $u_2 = 1$.
2. $B_{i,i+r}(u_1) = \mathbf{B}^{\mathbf{geo}}_{\mathbf{i,i+r}}(u_1, u_1)$
    end for
  end for
end for

return $B(u) = B_{0,n}(u)$

The generalized version of De Casteljau algorithm is given in Algorithm 5:

The resulting curve $([0,1], B)$ using Algorithm 5 is called **Bézier geodesic curve** which passes through P_0 at $u = 0$ and through P_n at $u = 1$.

9.4 Interpolation on $SO(m)$

Our problem formulation is well defined for any smooth Riemannian manifold. In practice, they can be applied to such manifolds if there is a way to compute the geodesic distance. The orthogonal group $SO(m)$ is such an m-dimensional manifold of great practical interest. Thus, to apply the generalized Algorithms 4 and 5 introduced in the previous section we need the geodesic distance [1]. In particular, $SO(m)$ is a naturally occurring example of a manifold that admits analytical expressions for computing geodesics between any two arbitrary points.

Recall that for A, $B \in SO(m)$, the geodesic expression $([0,1], \alpha)$ on $SO(m)$ joining A at time $t = 0$ and B at $t = 1$ is:

$$\alpha(t) = A \exp\left(t \log\left[A^t B\right]\right) \tag{9.2}$$

Then, the expression of the geodesic $([t_0, t_n], \alpha)$ on $SO(m)$ joining A at $t = t_i$ and B at $t = t_{i+r}$ is:

$$\alpha(t) = A \exp\left(\frac{t - t_i}{t_{i+r} - t_i} \log\left[A^t B\right]\right) \tag{9.3}$$

We will use both formulas (9.2) and (9.3) in algorithms 4 and 5 to construct Lagrange and Bézier curves passing through or as close as possible to the control points $P_0, \ldots, P_n \in SO(m)$.

9.4.1 Aitken–Neville Algorithm on $SO(m)$

Given a set of $n+1$ matrices $P_0, \ldots, P_n$ on $SO(m)$ at different instants of time $t_0 < \ldots < t_n$, using recursive algorithms described in the previous section, we can write: for

$0 \leq i \leq n-r, 1 \leq r \leq n$ and u_1 in $[t_0, t_n]$:

$$\begin{aligned}
\mathbf{L}^{\mathbf{geo}}_{\mathbf{i,i+r}}(u_1, u_2) &= L_{i,i+(r-1)}(u_1) \exp\left(\frac{u_2 - t_i}{t_{i+r} - t_i} \log\left[L^t_{i,i+(r-1)}(u_1) L_{i+1,i+r}(u_1)\right]\right), \ u_2 \in [t_0, t_n] \\
L_{i,i+r}(u) &= \mathbf{L}^{\mathbf{geo}}_{\mathbf{i,i+r}}(u, u)
\end{aligned}$$

where $L_{i,i}(u) = P_i$ in $[t_0, t_n]$, $0 \leq i \leq n$

Note that the expression of $L_{i,i+r}$ can be computed directly without determining the geodesic $\mathbf{L}^{\mathbf{geo}}_{\mathbf{i,i+r}}(u_1, .)$. Thus, we have

$$L_{i,i+r}(u) = L_{i,i+(r-1)}(u) \exp\left(\frac{u - t_i}{t_{i+r} - t_i} \log\left[L^t_{i,i+(r-1)}(u) L_{i+1,i+r}(u)\right]\right), \ u \in [t_0, t_n]$$

due to the fact that we have an explicit expression of the geodesic on $SO(m)$. Algorithm 6 gives the Aitken–Neville algorithm to construct Lagrange curves on $SO(m)$:

Algorithm 6 Aitken–Neville algorithm on $SO(m)$

Require: $P_0, \ldots, P_n \in SO(m)$, $t_0 < \ldots < t_n \in \mathbb{R}$, $\mathcal{D}$ a discretization of $[t_0, t_n]$
  for i=0 to $n-1$ **do**
    for $t \in \mathcal{D}$ **do**
      $L_{i,i+1}(t) = P_i \exp\left(\frac{t-t_i}{t_{i+1}-t_i} \log\left[P_i^t P_{i+1}\right]\right)$
    end for
  end for
  for r=1 to n **do**
    for i=0 to $n-r$ **do**
      for $t \in \mathcal{D}$ **do**
        $L_{i,i+r}(t) = L_{i,i+r-1}(t) \exp\left(\frac{t-t_i}{t_{i+r}-t_i} \log\left[L^t_{i,i+r-1}(t) L_{i+1,i+r}(t)\right]\right)$
      end for
    end for
  end for
  return $L(t) = L_{0,n}(t)$

9.4.2 De Casteljau Algorithm on $SO(m)$

Given a set of $n+1$ matrices $P_0, \ldots, P_n$ on $SO(m)$ at different instants of time $t_0 < \ldots < t_n$, using recursive algorithms described in the previous section 9.3.2, we have for $0 \leq i \leq n-r, 1 \leq r \leq n$ and u_1 in $[t_0, t_n]$:

$$\begin{aligned}
\mathbf{B}^{\mathbf{geo}}_{\mathbf{i,i+r}}(u_1, u_2) &= B_{i,i+(r-1)}(u_1) \exp\left(u_2 \log\left[B^t_{i,i+(r-1)}(u_1) B_{i+1,i+r}(u_1)\right]\right), \ u_2 \in [0, 1] \\
B_{i,i+r}(u) &= \mathbf{B}^{\mathbf{geo}}_{\mathbf{i,i+r}}(u, u)
\end{aligned}$$

Where $B_{i,i}(u) = P_i$ on $[0, 1]$, $0 \leq i \leq n$. Algorithm 7 gives the De Casteljau algorithm to construct Bézier curves on $SO(m)$:

9.4.3 Example of Fitting Curves on $SO(3)$

In this section we show results using Algorithms 6 and 7 on $SO(3)$. In order to have a visual example of rotations in space we will represent a rotation matrix by the trihedron

Algorithm 7 De Casteljau algorithm on $SO(m)$

Require: $P_0, \ldots, P_n \in SO(m)$, $\mathcal{D}$ a discretization of $[0,1]$

```
for i=0 to n do
  for t ∈ D do
    B_{i,i}(t) = P_i
  end for
end for
for r=1 to n do
  for i=0 to n − r do
    for t ∈ D do
      B_{i,i+r}(t) = B_{i,i+r−1}(t) exp (t log (B^t_{i,i+r−1}(t) B_{i+1,i+r}(t)))
    end for
  end for
end for
return B_{0,n}(t)
```

composed of its three column vectors. For example, the identity matrix will be represented by a trihedron of the unit vectors $e_1 = (1, 0, 0)$, $e_2 = (0, 1, 0)$, and $e_3 = (0, 0, 1)$ as shown in Figure 9.3.

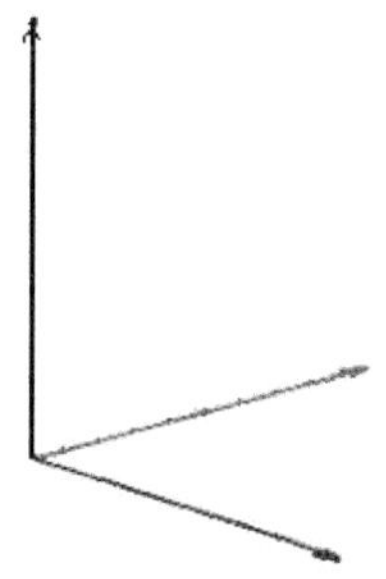

Figure 9.2: The three column vectors of a 3×3 rotation matrix represented as a trihedron.

For the following examples, we generated 3 orthogonal matrices as given data as shown in Figure 9.3 at time instants $t_i = i$ for $i = 0, 1, 2$. The first control point is the identity matrix, the second and the third are obtained by rotating the first one. A discrete version of the resulting Lagrange curve is shown in Figure 9.4 and Bézier curve is shown in Figure 9.5. To have a comparison with different interpolations we also show an interpolation using piecewise-geodesic in Figure 9.6.

9.5 Application: The Motion of a Rigid Object in Space

In this section we will consider the problem of fitting curves to a finite set of control points derived from a continuous observation of a rigid transformation of a 3D object in space. A similar idea was applied in [8] to build a trajectory of a satellite in space, where only rotations and translations were considered. The 3D object will be represented by its center of mass and a local trihedron as shwon in Figure 9.7.

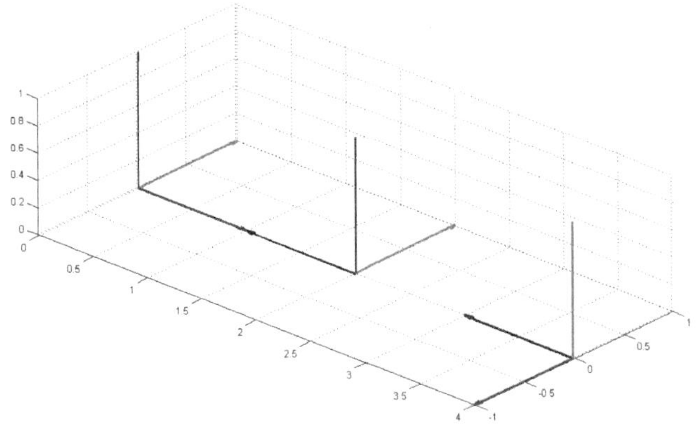

Figure 9.3: 3×3 rotation matrices used as given data on $SO(3)$.

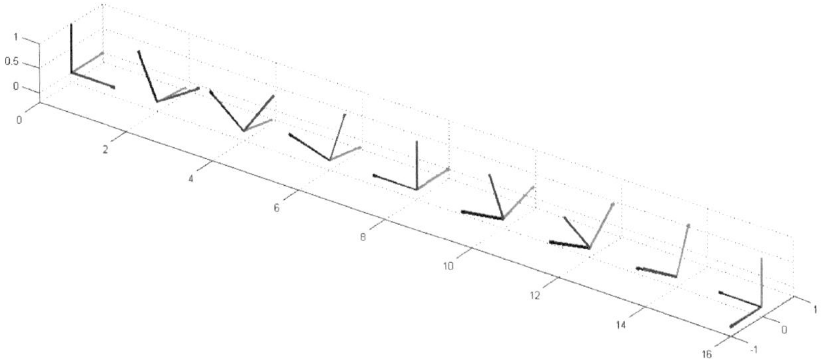

Figure 9.4: Lagrange interpolation on $SO(3)$ using matrices shown in Figure 9.3.

We are given a finite set of positions as 3D coordinates in $\mathbb{R}^3$ and a finite set of rotations at different instants of time. The goal is to interpolate between the given set of points in such a way that the object will pass through or as close as possible to the given positions, and will rotate by the given rotations, at the given instants. A key idea is to interpolate data in $SO(3) \times \mathbb{R}^3$ using Algorithms 4 and 5 in both $SO(m)$ and $\mathbb{R}^m$ with $m = 3$.

In order to visualize the end effect, Figure 9.8 and Figure 9.9 show the motion of the rigid body where position is given by the curve in $\mathbb{R}^3$ and rotation is displayed by rotating axes. The same idea is applied in Figure 9.10 where the interpolation is obtained by a piecewise geodesic. From resulting curves we observe that Lagrange construction yields a significant smoother interpolation.

Another example is shown in Figure 9.11 where we obtain different interpolating curves using Bézier in Figure 9.11(a) and Lagrange in Figure 9.11(b).

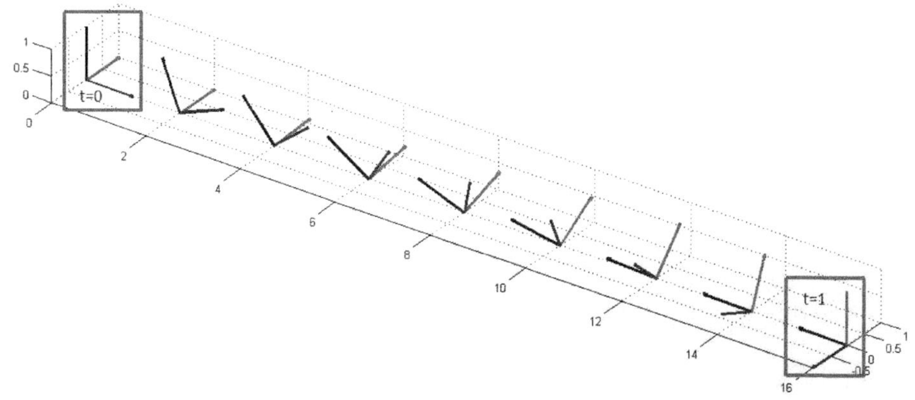

Figure 9.5: Bézier curve on $SO(3)$ using matrices shown in Figure 9.3.

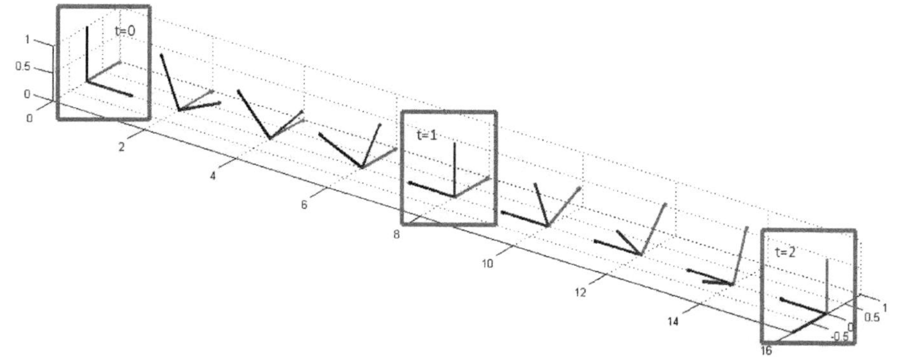

Figure 9.6: Piecewise geodesic on $SO(3)$ using matrices shown in Figure 9.3.

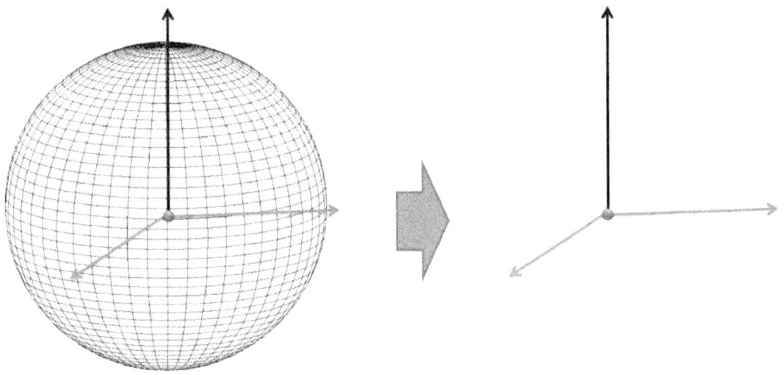

Figure 9.7: An example of a 3D object represented by its center of mass and a local trihedron.

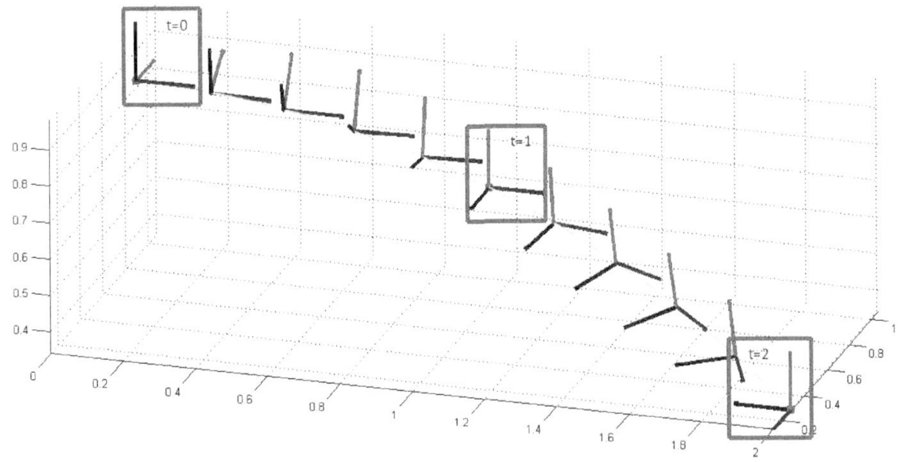

Figure 9.8: Interpolation on $SO(3) \times \mathbb{R}^3$ as a Lagrange curve.

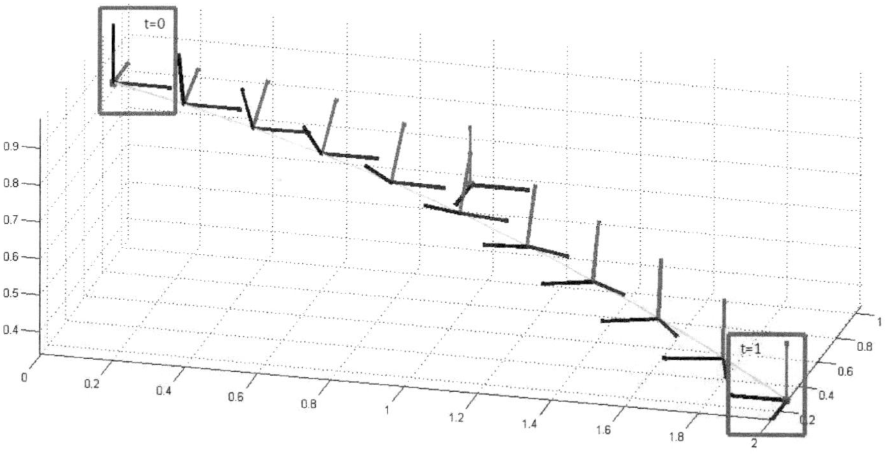

Figure 9.9: Interpolation on $SO(3) \times \mathbb{R}^3$ as a Bézier curve.

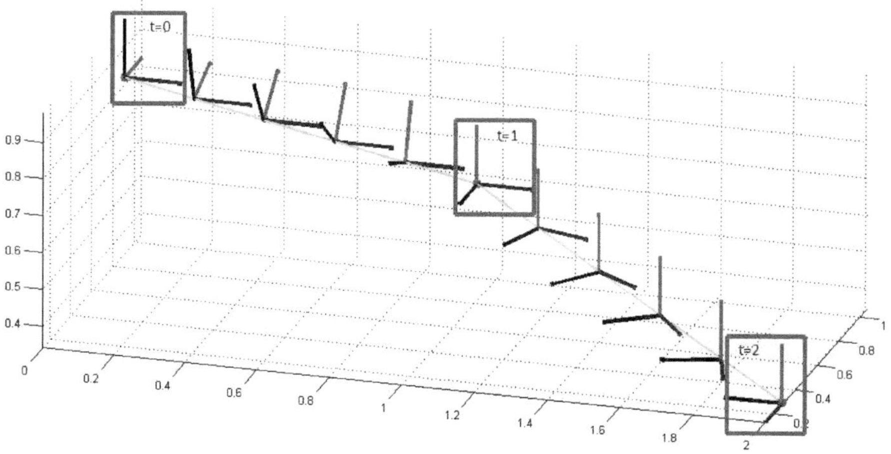

Figure 9.10: Interpolation on $SO(3) \times \mathbb{R}^3$ as a piecewise geodesic curve.

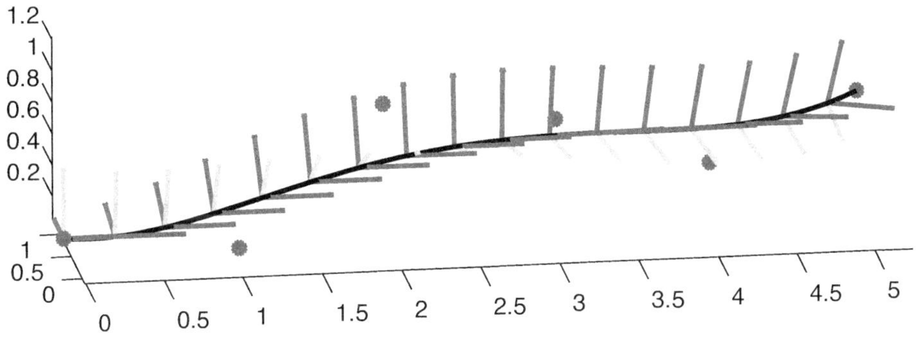

(a)

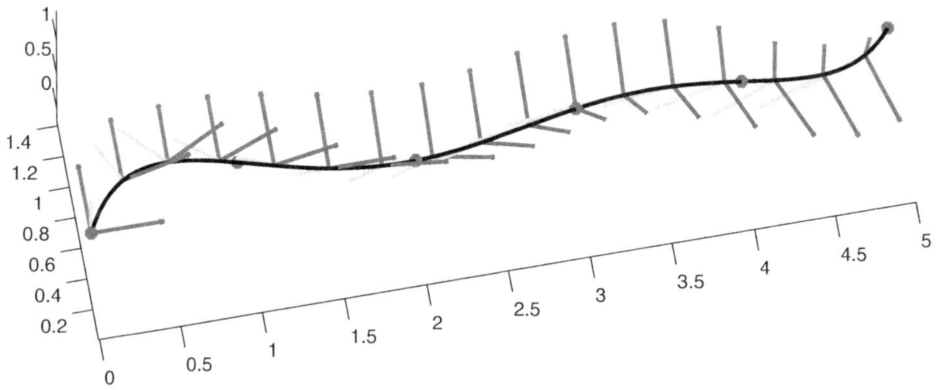

((b)

Figure 9.11: (a): Bézier curve using the De Casteljau algorithm and (b): Lagrange curve on using the Aitken–Neville algorithm.

9.6 Interpolation on Shape Manifold

9.6.1 Geodesic between 2D Shapes

Here we adopt the approach presented in Joshi et al. [9] because it greatly simplifies the elastic shape analysis. The main steps are: (i) defining a space of closed curves of interest, (ii) imposing a Riemannian structure on it using the elastic metric, and (iii) computing geodesic paths under this metric. These geodesic paths can then be interpreted as optimal elastic deformations of curves.

For the interval $I \equiv [0, 2\pi]$, let $\beta : I \to \mathbb{R}^3$ be a parameterized curve with a non-vanishing derivative everywhere. We represent its shape by the function:

$$q : I \to \mathbb{R}^3,\ q(s) = \frac{\dot{\beta}(s)}{\sqrt{||\dot{\beta}(s)||}} \in \mathbb{R}^3.$$

Where $||\cdot|| \equiv \sqrt{(\cdot,\cdot)_{\mathbb{R}^3}}$, and $(\cdot,\cdot)_{\mathbb{R}^3}$ is taken to be the standard Euclidean inner product in $\mathbb{R}^3$. The quantity $||q(s)||$ is the square-root of the instantaneous speed of the curve β, whereas the ratio $\frac{q(s)}{||q(s)||}$ is the direction for each $s \in [0, 2\pi)$ along the curve. Let $\mathcal{Q}$ be the space of all square integrable functions in $\mathbb{R}^3$,

$$\mathcal{Q} \equiv \{q = (q_1, q_2, q_3) | q(s) : I \to \mathbb{R}^3, q(s) \neq 0, \forall s\}.$$

The closure condition for a curve β requires that $\int_I \dot{\beta}(s)ds = 0$, which translates to $\int_I ||q(s)||q(s)\ ds = 0$. The space of all closed curves of unit length, and this representation is invariant under translation and scaling. $\mathcal{C}$ is endowed with a Riemannian structure using the metric: for any two tangent vectors $v_1, v_2 \in T_q(\mathcal{C})$,

$$\langle v_1 \rangle\, v_2 = \int_I (v_1(s), v_2(s))_{\mathbb{R}^3} ds\ . \tag{9.4}$$

Next, we want a tool to compute geodesic paths between arbitrary elements of $\mathcal{C}$. There have been two prominent numerical approaches for computing geodesic paths on nonlinear manifolds. One approach uses the shooting method [11] where, given a pair of shapes, one finds a tangent direction at the first shape such that its image under the exponential map reaches as close to the second shape as possible. We will use another, more stable approach that uses path-straightening flows to find a geodesic between two shapes. In this approach, the given pair of shapes is connected by an initial arbitrary path that is iteratively "straightened" so as to minimize its length. The path-straightening method, proposed by Klassen et al [10], overcomes some of the practical limitations in the shooting method. Other authors, including Schmidt et al. [19] and Glaunes et al [6], have also presented other variational techniques for finding optimal matches. Given two curves, represented by q_0 and q_1, our goal is to find a geodesic between them in $\mathcal{C}$. Let $\alpha : [0, 1] \to \mathcal{C}$ be any path connecting q_0, q_1 in $\mathcal{C}$, i.e. $\alpha(0) = q_0$ and $\alpha(1) = q_1$. Then, the critical points of the energy

$$E[\alpha] = \frac{1}{2} \int_0^1 \langle \dot{\alpha}(t), \dot{\alpha}(t) \rangle\ dt\ , \tag{9.5}$$

with the inner product defined in Eqn. 9.4, are geodesics in $\mathcal{C}$ (this result is true on a general manifold [20]). As described by Klassen et al. [10] (for general shape manifolds), one can use a gradient approach to find a critical point of E and reach a geodesic. The distance between the two curves q_0 and q_0 is given by the length of the geodesic α:

$$d_c(q_1, q_2) = \int_0^1 (\langle \alpha'(t) \rangle\, \alpha'(t))^{1/2} dt\ .$$

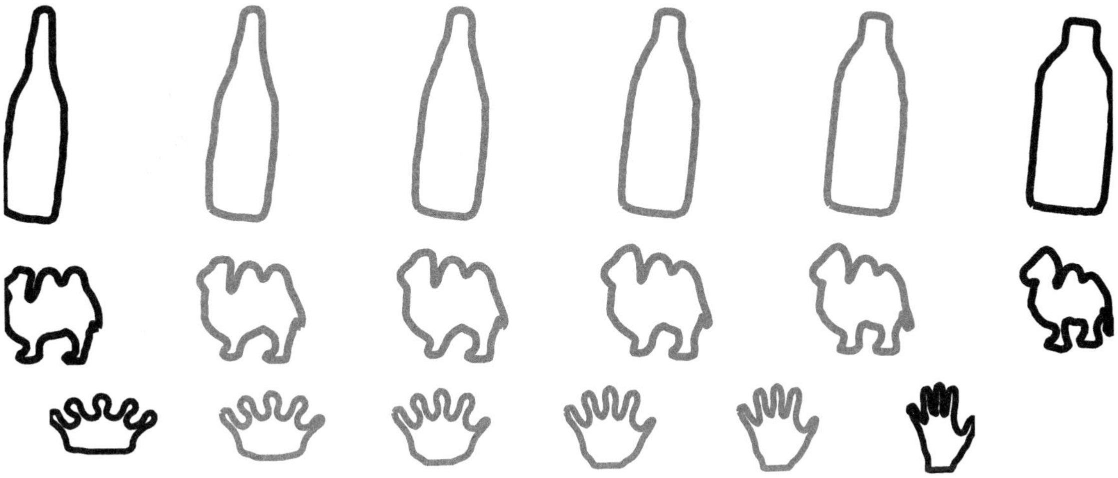

Figure 9.12: Elastic deformation as a geodesic between 2D shapes from shape database.

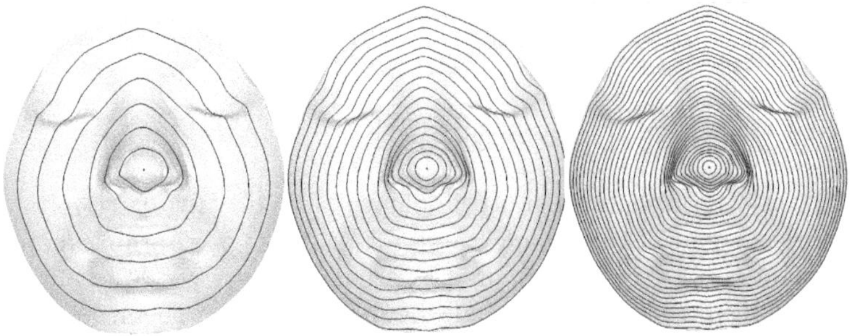

Figure 9.13: Representation of facial surfaces as indexed collection of closed curves in $\mathbb{R}^3$.

We call this the **elastic** distance in deforming the curve represented by q_0 to the curve represented by q_1.

We will illustrate these ideas using some examples. Firstly, we present some examples of elastic matching between planar shapes in Figure 9.12. Nonlinearity of matching between points across the two shapes emphasizes the elastic nature of this matching. One can also view these paths as optimal elastic deformations of one curve to another.

9.6.2 Geodesic between 3D Shapes

To analyze the morphing of a surface is much more complicated due to the corresponding difficulty in analyzing shapes of surfaces. The space of parameterizations of a surface is much larger than that of a curve, and this hinders an analysis of deformation in a way that is invariant to parametrization. One solution is to restrict to a family of parameterizations and perform shape analysis over that space. Although this cannot be done for all surfaces, it is natural for certain surfaces such as the facial surfaces as described next.

Using the approach of Samir et al. [17], we can represent a facial surface S as an indexed collection of facial curves, as shown in Figure 9.13. Each facial curve, denoted by c_λ, is obtained as a level set of the distance function from the tip of the nose; it is a

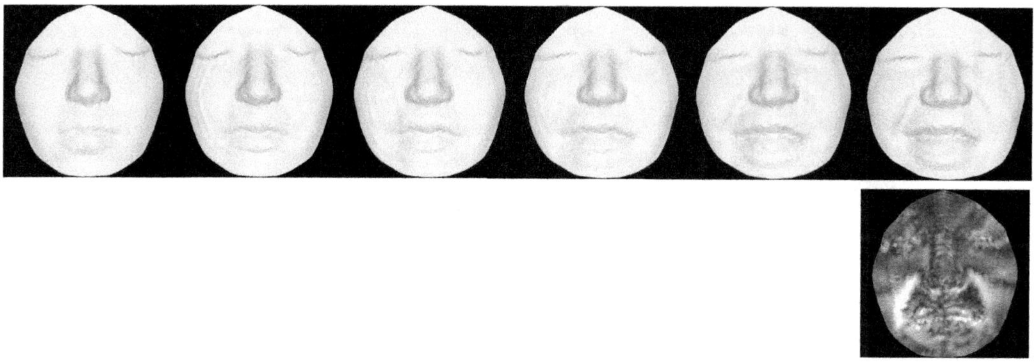

Figure 9.14: Geodesic path between the starting and the ending 3D faces in the first row, and the corresponding magnitude of deformation in the second row.

closed curve in $\mathbb{R}^3$. As earlier, let d_c denote the geodesic distance between closed curves in $\mathbb{R}^3$, when computed on the shape space $\mathcal{S} = \mathcal{C}/(SO(3) \times \Gamma)$, where $\mathcal{C}$ is the same as defined in the previous section except this time it is for curves in $\mathbb{R}^3$, and Γ is the set of all parameterizations. A surface S is represented as a collection $\cup_\lambda c_\lambda$ with $\lambda \in [0, 1]$ and the elastic distance between any two facial surfaces is given by: $d_s(S_1, S_2) = \sum_\lambda d_c(\lambda)$, where

$$d_c(\lambda) = \inf_{O \in SO(3), \gamma \in \Gamma} d_c(q_\lambda^1, \sqrt{\dot{\gamma}} O q_\lambda^2(\gamma)) \ . \tag{9.6}$$

Here q_λ^1 and q_λ^2 are q representations of the curves c_λ^1 and c_λ^2, respectively. According to this equation, for each pair of curves in S_1 and S_2, c_λ^1 and c_λ^2, we obtain an optimal rotation and re-parametrization of the second curve. To put together geodesic paths between full facial surfaces, we need a single rotational alignment between them, not individually for each curve as we have now. Thus we compute an average rotation:

$$\hat{O} = \text{average}\{O_\lambda\} \ ,$$

using a standard approach, and apply $\hat{O}$ to S_2 to align it with S_1. This global rotation, along with optimal re-parameterizations for each λ, provides an optimal alignment between individual facial curves and results in the shortest geodesic paths between them. Combining these geodesic paths, for all λs, one obtains geodesic paths between the original facial surfaces as shown in Figure 9.14.

9.7 Examples of Fitting Curves on Shape Manifolds

In this section, we present some examples and discuss the effectiveness of our method. At each step of the interpolation, optimal deformations between two shapes are computed using geodesics on a shape manifold, as segments were used in the classical De Casteljau algorithm on the Euclidean space. Note that all shapes are extracted from real data and are generated fully automatically without any user interaction.

9.7.1 2D Curves Morphing

In the first example (see Figure 9.15), curves are derived from a public shape database. In the second example (see Figure 9.16), curves are extracted from a video sequence of a human walk. In each example, only four key frames are selected to be used as control

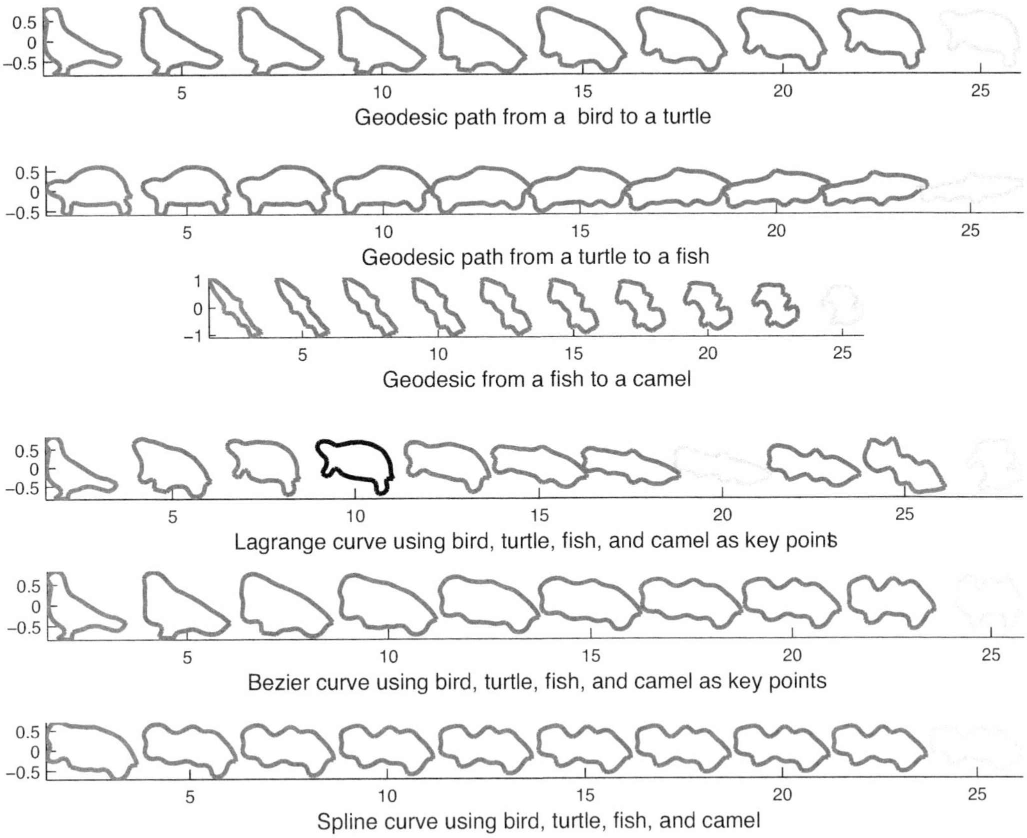

Figure 9.15: First three rows: geodesic between the ending shapes. Fourth row: Lagrange interpolation between four control points, ending points in previous rows. The fifth row shows Bézier interpolation, and the last row shows spline interpolation using the same control points.

points for interpolation. Curves are then extracted and represented as a vector of 100 points. Recall that shapes are invariant under rotation, translation, and re-parametrization. Thus, the alignment between the given curves is implicit in geodesics which makes the morphing process fully automatic. In Figure 9.15 and Figure 9.16, the first three rows show optimal deformations between ending shapes and the morphing sequences are shown in the last three rows. Thus, the fourth row shows Lagrange interpolation, the fifth row shows Bézier interpolation, and the last row shows spline interpolation. It is clear from Figure 9.15 and Figure 9.16 that Lagrange interpolation gives at least visually a good morphing and passes through the given data.

9.7.2 3D Face Morphing

In this example we show how to build an animation of 3D faces using different facial surfaces that represent the same person under different facial expressions. Despite some previous methods that show morphing between faces as a deformation from one face to another, which could be obtained here by a geodesic between two faces, our goal is to provide a way to build a morphing that includes a finite set of faces. So, as shown in Figure 9.17, we can

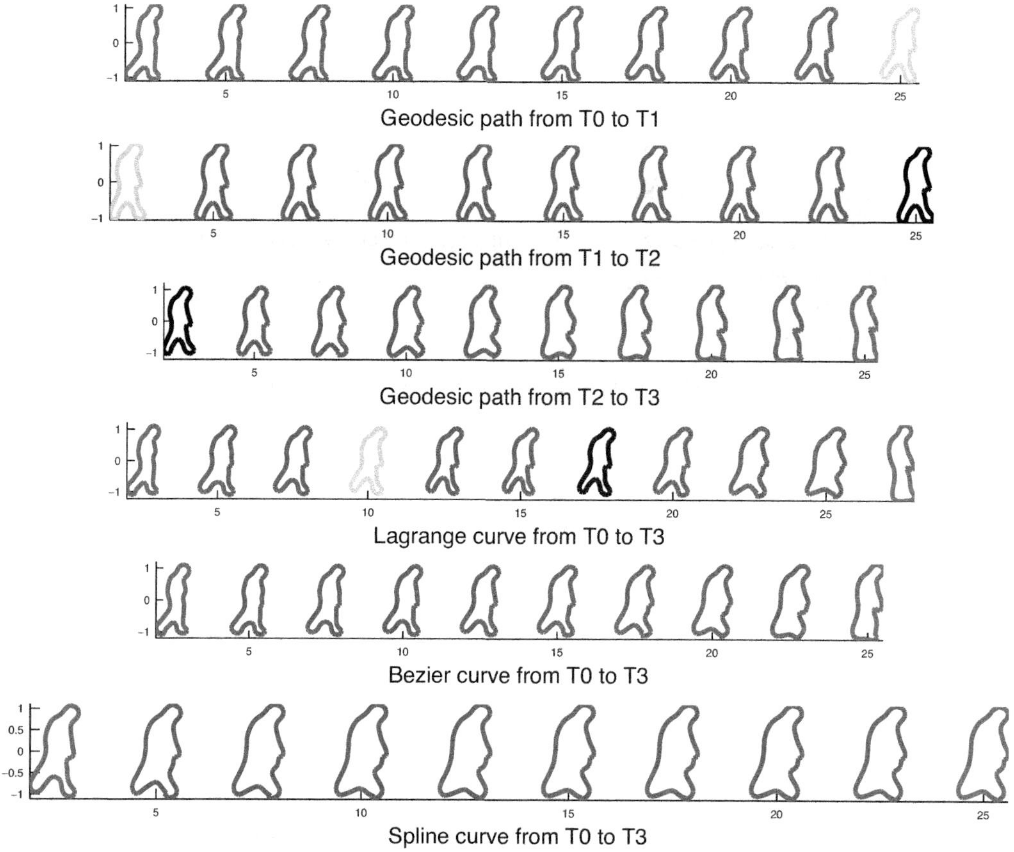

Figure 9.16: First three rows: geodesic between the ending shapes as human silhouettes from gait. Fourth row: Lagrange interpolation between four control points. The fifth row shows Bézier interpolation, and the last row shows spline interpolation using the same control points.

Figure 9.17: Morphing 3D faces by applying Lagrange interpolation on four different facial expressions of the same person.

use different facial expressions (four in the figure) and make the animation start from one face and pass through different facial expressions using Lagrange interpolation on a shape manifold. As mentioned above, no manual alignment is needed. Thus, the animation is fully automatic. In this experiment, we represent a face as a collection of 17 curves, and each curve is represented as a vector of 100 points. The method proposed in this chapter can be applied to more general surfaces if there is a natural way of representing them as indexed collections of closed curves. For more details, we refer the reader to [17].

9.8 Summary

The chapter presents a framework and algorithms for discrete interpolations on Riemannian manifolds and demonstrated it on $\mathbb{R}^2$ and $SO(3)$. Among many other applications, the method allows 2D and 3D shape metamorphosis based on Bézier, Lagrange, and spline interpolations on a shape manifold; thus, a fully automatic method to morph a shape passing through or as close as possible to a given finite set of other shapes. We then showed some examples using 2D curves from a walk-observation shape database, and a Lagrange interpolation between 3D faces to demonstrate the effectiveness of this framework.

Finally we note that the morphing algorithms presented in this chapter could be easily extended to other object-parameterizations if there is a way to compute geodesic between them.

Bibliography

[1] C. Altafini. The de casteljau algorithm on $se(3)$. In *Book chapter, Nonlinear control in the Year 2000*, pages 23–34, 2000.

[2] T. Beier and S. Neely. Feature-based image metamorphosis. In *In Computer Graphics (SIGGRAPH'92)*, pages 35–42, 1992.

[3] V. Blanz and T. Vetter. Face recognition based on fitting a 3d morphable model. *IEEE Transactions on Pattern Analysis and Machine Intelligence*, 25(9):1063–1074, 2003.

[4] C. Samir, D. Laurent, D. Laurent, K. A. Gallivan, C. Samir, P. Van Dooren, and P. A. Absil. Elastic morphing of 2d and 3d objects. In *Image Analysis and recognition*, volume 5627, pages 563–572, 2009.

[5] P. Crouch, G. Kun, and F. S. Leite. The de casteljau algorithm on the lie group and spheres. In *Journal of Dynamical and Control Systems*, volume 5, pages 397–429, 1999.

[6] J. Glaunes, A. Qiu, M. Miller, and L. Younes. Large deformation diffeomorphic metric curve mapping. In *International Journal of Computer Vision*, volume 80, pages 317–336, 2008.

[7] J. F. Hughes. Scheduled fourier volume morphing. In *Computer Graphics (SIGGRAPH'92)*, pages 43–46, 1992.

[8] J. Jakubiak, F. S. Leite, and R.C. Rodrigues. A two-step algorithm of smooth spline generation on Riemannian manifolds. In *Journal of Computational and Applied Mathematics*, pages 177–191, 2006.

[9] Shantanu H. Joshi, Eric Klassen, Anuj Srivastava, and Ian Jermyn. A novel representation for riemannian analysis of elastic curves in R^{n}. In *CVPR*, 2007.

[10] E. Klassen and A. Srivastava. Geodesic between 3D closed curves using path straightening. In A. Leonardis, H. Bischof, and A. Pinz, editors, *European Conference on Computer Vision*, pages 95–106, 2006.

[11] E. Klassen, A. Srivastava, W. Mio, and S. Joshi. Analysis of planar shapes using geodesic paths on shape spaces. *IEEE Patt. Analysis and Machine Intell.*, 26(3):372–383, March, 2004.

[12] A. Kume, I. L. Dryden, H. Le, and A. T. A. Wood. Fitting cubic splines to data in shape spaces of planar configurations. In *Proceedings in Statistics of Large Datasets, LASR, 119–122*, 2002.

[13] P. Lancaster and K. Salkauskas. Curve and surface fitting. In *Academic Press*, 1986.

[14] F. Lazarus and A. Verroust. Three-dimensional metamorphosis: a survey. In *The Visual Computer*, pages 373–389, 1998.

[15] Achan Lin and Marshall Walker. CAGD techniques for differentiable manifolds. In *Proceedings of the 2001 International Symposium Algorithms for Approximation IV*, 2001.

[16] T. Popeil and L. Noakes. Bézier curves and c^2 interpolation in Riemannian manifolds. In *Journal of Approximation Theory*, pages 111–127, 2007.

[17] C. Samir, A. Srivastava, M. Daoudi, and E. Klassen. An intrinsic framework for analysis of facial surfaces. *International Journal of Computer Vision*, volume 82, pages 80–95, 2009.

[18] Chafik Samir, P.-A. Absil, Anuj Srivastava, and Eric Klassen. A gradient-descent method for curve fitting on Riemannian manifolds, 2011. Accepted for publication in Foundations of Computational Mathematics.

[19] F. R. Schmidt, M. Clausen, and D. Cremers. Shape matching by variational computation of geodesics on a manifold. In *Pattern Recognition (Proc. DAGM)*, volume 4174 of *LNCS*, pages 142–151, Berlin, Germany, September 2006. Springer.

[20] Michael Spivak. *A Comprehensive Introduction to Differential Geometry, Vol I & II.* Publish or Perish, Inc., Berkeley, 1979.

[21] R. Whitaker and D. Breen. Level-set models for the deformation of solid object. In *Third International Workshop on Implicit Surfaces*, pages 19–35, 1998.

[22] G. Wolberg. Digital image warping. In *IEEE Computer Society Press*, 1990.

[23] H. Yang and B. Juttler. 3d shape metamorphosis based on t-spline level sets. In *The Visual Computer*, pages 1015–1025, 2007.

Chapter 10

Learning Image Manifolds from Local Features

Ahmed Elgammal and Marwan Torki

10.1 Introduction

Visual recognition is a fundamental yet challenging computer vision task. In recent years there has been tremendous interest in investigating the use of local features and parts in generic object recognition-related problems, such as object categorization, localization, discovering object categories, recognizing objects from different views, *etc.* In this chapter we present a framework for visual recognition that emphasizes the role of local features, the role of geometry, and the role of manifold learning. The framework learns an image manifold embedding from local features and their spatial arrangement. Based on that embedding several recognition-related problems can be solved, such as object categorization, category discovery, feature matching, regression, etc. We start by discussing the role of local features, geometry and manifold learning, and follow that by discussing the challenges in learning image manifolds from local features.

1) The Role of Local Features: Object recognition based on local image features has shown a lot of success recently for objects with large within-class variability in shape and appearance [23, 39, 51, 69, 2, 8, 20, 60, 21]. In such approaches, objects are modeled as a collection of parts or local features and the recognition is based on inferring the class of the object based on parts' appearance and (possibly) their spatial arrangement. Typically, such approaches find interest points using some operator such as corners [27] and then extract local image descriptors around such interest points. Several local image descriptors have been suggested and evaluated [41], such as Lowe's scale invariant features (SIFT) [39], Geometric Blur [7], and many others (see Section 10.7). Such highly discriminative local appearance features have been successfully used for recognition even without any shape (structure) information, *e.g.*, bag-of-words like approaches [71, 54, 41].

2) The Role of Geometry: The spatial structure, or the arrangement of the local features plays an essential role in perception since it encodes the shape. There are no better examples to show the importance of the shape in recognition over the appearance of local parts than the paintings of the Italian painter Giuseppe Arcimboldo (1527–1593). Arcimboldo is famous for painting portraits that are made of parts of different objects such as flowers, vegetables, fruits, fish, etc. Examples are shown in Figure 10.1. Human perception has no

Figure 10.1: Example painting of Giuseppe Arcimboldo (1527–1593). Faces are composed of parts of irrelevant objects.

problem recognizing the faces in the paintings mainly from the shape, i.e., the arrangement of parts, rather than from the appearance of the local parts. There are many other examples that can show such a point. One argument might be that it is a matter of scale: At the right scale the local parts can become discriminative. On the contrary, we believe that at the right scale the arrangement of the local features would become discriminative and not the local feature appearance.

There is a fundamental trade-off in part-structure approaches in general: The more discriminative and/or invariant a feature is, the sparser this feature becomes. Sparse features result in losing the spatial structure. For example, a corner detector results in dense but indiscriminative features while an affine invariant feature detector like SIFT will result in sparse features that do not necessarily capture the spatial arrangement. The above trade-off shapes the research in object recognition and matching. At one extreme are approaches such as bag-of-feature approaches [71, 54] that depend on highly discriminative features and end up with sparse features that do not represent the shape of the object. Therefore, such approaches tend to depend heavily on the feature distribution in recognition. Many researches recently have tried to include the spatial information of features, *e.g.* , by spatial partitioning and spatial histograms, *e.g.* [40, 32, 25, 55]. On the other end of the tradeoff are approaches that focus on the spatial arrangement for recognition. They tend to use very abstract and primitive feature detectors like corner detectors, which result in dense binary or oriented features. In such cases, the correspondences between features are established on the spatial arrangement level, typically through formulating the problem as a graph matching problem, e.g., [5, 61].

3) The Role of Manifold: Learning image manifolds has been shown to be quite useful in recognition, for example for learning appearance manifolds from different views [44], learning activity and pose manifolds for activity recognition and tracking [17, 65], *etc.*. Almost all the prior applications of image manifold learning, whether linear or nonlinear, have been based on holistic image representations where images are represented as vectors, *e.g.*, the seminal work of Murase and Nayar [44], or by establishing a correspondence framework between features or landmarks, e.g., [11].

The Manifold of Local Features:

Consider collections of images from any of the following cases or combinations of them:

- Different instances of an object class (within-class variations);
- Different views of an object;
- Articulation and deformation of an object;

- Different objects across-classes or within-class sharing a certain attribute.

Each image is represented as a collection of local features. In all these cases, both the features' appearance and their spatial arrangement will change as a function of all the above-mentioned factors. Whether a feature appears in a given frame and where, relative to other features, are functions of the viewpoint of the object and/or the articulation of the object and/or the object instance structure and/or a latent attribute.

Consider, in particular, the case of different views of the same object. There is an underlying manifold (or a subspace) where the spatial arrangement of the features should follow. For example, if the object is viewed from a view circle, which constitutes a one-dimensional view manifold, there should be a representation where the features and their spatial arrangement are expected to be evolving on a manifold of dimensionality, at most one (assuming we can factor out all other nuisance factors). Similarly, if we consider a full view sphere, a two-dimensional manifold, the features and their spatial arrangement should be evolving on a manifold of dimensionality, at most two. *The fundamental question is what is such representation that reveals the underlying manifold topology.* The same argument holds for the cases of within-class variability, articulation, and deformation, and across-class attributes; but in such cases, the underlying manifold dimensionality might not be known.

A central challenging question is how can we learn image manifolds from a bunch of local features in a smooth way such that we can capture the feature similarity and spatial arrangement variability between images. If we can answer this question, that will open the door for explicit modeling of within-class variability manifolds, objects' view manifolds, activity manifolds, and attribute manifolds, all from local features.

Why manifold learning from local features is challenging:

There are different ways researchers have approached the study of image manifolds, which are not applicable here. This points out the challenges for the case of learning from local features.

1. *Image vectorization based analysis:* Manifold analysis requires a representation of images in a vector space or in a metric space. Therefore, almost all the prior applications for image manifold learning, whether linear or nonlinear, have been based on wholistic image representations where images are represented as vectors [44, 57, 66, 17]. Such wholisitic image representation provides a vector space representation and a correspondence frame between pixels in images.

2. *Histogram based analysis:* On the other hand, vectorized representations of local features based on histograms, *e.g.* bag-of-words alike representations, cannot be used for learning image manifolds since theoretically histograms are not vector spaces. Histograms do not provide smooth transition between different images with the change in the feature-spatial structure. Extensions to the bag-of-words approach, where the spatial information is encoded in a histogram structure, e.g., [40, 32, 55], cannot be used, for the same reasons.

3. *Landmark based analysis:* Alternatively, manifold learning can be done on local features if we can establish full correspondences between these features in all images, which explicitly establish a vector representation of all the features. For example, Active Shape Models (ASM) [11] and similar algorithms use specific landmarks that can be matched in all images. Obviously it is not possible to establish such full correspondences between all features, since the same local features are not expected to be visible in all images. This is a challenge in the context of generic object recognition, given the large within-class variability. Establishing a full correspondence frame between features is also not feasible between different views of an object or different

frames of an articulated motion because of self occlusion or between different objects sharing a common attribute.

4. *Kernel-based analysis:* Another alternative for learning image manifolds is to learn the manifold in a metric space, where we can learn a similarity metric between images (from local features). Once such a similarity metric is defined, any manifold learning technique can be used. Since we are interested in problems such as learning within-class variability manifolds, view manifolds, and activity manifolds, the similarity kernel should reflect both the appearance affinity of local features and the spatial structure similarity in a *smooth* way to be able to capture the topology of the underlying image manifold without distorting it. Such a similarity kernel should also be robust to clutter. There have been a variety of similarity kernels based on local features, e.g., pyramid matching kernel [25], string kernels [14], *etc.*. However, to the best of our knowledge, none of these existing similarity measures were shown to be able to learn a smooth manifold representation.

Framework Overview: In the following sections we present a framework for learning an image manifold representation from collections of local features in images. Section 10.2 shows how to learn a feature embedding representation that preserves both the local appearance similarity as well as the spatial structure of the features. Section 10.3 shows how to embed features from a new image by introducing a solution for the out-of-sample that is suitable for this context. By solving these two problems and defining a proper distance measure in the feature embedding space, an image manifold embedding space can be obtained. Section 10.5 illustrates several applications of the framework for object categorization, localization, category discovery, and feature matching.

10.2 Joint Feature–Spatial Embedding

We are given K images; each is represented with a set of feature points. Let us denote such sets by, $X^1, X^2, \cdots X^K$ where $X^k = \{(x_1^k, f_1^k), \cdots, (x_{N_k}^k, f_{N_k}^k)\}$. Each feature point (x_i^k, f_i^k) is defined by its spatial location, $x_i^k \in \mathbb{R}^2$, in its image plane and its appearance descriptor $f_i^k \in \mathbb{R}^D$, where D is the dimensionality of the feature descriptor space. Throughout this chapter, we will use superscripts to indicate an image and subscripts to indicate point index within that image, i.e., $\boldsymbol{x}_i^k$ denotes the location of feature i in the k-th image. For example, the feature descriptor can be a SIFT [38], GB [7], etc. Notice that the number of features in each image might be different. We use N_k to denote the number of feature points in the k-th image. Let N be the total number of points in all sets, i.e., $N = \sum_{k=1}^{K} N_k$.

We are looking for an embedding for all the feature points into a common embedding space. Let $y_i^k \in \mathbb{R}^d$ denote the embedding coordinate of point (x_i^k, f_i^k), where d is the dimensionality of the embedding space, i.e., we are seeking a set of embedded point coordinates $Y^k = \{y_1^k, \cdots, y_{N_k}^k\}$ for each input feature set X^k. The embedding should satisfy the following two constraints:

- The feature points from different point sets with high feature similarity should become close to each other in the resulting embedding as long as they do not violate the spatial structure.
- The spatial structure of each point set should be preserved in the embedding space.

To achieve a model that preserves these two constraints we use two data kernels based on the affinities in the spatial and descriptor domains separately. The spatial affinity (structure) is computed within each image and is represented by a weight matrix $\mathbf{S}^k$ where

$\mathbf{S}^k_{ij} = K_s(x^k_i, x^k_j)$ and $K_s(\cdot, \cdot)$ is a spatial kernel local to the k-th image that measures the spatial proximity. Notice that we only measure intra-image spatial affinity; no geometric similarity is measured across images. The feature affinity between images p and q is represented by the weight matrix $\mathbf{U}^{pq}$ where $\mathbf{U}^{pq}_{ij} = K_f(f^p_i, f^q_j)$ and $K_f(\cdot, \cdot)$ is a feature kernel that measures the similarity in the descriptor domain between the i-th feature in image p and the j-th feature in image q. Here we describe the framework given any spatial and feature weights in general and later in this section we will give specific details on which kernels we use.

Let us jump ahead and assume an embedding can be achieved satisfying the aforementioned spatial structure and the feature similarity constraints. Such an embedding space represents a new Euclidean "Feature" space that encodes both the features' appearance and the spatial structure information. Given such an embedding, the similarity between two sets of features from two images can be computed within that Euclidean space with any suitable set similarity kernel. Moreover, unsupervised clustering can also be achieved in this space.

10.2.1 Objective Function

Given the above stated goals, we reach the following objective function on the embedded points Y, which need to be minimized

$$\Phi(Y) = \sum_k \sum_{i,j} \|y^k_i - y^k_j\|^2 \mathbf{S}^k_{ij} + \sum_{p,q} \sum_{i,j} \|y^p_i - y^q_j\|^2 \mathbf{U}^{pq}_{ij}, \tag{10.1}$$

where k, p and $q = 1, \cdots, K$, $p \neq q$, and $\|\cdot\|$ is the L2 Norm. The objective function is intuitive; the first term preserves the spatial arrangement within each set, since it tries to keep the embedding coordinates y^k_i and y^k_j of any two points x^k_i and x^k_j in a given point set close to each other based on their spatial kernel weight $\mathbf{S}^k_{ij}$. The second term of the objective function tries to bring close the embedded points y^p_i and y^q_j if their feature similarity kernel $\mathbf{U}^{pq}_{ij}$ is high.

This objective function can be rewritten using one set of weights defined on the whole set of input points as:

$$\Phi(Y) = \sum_{p,q} \sum_{i,j} \|y^p_i - y^q_j\|^2 \mathbf{A}^{pq}_{ij}, \tag{10.2}$$

where the matrix $\mathbf{A}$ is defined as

$$\mathbf{A}^{pq}_{ij} = \begin{cases} \mathbf{S}^k_{ij} & p = q = k \\ \mathbf{U}^{pq}_{ij} & p \neq q \end{cases} \tag{10.3}$$

where $\mathbf{A}^{pq}$ is the pq block of $\mathbf{A}$.

The matrix $\mathbf{A}$ is an $N \times N$ weight matrix with $K \times K$ blocks where the pq block is of size $N_p \times N_q$. The k-th diagonal block is the spatial structure kernel $\mathbf{S}^k$ for the k-th set. The off-diagonal pq block is the descriptor similarity kernels $\mathbf{U}^{pq}$. The matrix $\mathbf{A}$ is symmetric by definition since diagonal blocks are symmetric and since $\mathbf{U}^{pq} = \mathbf{U}^{qp^T}$. The matrix $\mathbf{A}$ can be interpreted as a weight matrix between points on a large point set where all the input points are involved in this point set. Points from a given image are linked by weights representing their spatial structure $\mathbf{S}^k$, while nodes across different data sets are linked by suitable weights representing their feature similarity kernel $\mathbf{U}^{pq}$. Notice that the size of the matrix $\mathbf{A}$ is linear in the number of input points.

We can see that the objective function Equation 10.2 reduces to the problem of Laplacian embedding [45] of the point set defined by the weight matrix $\mathbf{A}$. Therefore the objective function reduces to

$$\mathbf{Y}^* = \arg \min_{\mathbf{Y}^T\mathbf{DY}=I} tr(\mathbf{Y}^T\mathbf{LY}), \tag{10.4}$$

where $\mathbf{L}$ is the Laplacian of the matrix $\mathbf{A}$, i.e., $\mathbf{L} = \mathbf{D} - \mathbf{A}$, where $\mathbf{D}$ is the diagonal matrix defined as $\mathbf{D}_{ii} = \sum_j \mathbf{A}_{ij}$. The $N \times d$ matrix $\mathbf{Y}$ is the stacking of the desired embedding coordinates such that,

$$\mathbf{Y} = \left[y_1^1, \ldots, y_{N_1}^1, y_1^2, \ldots, y_{N_2}^2, \ldots y_1^K, \ldots, y_{N_K}^K\right]^T.$$

The constraint $\mathbf{Y}^T\mathbf{DY} = \mathbf{I}$ removes the arbitrary scaling and avoids degenerate solutions [45]. Minimizing this objective function is a straightforward generalized eigenvector problem: $\mathbf{L}y = \lambda\mathbf{D}y$. The optimal solution can be obtained by the bottom d nonzero eigenvectors. The required N embedding points Y are stacked in the d vectors in such a way that the embedding of the points of the first point set will be the first N_1 rows followed by the N_2 points of the second point set, and so on.

10.2.2 Intra-Image Spatial Structure

The spatial structure weight matrix $\mathbf{S}^k$ should reflect the spatial arrangement of the features in each image k. In general, it is desired that the spatial weight kernel be invariant to geometric transformations. However, this is not always achievable.

One obvious choice is a kernel based on the Euclidean distances between features in the image space, which would be invariant to translation and rotation. Instead we use an affine invariant kernel based on subspace invariance [68]. Given a set of feature points from an image at locations $\{x_i \in \mathbb{R}^2, i = 1, \cdots, N\}$, we can construct a configuration matrix

$$\mathbf{X} = [\mathbf{x}_1\mathbf{x}_2 \cdots \mathbf{x}_N] \in \mathbb{R}^{N\times 3}$$

where $\mathbf{x}_i$ is the homogeneous coordinate of point x_i. The range space of such a configuration matrix is invariant under affine transformation. It was shown in [68] that an affine representation can be achieved by QR decomposition of the projection matrix of $\mathbf{X}$, *i.e.*

$$\mathbf{QR} = \mathbf{X}(\mathbf{X}^T\mathbf{X})^{-1}\mathbf{X}^T$$

The first three columns of $\mathbf{Q}$, denoted by $\mathbf{Q}'$, gives an affine invariant representation of the points. We use a Gaussian kernel based on the Euclidean distance in this affine invariant space, *i.e.*,

$$K_s(x_i, x_j) = e^{-\|q_i - q_j\|^2/2\sigma^2}$$

where q_i, q_j are the i-th and j-th rows of $\mathbf{Q}'$.

10.2.3 Inter-Image Feature Affinity

The feature weight matrix $\mathbf{U}^{pq}$ should reflect the feature-to-feature similarity in the descriptor space between the p-th and q-th sets. An obvious choice is the widely used affinity based on a Gaussian kernel on the squared Euclidean distance in the feature space, i.e.,

$$\mathbf{G}_{ij}^{pq} = e^{-\left\|f_i^p - f_j^q\right\|^2/2\sigma^2}$$

given a scale σ. Another possible choice is a soft correspondence kernel that enforces the exclusion principle based on the Scott and Longuet-Higgins algorithm [52]. This is particularly useful for feature matching application [58] as will be discussed in section 10.5.6.

10.3 Solving the Out-of-Sample Problem

Given the feature embedding space learned from a collection of training images and given a new image represented with a set of features $X^\nu = \{(x_i^\nu, f_i^\nu)\}$, it is desired to find the coordinates of these new feature points in the embedding space. This is an out-of-sample problem; however, it is quite challenging. Most of out-of-sample solutions [6] depends on learning a nonlinear mapping function between the input space and the embedding space. This is not applicable here since the input is not a vector space, rather a collection of points. Moreover, the embedding coordinate of a given feature depends on all the features in the new image (because of the spatial kernel). The solution we introduce here is inspired by the formulation in [72][1]. For clarity, we show how to solve for the coordinates of the new features of a single new image. The solution can be extended to embed any number of new images in batches in a straightforward way.

We can measure the feature affinity in the descriptor space between the features of the new image and the training data descriptors using the feature affinity kernel defined in Section 10.2. The feature affinity between image p and the new image is represented by the weight matrix $\mathbf{U}^{\nu,p}$ where $\mathbf{U}_{ij}^{\nu,p} = K_f(f_i^\nu, f_j^p)$. Similarly, the spatial affinity (structure) within the new image can be encoded with the spatial affinity kernel. The spatial affinity (structure) of the new image's features is represented by a weight matrix $\mathbf{S}^\nu$ where $\mathbf{S}_{ij}^\nu = K_s(x_i^\nu, x_j^\nu)$. Notice that, consistently, we do not measure any inter geometric similarity between images, we only encode intra-geometric constraints within each image.

We have a new embedding problem in hand. Given the sets $X^1, X^2, \cdots X^K, X^\nu$ where the first K sets are the training data and X^ν is the new set, we need to find embedding coordinates for all the features in all the sets, i.e., we need to find $\{y_i^k\} \cup \{y_j^\nu\}$, $i = 1, \cdots, N_k$ and $k = 1, \cdots, K$, $j = 1, \cdots, N_\nu$ using the same objective function in Equation 10.1; in this case the indices k, p, and $q = 1, \cdots K+1$, to include the new set. *However, we need to preserve the coordinates of the already embedded points.* Let $\hat{y}_i^k$ be the original embedding coordinates of the training data. We now have a new constraint that we need to satisfy

$$y_i^k = \hat{y}_i^k, \text{for } i = 1, \cdots, N_k, k = 1, \cdots, K$$

Following the same derivation in Section 10.2, and adding the new constraint, we reach the following optimization problem in $\mathbf{Y}$

$$\begin{array}{ll} \min & tr(\mathbf{Y}^T \mathbf{L} \mathbf{Y}) \\ s.t. & y_i^k = \hat{y}_i^k, i = 1, \cdots, N_k, k = 1, \cdots, K \end{array} \tag{10.5}$$

where

$$\mathbf{Y} = \left[y_1^1, \ldots, y_{N_1}^1, \cdots y_1^K, \ldots, y_{N_K}^K, y_1^\nu, \ldots, y_{N_\nu}^\nu\right]^T$$

where $\mathbf{L}$ is the laplacian of the $(N + N_\nu) \times (N + N_\nu)$ matrix $\mathbf{A}$ is defined as

$$\mathbf{A} = \begin{pmatrix} \mathbf{A}^{train} & \mathbf{U}^{\nu^T} \\ \mathbf{U}^\nu & \mathbf{S}^\nu \end{pmatrix} \tag{10.6}$$

where $\mathbf{A}^{train}$ is defined in Equation 10.3 and $\mathbf{U}^\nu = [\mathbf{U}^{\nu,1} \cdots \mathbf{U}^{\nu,K}]$. Notice that the constraint $\mathbf{Y}^T \mathbf{D} \mathbf{Y} = \mathbf{I}$, which was used in Equation 10.4, is not needed anymore since the equality constraints avoid the degenerate solution.

The out-of-sample solution described earlier is used to obtain such a function. We can achieve a closed form solution for this function given the spatial and feature affinity matrices $\mathbf{S}^\nu$, $\mathbf{U}^\nu$

$$\mathbf{Y}^\nu = (\mathbf{L}^\nu)^{-1} \mathbf{U}^\nu \mathbf{Y}^\tau \tag{10.7}$$

[1]We are not using the approach in [72] for coordinate propagation; we are only using a similar optimization formulation.

where $\mathbf{Y}^\top$ is an $N \times d$ matrix stacking of the embedding coordinate of the training features and $\mathbf{L}^\nu$ is a block of the Laplacian $\mathbf{L}$ corresponding to the spatial affinity block $\mathbf{S}^\nu$.

10.3.1 Populating the Embedding Space

The out-of-sample framework is essential not only to be able to embed features from a new image for classification purposes, but also to be able to embed a large number of images with a large number of features. The feature embedding framework in Section 10.2 solves an Eigenvalue problem on a matrix of size $N \times N$ where N is the total number of features in all training data. Therefore, there is a computational limitation on the number of training images and the number of features per image that can be used. Given a large training data, we use a two-step procedure to establish a comprehensive feature embedding space:

1. *Initial Embedding:* Given a small subset of training data with a small number of features per image, solve for an initial embedding using Equation 10.4.
2. *Populate Embedding:* Embed the whole training data with a larger number of features per image, one image at a time, by solving the out-of-sample problem in Equation 10.5.

10.4 From Feature Embedding to Image Embedding

The embedding achieved in Section 10.2 is an embedding of the features where each image is represented by a set of coordinates in that space. This Euclidean space can be the basis to study image manifolds. All we need is a measure of similarity between two images in that space. There are a variety of similarity measures that can be used. For robustness, we chose to use a percentile-based Hausdorff-based distance to measure the distance between two sets of features from two images, defined as

$$H_l(X^p, X^q) = \max\{\overset{l\%}{\max_j} \min_i \|y_i^p - y_j^q\|, \overset{l\%}{\max_i} \min_j \|y_i^p - y_j^q\|\} \tag{10.8}$$

where l is the percentile used. In all the experiments we set the percentile to 50%, i.e., the median. Since this distance is measured in the feature embedding space, it reflects both feature similarity and shape similarity. However, one problem with this distance is that it is not a metric and does not guarantee a positive semi-definite kernel. Therefore, we use this measure to compute a positive definite matrix $\mathbf{H}^+$ by computing the eigenvectors corresponding to the positive eigenvalues of the original $\mathbf{H}_{pq} = H_l(X^p, X^q)$.

Once a distance measure between images is defined, any manifold embedding techniques, such as MDS [13], LLE [48], Laplacian Eigen maps [45], etc., can be used to achieve an embedding of the image manifold where each image is represented as a point in that space. We call this space "Image-Embedding" space and denote its dimensionality by d_I to disambiguate it from the "Feature-Embedding" space with dimensionality d.

10.5 Applications

10.5.1 Visualizing Objects View Manifold

The COIL data set [44] has been widely used in holistic recognition approaches where images are represented by vectors [44]. This is a relatively easy data set where the view manifold of an object can be embedded using PCA, using the whole image as a vector representation [44]. It has also been used extensively in manifold learning literature, also using the whole image as a vector representation. We use this data to validate that our approach can really achieve an embedding that is topologically correct using local features and the proposed framework. Figure 10.2 shows two examples of the resulting view manifold embedding. In this example we used 36 images with 60 GB features [7] per image. The figure clearly shows an embedding of a closed one-dimensional manifold in a two-dimensional embedding space.

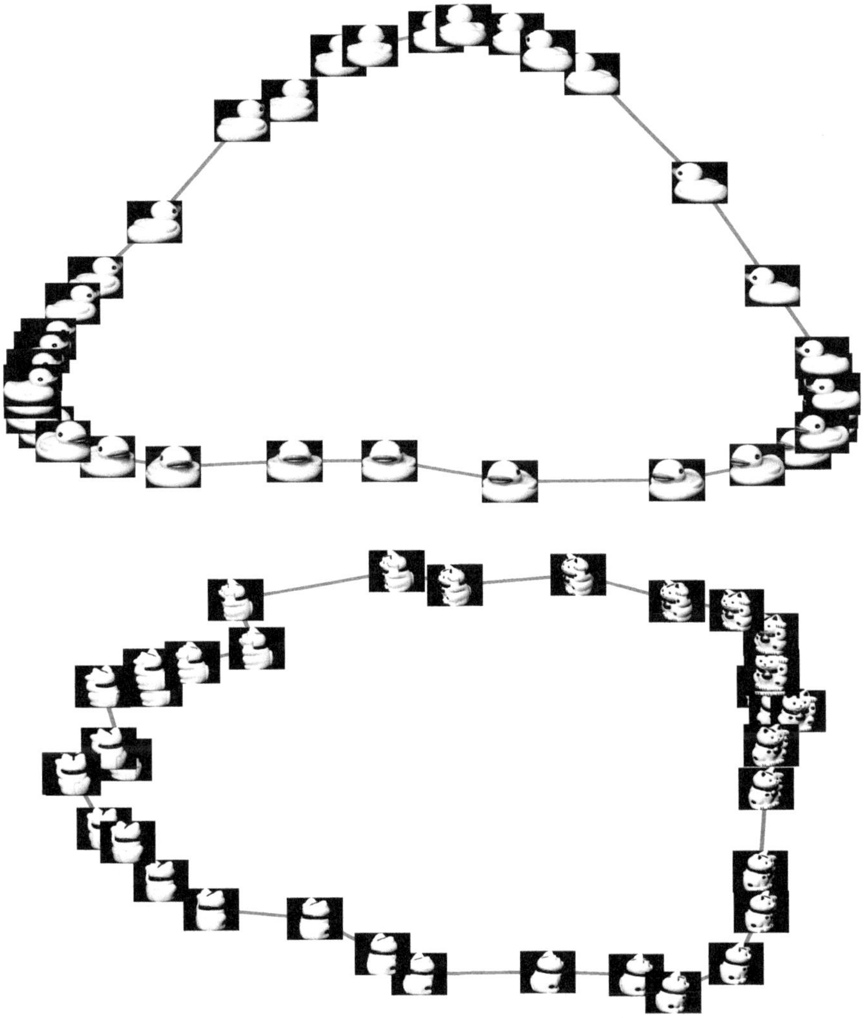

Figure 10.2: Examples of view manifolds learned from local features.

10.5.2 What the Image Embedding Captures

Figure 10.3 shows the resulting image embedding space (the first two dimensions are shown) of images from the "Shape" dataset [55]. The Shape dataset contains 10 classes (cup, fork, hammer, knife, mug, pan, pliers, pot, sauce pan and scissors), with a total of 724 images. The dataset exhibits large within-class variation and moreover there is similarity

Figure 10.3: Manifold embedding for 60 samples from shape dataset using 60 GB local features per image.

between classes, e.g., mugs and cups; saucepans and pots. We used 60 local features per image. Sixty images were used to learn the initial feature embedding of dimensionality 60 (6 samples per class chosen randomly). Each image is represented using 60 randomly chosen geometric blur local feature descriptors [7]. The initial feature embedding is then expanded using the out-of-sample solution to include all the training images with 120 features per images. We can notice how different objects are clustered in the space. It is clear that the embedding captures the objects global shape from the local feature arrangement, i.e., the global spatial arrangement is captured. There are many interesting semantics that we can notice in the embedding. There are many interesting structures we can notice. We can notice that objects with similar semantic attributes are grouped together. For example, elongated objects (e.g., forks and knifes) are to the left, cylindrical objects (e.g., mugs) are to the top right, circular objects (e.g., pans) are to the bottom right, i.e., the embedding captures shape attributes. Beyond shape, we can also notice that other semantic attributes are captured, e.g., metal forks, knives and other metal objects with black handles, mugs with texture, metal pots and pans, notice that this is a two-dimensional projection of the embedding, the dimensionality of the embedding space itself is much higher. This points out that this embedding space captures different global semantic similarities between images only based on local appearance and arrangement information.

Figure 10.4-top shows an example embedding of sample images from four classes of the Caltech-101 dataset [37] where the manifold was learned from local features detected on each image. As can be noticed, the images contain a significant amount of clutter, yet the embedding clearly reflects the perceptual similarity between images as we might expect. This obviously cannot be achieved using holistic image vectorization, as can be seen in Figure 10.4-bottom, where the embedding is dominated by similarity in image intensity.

Figure 10.5 shows an embedding of four classes in Caltech-4 [37] (2880 images of faces,

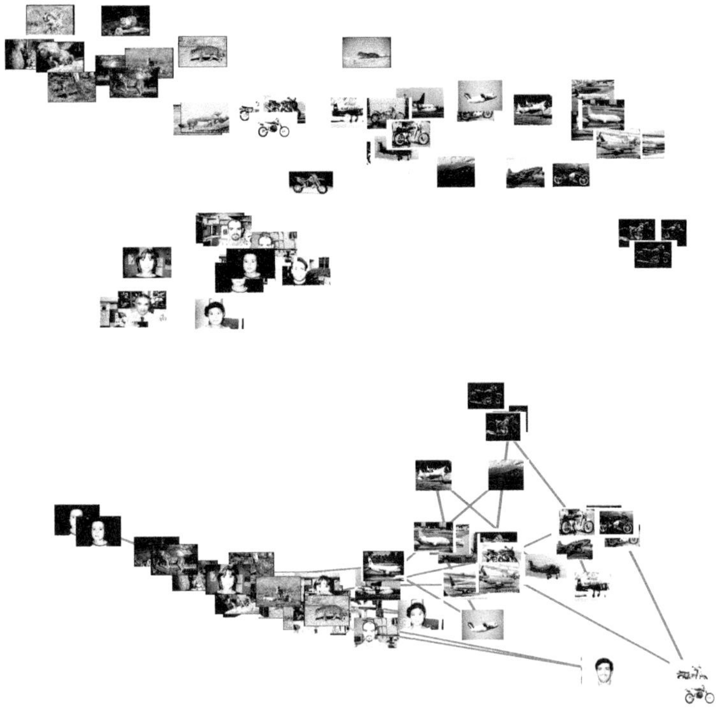

Figure 10.4: Example embedding result of samples from four classes of Caltech-101. Top: Embedding using our framework using 60 Geometric Blur local features per image. The embedding reflects the perceptual similarity between the images. Bottom: Embedding based on Euclidean image distance (no local features, image as a vector representation). Notice that Euclidean image distance based embedding is dominated by image intensity, i.e., darker images are clustered together and brighter images are clustered.

airplanes, motorbikes, cars-rear). We can notice that the classes are well clustered in the space, even though only the first two dimensions' embedding are shown.

10.5.3 Object Categorization

We describe the object categorization problem as an application of learning the image manifold from local features with their spatial arrangement. The goal is to achieve an embedding of a collection of images to facilitate the categorization task. The resulting embedding captures both appearance and shape similarities. Using such an embedding gives more accurate results when contrasted with other state-of-the art methods.

In [59] the "Shape" dataset [55] was used to evaluate the application of the framework for object categorization based on both the feature embedding and image embedding. Different training/testing random splits were used for training with 1/5, 1/3, 1/2, and 2/3 splits and 10 times cross validation and average accuracies were reported. Four different classifiers were evaluated based on the proposed representation: 1) Feature-embedding with SVM, 2) Image embedding with SVM, 3) Feature embedding with 1-NN classifier, 4) Image-embedding with 1-NN classifier. Table 10.1 shows the results for the four different classifier settings. We can clearly notice that image manifold-based classifiers enhance the results over feature embedding-based classifiers. In [59] several other data sets were used to evaluate the

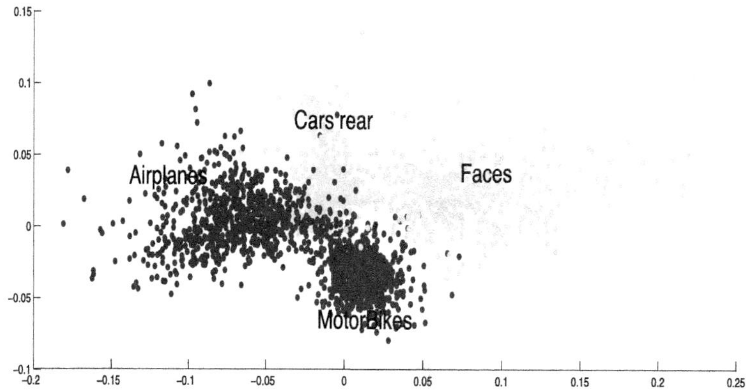

Figure 10.5: (See Color Insert.) Manifold embedding for all images in Caltech-4-II. Only first two dimensions are shown.

Table 10.1: Shape dataset: Average accuracy for different classifier settings based on the proposed representation.

	traning/test splits			
Classifier	1/5	1/3	1/2	2/3
Feature embedding - SVM	74.25	80.29	82.85	87.02
Image Manifold - SVM	80.85	84.96	88.37	91.27
Feature embedding - 1-NN	70.90	74.13	77.49	79.63
Image Manifold - 1-NN	71.93	75.29	78.26	79.34

performance with a similar conclusion. The evaluation also showed very good recognition rates (above 90%) even with as low as 5 training images.

In [55] the Shape dataset was used to compare the effect of modeling feature geometry by dividing the object's bounding box to 9 grid cells (localized bag of words) in comparison to geometry-free bag of words. Results were reported using SIFT [38], GB [7], and KAS [22] features. Table 10.2 shows the reported accuracy in [55] for comparison. All reported results are based on 2:1 ratio for training/testing split. Unlike [55] where bounding boxes are used both in training and testing, we do not use any bounding box information since our approach does not assume a bounding box for the object to encode the geometry and yet gets better result.

10.5.4 Object Localization

Many approaches that encode feature geometry are based on a bounding box, e.g., [55, 25]. Our approach does not require such a constraint and it is robust to the existence of heavy

Table 10.2: Shape dataset: Comparison with reported results.

Accuracy %			
Feature used	SIFT	GB	KAS
Our approach	-	91.27	-
bag of words (reported by [55])	75	69	65
Localized bag of words ([55])	88	86	85

Table 10.3: Object localization results.

Class	TPR	FPR	BBHR	BBMR
Airplanes	98.08%	1.92%	100%	0/800
Faces	68.43%	31.57%	96.32%	16/435
Leopards	76.81%	23.19%	98%	4/200
Motorbikes	99.63%	0.37%	100%	0/798

Table 10.4: Caltech-4, 5, and 6: Average clustering accuracy, best results are shown in bold.

Categories	FE Clustering	Baseline	[30]	[34]	[26]	Baseline [34]
Caltech-4	**99.54**±0.31	96.43	98.55	98.03	86	87.37
Caltech-5	**98.59**±0.47	96.28	97.30	96.92	NA	83.78
Caltech-6	**97.48**±0.57	94.03	95.42	96.15	NA	83.53

visual clutter. Therefore, it can be used in localization as well as recognition. We used Caltech-4 data {Airplane, Leopards, Faces, Motorbikes} for evaluation. In this case we learned the feature embedding from all the four classes, using only 12 images per class. For evaluation we used 120 features in each query image and embed them by out-of-sample. The object is localized by finding the top 20% features closer to the training data (by computing feature distances in the feature embedding space.) Table 10.3 shows the results in terms of the True Positive Ratio (TPR): the percentage of localized features inside the bounding box, and False Positive Ratio (FPR), Bounding Box Hit Ratio (BBHR), the percentage of images with more than 5 features localized (a metric defined in [30]), and Bounding Box Miss Ratio (BBMR).

10.5.5 Unsupervised Category Discovery

Another intersting application for framework is unsupervised categroy discovery. We tested the approach for unsupervised category dicovery by following the setup by [26, 30, 34] on the same benchmark subsets of Caltech-101 dataset. Namely we use the {Airplane, Cars-rear, Faces, Motorbikes} for Caltech-4. We add the class {Watches} for Caltech-5 and the class {Ketches} for Caltech-6. The output is the classification of images according to object category. We use the clustering accuracy as our measure to evaluate the categorization process. We report the average accuracy over 40 runs.

We use NCUT spectral clustering algorithm to compute the desired clustering. Using the $\mathbf{H}^+$ matrix, we compute a weight matrix $\mathbf{W}$ as an input to the clustering algorithm. We further use the K-nearest neighbor graph on the weight matrix $\mathbf{W}$, where K is $O(log(M))$ and M is number of images in the dataset.

We randomly select $12 \times C$ random samples to form an initial embedding that is used to generate initially the common feature embedding of all features. We select 120 features per image for initial embedding and we out-of-sample 420 features (at the most) per image. This results in a common feature embedding that has $100C \times 420$ features. We chose dimensionality of the common feature embedding = 120. Table 10.4 shows comparative evaluation, the state of the art results in [30, 34]. We also show the results by using the baseline that uses feature descriptor similarity to compute $\mathbf{H}_{descriptor}$; in other words there is no spatial arrangement proximity in this $\mathbf{H}_{descriptor}$. The results show that our method is doing an extremely excellent job for all the subsets Caltech-4, 5, and 6. We infer from these results that the approaches that use explicit spatially consistent matching steps like [30, 34] can be outperformed by using a common feature embedding space that encodes the spatial proximity and appearance similarity at the same time, which is done without an explicit matching step.

10.5.6 Multiple Set Feature Matching

Finding correspondences between features in different images plays an important role in many computer vision tasks, including stereoscopic vision, object recognition, image registration, mosaicing, structure from motion, motion segmentation, tracking, etc. [43]. Several robust and optimal approaches have been developed for finding consistent matches for rigid objects by exploiting a prior geometric constraint [63]. The problem becomes more challenging in a general setting, e.g., matching features on an articulated object, deformable object, or matching between two instances (or a model to an instance) of the same object class for recognition and localization. For such problems, many researchers have recently tended to use high-dimensional descriptors encoding the local appearance (e.g., SIFT features [38]). Using such highly discriminative features makes it possible to solve for correspondences without much structure information or avoid solving for correspondences all together, which is quite a popular trend in object categorization [49]. This is also motivated by avoiding the high complexity of solving for spatially consistent matches.

The framework for the joint feature-spatial embedding presented in this chapter provides a way to find consistent matches between *multiple* sets of features where both the feature descriptor similarity and the spatial arrangement of the features need to be enforced. However, the spatial arrangement of the features needs to be encoded and enforced in a relaxed manner to be able to deal with non-rigidity, articulation, deformation, and within class variation.

The problem of matching appearance features between two images in a spatially consistent way has been addressed recently (e.g., [36, 12, 10, 61]). Typically this problem is formulated as an attributed graph matching problem where graph nodes represent the feature descriptors and edges represent the spatial relations between features. Enforcing consistency between the matches led researchers to formulate this problem as a quadratic assignment problem where a linear term is used for node compatibility and a quadratic term is used for edge compatibility. This yields an NP-hard problem [10]. Even though some efficient solutions (e.g., linear complexity in the problem description length) have been proposed for such a problem [12] the problem description itself remains quadratic, since consistency has to be modeled between every pair of edges in the two graphs. This puts a huge limitation on the applicability of such approaches to handle a large number of features; for example, for matching n features in two images, an edge compatibility matrix of size $n^2 \times n^2$, i.e., $O(n^4)$, needs to be computed and manipulated to encode the edge compatibility constraints. Obviously this is prohibitively complex and does not scale to handle a large number of features.

The problem of consistent matching can be formulated as an embedding problem [58] where the goal is to embed all the features in a Euclidean embedding space where the locations of the features in that space reflect both the descriptor similarity and the spatial arrangement. This is achieved through minimizing the same objective function in Equation 10.1 enforcing both the feature similarity and the spatial arrangement. A soft correspondence kernel that enforces the exclusion principle based on the Scott and Longuet-Higgins algorithm [52] is advantageous for such application. The embedding space acts as a new unified feature space (encoding both the descriptor and spatial constraints) where the matching can be easily solved. This embeding-based matching framework directly generalizes to matching multiple sets of features in one shot through solving one eigenvalue problem. An interesting point about this formulation is that the spatial arrangement for each set is only encoded within that set itself, i.e., in a graph matching context no com-

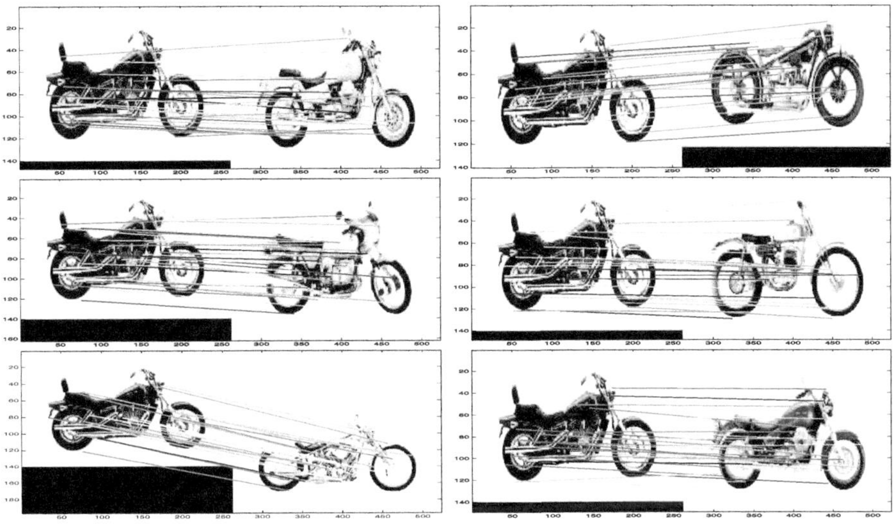

Figure 10.6: Sample matching results on Caltech-101 motorbike images.

patibility needs to be computed between the edges (no quadratic terms or higher order terms), yet we can enforce spatial consistency. Therefore, this approach is scalable and can deal with hundreds and thousands of features. Minimizing the objective function in the proposed framework can be done by solving an eigenvalue problem *which size is linear in the number of features in all images.*

Figure 10.6 shows sample matches on motorbike images from Caltech-101 [37]. Eight images were used to achieve a unified feature embedding and then pairwise matching was performed in the embedding space using the Scott and Longuet-Higgins (SLH) algorithm [52]. Extensive evaluation of the feature matching application of the framework can be found in [58]

10.6 Summary

In this chapter we presented a framework that enables the study of image manifolds from local features. We introduced an approach to embed local features based on their inter-image similarity and their intra-image structure. We also introduced a relevant solution for the out-of-sample problem, which is essential to be able to embed large data sets. Given these two components we showed that we can embed image manifolds from local features in a way that reflects the perceptual similarity and preserves the topology of the manifold. Experimental results showed that the framework can achieve superior results in recognition and localization. Computationally, the approach is very efficient. The initial embedding is achieved by solving an eigenvalue problem which is done offline. Incremental addition of images, as well as solving out-of-sample for a query image is done in a time that is negligible to the time needed by the feature detector per image.

10.7 Bibliographical and Historical Remarks

The use of local features and parts for visual recognition has been rooted in the computer vision literature for long time, e.g., [23], however such a paradigm received extensive interest in the last decade, e.g., [39, 51, 69, 2, 8, 20, 60, 21], and others. Several local feature descriptors have been proposed and widely used, such as Lowe's scale invariant features (SIFT) [39], entropy-based scale invariant features [29, 20], Geometric Blur [7], contour based features (kAS) [22], and other local features that exhibit affine invariance, such as [3, 62, 50].

Modeling the spatial structure of an object varies dramatically in the literature of object classification. At the extreme are approaches that totally ignore the structure and classify the object only based on the statistics of the features (parts) as an unordered set, e.g., bag-of-features approaches [71, 54]. Generalized Hough-transform-like approaches provide a way to encode spatial structure in a loose manner [35, 46]. A similar idea was used earlier in the constellation model of Weber et al. [69] where part locations were modeled statistically given a central coordinate system, also in [20]. Pairwise distances and directions between parts have also been used to encode the spatial structure, e.g., [1]. Felzenszwalb and Huttenlocher's Pictorial structure [19] uses springlike constraints between pairs of parts as well to encode global structure. The constellation model of Weber et al. [69] constrains the part locations given a central coordinate system.

The seminal work of Murase and Nayar [44] showed how linear dimensionality reduction using PCA [28] can be used to establish a representation of an object's view and illumination manifolds. Using such representation, recognition of a query instance can be achieved by searching for the closest manifold. Such subspace analysis has been extended to decompose multiple orthogonal factors using bilinear models [57] and multi-linear tensor analysis [66].

The introduction of nonlinear dimensionality reduction techniques such as Local Linear Embedding (LLE) [48], Isometric Feature Mapping (Isomap) [56], and others [56, 48, 4, 9, 31, 70, 42] made it possible to represent complex manifolds in low-dimensional embedding spaces in ways that preserve the manifold topology. Such manifold learning approaches have been used successfully in human body pose estimation and tracking [17, 18, 65, 33].

There is a huge literature on formulating correspondence finding as a graph-matching problem. We refer the reader to [10] for an excellent survey on this subject. Matching two sets of features can be formulated as a bipartite graph matching in the descriptor space, e.g., [5], and the matches can be computed using combinatorial optimization, e.g., the Hungarian algorithm [47]. Alternatively, spectral decomposition of the cost matrix can yield an approximate relaxed solution, e.g., [52, 15], which solves for an orthonormal matrix approximation for the permutation matrix. Alternatively, matching can be formulated as a graph isomorphism problem between two weighted or unweighted graphs to enforce edge compatibility, e.g., [64, 53, 67]. The intuition behind such approaches is that the spectrum of a graph is invariant under node permutation and, hence, two isomorphic graphs should have the same spectrum, but the converse does not hold. Several approaches formulated matching as a quadratic assignment problem and introduced efficient ways to solve it, e.g., [24, 7, 12, 36, 61]. Such formulation enforces edgewise consistency on the matching; however, that limits the scalability of such approaches to a large number of features. Even higher order consistency terms have been introduced [16]. In [10] an approach was introduced to learn the compatibility functions from examples and it was found that linear assignment with such a learning scheme outperforms quadratic assignment solutions such as [12]. In [58] the approach described in this chapter was also shown to outperform quadratic assignment and without the need to resort to edge compatibilities.

Acknowledgments: This research is partially funded by NSF CAREER award IIS-0546372.

Bibliography

[1] S. Agarwal, A. Awan, and D. Roth. Learning to detect objects in images via a sparse, part-based representation. *TPAMI*, 26(11):1475–1490, 2004.

[2] S. Agarwal and D. Roth. Learning a sparse representation for object detection. In *ECCV*, pages 113–130, 2002.

[3] A. Baumberg. Reliable feature matching across widely separated views. In *CVPR*, pages 774–781, 2004.

[4] M. Belkin and P. Niyogi. Laplacian eigenmaps for dimensionality reduction and data representation. *Neural Comput.*, 15(6):1373–1396, 2003.

[5] S. Belongie, J. Malik, and J. Puzicha. Shape matching and object recognition using shape contexts. *TPAMI*, 2002.

[6] Y. Bengio, J.-F. Paiement, P. Vincent, O. Delalleau, N. L. Roux, and M. Ouimet. Out-of-sample extensions for lle, isomap, mds, eigenmaps, and spectral clustering. In *NIPS 16*, 2004.

[7] A. C. Berg. *Shape Matching and Object Recognition.* PhD thesis, University of California, Berkeley, 2005.

[8] E. Borenstein and S. Ullman. Class-specific, top-down segmentation. In *ECCV*, pages 109–124, 2002.

[9] M. Brand and K. Huang. A unifying theorem for spectral embedding and clustering. In *Proc. of the Ninth International Workshop on AI and Statistics*, 2003.

[10] T. S. Caetano, J. J. McAuley, L. Cheng, Q. V. Le, and A. J. Smola. Learning graph matching. *TPAMI*, 2009.

[11] T. F. Cootes, C. J. Taylor, D. H. Cooper, and J. Graham. Active shape models: Their training and application. *CVIU*, 61(1):38–59, 1995.

[12] T. Cour, P. Srinivasan, and J. Shi. Balanced graph matching. *NIPS*, 2006.

[13] T. Cox and M. Cox. *Multidimensional scaling.* London: Chapman & Hall, 1994.

[14] M. Daliri, E. Delponte, A. Verri, and V. Torre. Shape categorization using string kernels. In *SSPR06*, pages 297–305, 2006.

[15] E. Delponte, F. Isgrò, F. Odone, and A. Verri. Svd-matching using sift features. *Graph. Models*, 2006.

[16] O. Duchenne, F. Bach, I. S. Kweon, and J. Ponce. A tensor-based algorithm for high-order graph matching. *CVPR*, 2009.

[17] A. Elgammal and C.-S. Lee. Inferring 3d body pose from silhouettes using activity manifold learning. In *CVPR*, volume 2, pages 681–688, 2004.

[18] A. Elgammal and C.-S. Lee. Separating style and content on a nonlinear manifold. In *CVPR*, volume 1, pages 478–485, 2004.

[19] P. F. Felzenszwalb and D. P. Huttenlocher. Pictorial structures for object recognition. *IJCV*, 61(1):55–79, 2005.

[20] R. Fergus, P. Perona, and A. Zisserman. Object class recognition by unsupervised scale-invariant learning. In *CVPR (2)*, pages 264–271, 2003.

[21] R. Fergus, P. Perona, and A. Zisserman. A sparse object category model for efficient learning and exhaustive recognition. In *CVPR*, 2005.

[22] V. Ferrari, L. Fevrier, F. Jurie, and C. Schmid. Groups of adjacent contour segments for object detection. *TPAMI*, 30(1):36–51, 2008.

[23] M. Fischler and R. Elschlager. The representation and matching of pictorial structures, 1973. IEEE Transaction on Computer c-22(1):67–92.

[24] S. Gold and A. Rangarajan. A graduated assignment algorithm for graph matching. *TPAMI*, 1996.

[25] K. Grauman and T. Darrell. The pyramid match kernel: discriminative classification with sets of image features. In *ICCV*, volume 2, pages 1458–1465 Vol. 2, October 2005.

[26] K. Grauman and T. Darrell. Unsupervised learning of categories from sets of partially matching image features. In *CVPR*, 2006.

[27] C. Harris and M. Stephens. A combined corner and edge detector. In *Proc. of The Fourth Alvey Vision Conference*, 1988.

[28] I. T. Jolliffe. *Principal Component Analysis*. Springer-Verlag, 1986.

[29] T. Kadir and M. Brady. Scale, saliency and image description. *IJCV*, 2001.

[30] G. Kim, C. Faloutsos, and M. Hebert. Unsupervised modeling of object categories using link analysis techniques. In *CVPR*, 2008.

[31] N. Lawrence. Gaussian process latent variable models for visualization of high dimensional data. In *NIPS*, 2003.

[32] S. Lazebnik, C. Schmid, and J. Ponce. Beyond bags of features: Spatial pyramid matching for recognizing natural scene categories, pages II: 2169–2178, 2006.

[33] C.-S. Lee and A. Elgammal. Coupled visual and kinematics manifold models for human motion analysis. *IJCV*, July 2009.

[34] Y. J. Lee and K. Grauman. Shape discovery from unlabeled image collections. In *CVPR*, 2009.

[35] B. Leibe, A. Leonardis, and B. Schiele. Combined object categorization and segmentation with an implicit shape model. In *ECCV workshop on statistical learning in computer vision*, pages 17–32, 2004.

[36] M. Leordeanu and M. Hebert. A spectral technique for correspondence problems using pairwise constraints. *ICCV*, 2005.

[37] F. Li, R. Fergus, and P. Perona. Learning generative visual models from few training examples: An incremental Bayesian approach tested on 101 object categories. *CVIU*, 106(1):59–70, April 2007.

[38] D. Lowe. Distinctive image features from scale-invariant keypoints. *IJCV*, 2004.

[39] D. G. Lowe. Object recognition from local scale-invariant features. In *ICCV*, pages 1150–1157, 1999.

[40] M. Marszaek and C. Schmid. Spatial weighting for bag-of-features. In *CVPR*, pages II: 2118–2125, 2006.

[41] K. Mikolajczyk and C. Schmid. A performance evaluation of local descriptors. *TPAMI*, 2005.

[42] P. Mordohai and G. Medioni. Unsupervised dimensionality estimation and manifold learning in high-dimensional spaces by tensor voting. In *Proc. of AJCAI*, 2005.

[43] P. Moreels and P. Perona. Evaluation of features detectors and descriptors based on 3D objects. *IJCV*, 2007.

[44] H. Murase and S. Nayar. Visual learning and recognition of 3D objects from appearance. *IJCV*, 14:5–24, 1995.

[45] P. Niyogi. Laplacian eigenmaps for dimensionality reduction and data representation. *Neural Computation*, 2003.

[46] A. Opelt, A. Pinz, and A. Zisserman. A boundary-fragment-model for object detection. In *ECCV*, 2006.

[47] C. Papadimitriou and K. Stieglitz. *Combinatorial Optimization Algorithms and Complexity*. Prentice Hall, 1982.

[48] S. Roweis and L. Saul. Nonlinear dimensionality reduction by locally linear embedding. *Sciene*, 290(5500):2323–2326, 2000.

[49] S. Savarese and L. Fei-Fei. 3D generic object categorization, localization and pose estimation. *ICCV*, 2007.

[50] F. Schaffalitzky and A. Zisserman. Multi-view matching for unordered image sets, or "how do i organize my holiday snaps?" In *ECCV (1)*, pages 414–431, 2002.

[51] C. Schmid and R. Mohr. Local grayvalue invariants for image retrieval. *TPAMI*, 19(5):530–535, 1997.

[52] G. Scott and H. Longuett-Higgins. An algorithm for associating the features of two images. *The Royal Society of London*, 1991.

[53] L. Shapiro and J. Brady. Feature-based correspondence: an eigenvector approach. *IVC*, 1992.

[54] J. Sivic, B. C. Russell, A. A. Efros, A. Zisserman, and W. T. Freeman. Discovering objects and their location in images. In *ICCV*, 2005.

[55] M. Stark and B. Schiele. How good are local features for classes of geometric objects. In *ICCV*, pages 1–8, Oct. 2007.

[56] J. Tenenbaum. Mapping a manifold of perceptual observations. In *NIPS*, volume 10, pages 682–688, 1998.

[57] J. Tenenbaum and W. T. Freeman. Separating style and content with bilinear models. *Neural Computation*, 12:1247–1283, 2000.

[58] M. Torki and A. Elgammal. One-shot multi-set non-rigid feature-spatial matching. In *CVPR*, 2010.

[59] M. Torki and A. Elgammal. Putting local features on a manifold. In *CVPR*, 2010.

[60] A. B. Torralba, K. P. Murphy, and W. T. Freeman. Sharing visual features for multiclass and multiview object detection. In *CVPR*, 2004.

[61] L. Torresani, V. Kolmogorov, and C. Rother. Feature correspondence via graph matching: Models and global optimization. *ECCV*, 2008.

[62] T. Tuytelaars and L. J. V. Gool. Wide baseline stereo matching based on local, affinely invariant regions. In *BMVC*, 2000.

[63] S. Ullman. Aligning pictorial descriptions: An approach to object recognition. *Cognition*, 1989.

[64] S. Umeyama. An eigen decomposition approach to weighted graph matching problems. *TPAMI*, 1988.

[65] R. Urtasun, D. J. Fleet, and P. Fua. 3D people tracking with Gaussian process dynamical models. In *CVPR*, pages 238–245, 2006.

[66] M. A. O. Vasilescu and D. Terzopoulos. Multilinear analysis of image ensembles: Tensorfaces. In *Proc. of ECCV, Copenhagen, Denmark*, pages 447–460, 2002.

[67] H. Wang and E. R. Hancock. Correspondence matching using kernel principal components analysis and label consistency constraints. *PR*, 2006.

[68] Z. Wang and H. Xiao. Dimension-free afne shape matching through subspace invariance. *CVPR*, 2009.

[69] M. Weber, M. Welling, and P. Perona. Unsupervised learning of models for recognition. In *ECCV (1)*, pages 18–32, 2000.

[70] K. W. Weinberger and L. K. Saul. Unsupervised learning of image manifolds by semidefinite programming. In *CVPR*, volume 2, pages 988–995, 2004.

[71] J. Willamowski, D. Arregui, G. Csurka, C. R. Dance, and L. Fan. Categorizing nine visual classes using local appearance descriptors. In *IWLAVS*, 2004.

[72] S. Xiang, F. Nie, Y. Song, C. Zhang, and C. Zhang. Embedding new data points for manifold learning via coordinate propagation. *Knowl. Inf. Syst.*, 19(2):159–184, 2009.

Chapter 11

Human Motion Analysis Applications of Manifold Learning

Ahmed Elgammal and Chan Su Lee

11.1 Introduction

The human body is an articulated object with high degrees of freedom. It moves through the three-dimensional world and such motion is constrained by body dynamics and projected by lenses to form the visual input we capture through our cameras. Therefore, the changes (deformation) in appearance (texture, contours, edges, etc.) in the visual input (image sequences) corresponding to performing certain actions, such as facial expression or gesturing, are well constrained by the 3D body structure and the dynamics of the action being performed. Such constraints are explicitly exploited to recover the body configuration and motion in model-based approaches [35, 31, 14, 74, 72, 26, 37, 83] through explicitly specifying articulated models of the body parts, joint angles, and their kinematics (or dynamics) as well as models for camera geometry and image formation. Recovering body configuration in these approaches involves searching high dimensional spaces (body configuration and geometric transformation), which is typically formulated deterministically as a nonlinear optimization problem, e.g., [71, 72], or probabilistically as a maximum likelihood problem, e.g., [83]. Such approaches achieve significant success when the search problem is constrained as in tracking context. However, initialization remains the most challenging problem, which can be partially alleviated by sampling approaches. The dimensionality of the initialization problem increases as we incorporate models for variations between individuals in physical body style, models for variations in action style, or models for clothing, etc. Partial recovery of body configuration can also be achieved through intermediate view-based representations (models) that may or may not be tied to specific body parts [20, 13, 101, 36, 6, 30, 102, 25, 84, 27]. In such a case, constancy of the local appearance of individual body parts is exploited. Alternative paradigms are appearance-based and motion-based approaches where the focus is to track and recognize human activities without full recovery of the 3D body pose [68, 63, 67, 69, 64, 86, 73, 7, 19].

Recently, there has been research for recovering body posture directly from the visual input by posing the problem as a learning problem through searching a pre-labelled database of body posture [60, 40, 81] or through learning regression models from input to output [32, 9, 76, 77, 75, 15, 70]. All these approaches pose the problem as a machine learning problem where the objective is to learn input-output mapping from input-output pairs of training data. Such approaches have great potential for solving the initialization problem for model-based vision. However, these approaches are challenged by the existence of a wide range of variability in the input domain.

Role of Manifold:

Despite the high dimensionality of the configuration space, many human motion activities lie intrinsically on low dimensional manifolds. This is true if we consider the body kinematics as well as if we consider the observed motion through image sequences. Let us consider the observed motion. For example, the shape of the human silhouette walking or performing a gesture is an example of a dynamic shape where the shape deforms over time based on the action performed. These deformations are constrained by the physical body constraints and the temporal constraints posed by the action being performed. If we consider these silhouettes through the walking cycle as points in a high dimensional visual input space, then, given the spatial and the temporal constraints, it is expected that these points will lay on a low dimensional manifold. Intuitively, the gait is a 1-dimensional manifold which is embedded in a high dimensional visual space. This was also shown in [8]. Such a manifold can be twisted, and self-intersect in such high dimensional visual space.

Similarly, the appearance of a face performing facial expressions is an example of dynamic appearance that lies on a low dimensional manifold in the visual input space. In fact if we consider certain classes of motion such as gait, or a single gesture, or a single facial expression, and if we factor out all other sources of variability, each of such motions lies on a one-dimensional manifold, i.e., a trajectory in the visual input space. Such manifolds are nonlinear and non-Euclidean.

Therefore, researchers have tried to exploit the manifold structure as a constraint in tasks such as tracking and activity recognition in an implicit way. Learning nonlinear deformation manifolds is typically performed in the visual input space or through intermediate representations. For example, Exemplar-based approaches such as [90] implicitly model nonlinear manifolds through points (exemplars) along the manifold. Such exemplars are represented in the visual input space. HMM models provide a probabilistic piecewise linear approximation which can be used to learn nonlinear manifolds as in [12] and in [9].

Although the intrinsic body configuration manifolds might be very low in dimensionality, the resulting appearance manifolds are challenging to model given various aspects that affect the appearance, such as the shape and appearance of the person performing the motion, or variation in the view point, or illumination. Such variability makes the task of learning visual manifold very challenging because we are dealing with data points that lie on multiple manifolds at the same time: body configuration manifold, view manifold, shape manifold, illumination manifold, etc.

Linear, Bilinear and Multi-linear Models:

Can we decompose the configuration using linear models? Linear models, such as PCA [34], have been widely used in appearance modeling to discover subspaces for variations. For example, PCA has been used extensively for face recognition such as in [61, 1, 17, 54] and to model the appearance and view manifolds for 3D object recognition as in [62]. Such subspace analysis can be further extended to decompose multiple orthogonal factors using bilinear models and multi-linear tensor analysis [88, 95]. The pioneering work of Tenenbaum and Freeman [88] formulated the separation of style and content using a bilinear model framework [55]. In that work, a bilinear model was used to decompose face appearance into two factors: head pose and different people as style and content interchangeably. They presented a computational framework for model fitting using SVD. Bilinear models have been used earlier in other contexts [55, 56]. In [95] multi-linear tensor analysis was used to decompose face images into orthogonal factors controlling the appearance of the face, including geometry (people), expressions, head pose, and illumination. They employed high order singular value decomposition (HOSVD) [41] to fit multi-linear models. Tensor representation of image data was used in [82] for video compression and in [94, 98] for motion analysis and synthesis. N-mode analysis of higher-order tensors was originally proposed and developed in [91, 38, 55] and others. Another extension is algebraic solution for subspace clustering through generalized-PCA [97, 96]

Figure 11.1: Twenty sample frames from a walking cycle from a side view. Each row represents half a cycle. Notice the similarity between the two half cycles. The right part shows the similarity matrix: each row and column corresponds to one sample. Darker means closer distance and brighter means larger distances. The two dark lines parallel to the diagonal show the similarity between the two half cycles.

In our case, the object is dynamic. So, can we decompose the configuration from the shape (appearance) using linear embedding? For our case, the shape temporally undergoes deformations and self-occlusion which result in the points lying on a nonlinear, twisted manifold. This can be illustrated if we consider the walking cycle in Figure 11.1. The two shapes in the middle of the two rows correspond to the farthest points in the walking cycle kinematically and are supposedly the farthest points on the manifold in terms of the geodesic distance along the manifold. In the Euclidean visual input space these two points are very close to each other as can be noticed from the distance plot on the right of Figure 11.1. Because of such nonlinearity, PCA will not be able to discover the underlying manifold. Simply, linear models will not be able to interpolate intermediate poses. For the same reason, multidimensional scaling (MDS) [18] also fails to recover such a manifold.

Nonlinear Dimensionality Reduction and Decomposition of Orthogonal Factors:

Recently some promising frameworks for nonlinear dimensionality reduction have been introduced, e.g., [87, 79, 2, 11, 43, 100, 59]. Such approaches can achieve embedding of nonlinear manifolds through changing the metric from the original space to the embedding space based on local structure of the manifold. While there are various such approaches, they mainly fall into two categories: Spectral-embedding approaches and Statistical approaches. Spectral embedding includes approaches such as isometric feature mapping (Isomap) [87], local linear embedding (LLE) [79], Laplacian eigenmaps [2], and manifold charting [11]. Spectral-embedding approaches, in general, construct an affinity matrix between data points using data dependent kernels, which reflect local manifold structure. Embedding is then achieved through solving an eigenvalue problem on such a matrix. It was shown in [3, 29] that these approaches are all instances of kernel-based learning, in particular kernel principle component analysis (KPCA) [80]. In [4] there is an approach for embedding out-of-sample points to complement such approaches. Along the same line, our work [24, 21] introduced a general framework for mapping between input and embedding spaces.

All these nonlinear embedding frameworks were shown to be able to embed nonlinear manifolds into low-dimensional Euclidean spaces for toy examples as well as for real images. Such approaches are able to embed image ensembles nonlinearly into low dimensional spaces where various orthogonal perceptual aspects can be shown to correspond to certain directions or clusters in the embedding spaces. In this sense, such nonlinear dimensionality reduction frameworks present an alternative solution to the decomposition problems. However, the application of such approaches is limited to embedding of a single manifold.

Biological Motivation:

While the role of manifold representations is still unclear in perception, it is clear that images of the same objects lie on a low dimensional manifold in the visual space defined by the retinal array. On the other hand, neurophysiologists have found that neural population activity firing is typically a function of a small number of variables, which implies that population activity also lies on low dimensional manifolds [33].

11.2 Learning a Simple Motion Manifold

11.2.1 Case Study: The Gait Manifold

In order to achieve a low dimensional embedding of the gait manifold, nonlinear dimensionality reduction techniques such as LLE [79], Isomap [87], and others can be used. Most of these techniques result in qualitatively similar manifold embedding. As a result of nonlinear dimensionality reduction we can reach an embedding of the gait manifold in a low dimension Euclidean space [21]. Figure 11.2 illustrates the resulting embedded manifold for a side view of the walker.[1] Figure 11.3 illustrates the embedded manifolds for five different view points of the walker. For a given view point, the walking cycle evolves along a closed curve in the embedded space, i.e., only one degree of freedom controls the walking cycle which corresponds to the constrained body pose as a function of the time. Such conclusion is conforming with the intuition that the gait manifold is one dimensional.

One important question is what is the least dimensional embedding space we can use to embed the walking cycle in a way that discriminates different poses through the whole cycle. The answer depends on the view point. The manifold twists in the embedding space given the different view points which impose different self occlusions. The least twisted manifold is the manifold for the back view as this is the least self occluding view (left most manifold in Figure 11.3). In this case the manifold can be embedded in a two dimensional space. For other views the curve starts to twist to be a three dimensional space curve. This is primarily because of the similarity imposed by the view point which attracts far away points on the manifold closer. The ultimate twist happens in the side view manifold where the curve twists to be a figure eight shape where each cycle of the eight (half eight) lies in a different plane. Each half of the "eight" figure corresponds to half a walking cycle. The cross point represents the body pose where it is totally ambiguous from the side view to determine from the shape of the contour which leg is in front as can be noticed in Figure 11.2. Therefore, in a side view, three-dimensional embedding space is the least we can use to discriminate different poses. Embedding a side view cycle in a two-dimensional embedding space results in an embedding similar to that shown in the top left of Figure 11.2 where the two half cycles lie over each other. Different people are expected to have different manifolds. However, such manifolds are all topologically equivalent. This can be noticed in Figure 11.4c. Such property will be exploited later in the chapter to learn unified representations from multiple manifolds.

[1]The data used are from the CMU Mobo gait data set which contains 25 people from six different view points. We used data sets of walking people from multiple views. Each data set consists of 300 frames and each contains about 8 to 11 walking cycles of the same person from certain view points. The walkers were using a treadmill which might result in different dynamics from the natural walking.

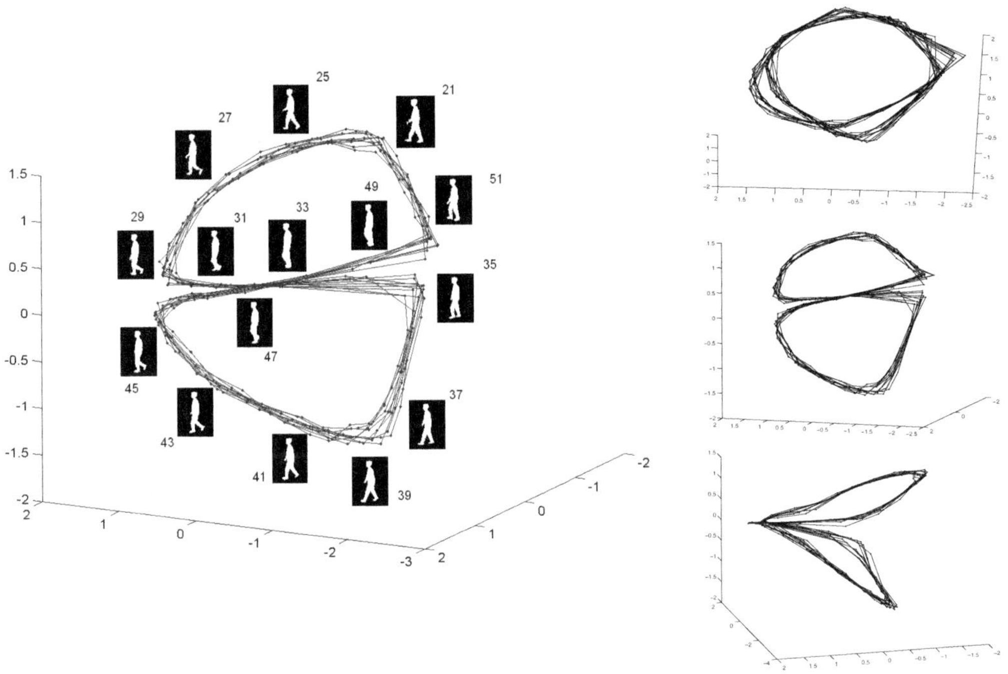

Figure 11.2: Embedded gait manifold for a side view of the walker. Left: sample frames from a walking cycle along the manifold with the frame numbers shown to indicate the order. Ten walking cycles are shown. Right: three different views of the manifold.

Figure 11.3: Embedded manifolds for different views of the walkers. Frontal view manifold is the rightmost one and back view manifold is the leftmost one. We choose the view of the manifold that best illustrates its shape in the 3D embedding space.

11.2.2 Learning the Visual Manifold: Generative Model

Given that we can achieve a low dimensional embedding of the visual manifold of dynamic shape data, such as the gait data shown above, the question is how to use this embedding to learn representations of moving (dynamic) objects that support tasks such as synthesis, pose recovery, reconstruction, and tracking. In the simplest form, assuming no other source of variability besides the intrinsic motion, we can think of a view-based generative model of the form

$$y_t = T_\alpha \gamma(x_t; a) \tag{11.1}$$

where the shape (appearance), y_t, at time t is an instance driven from a generative model where the function γ is a mapping function that maps body configuration x_t at time t into the image space. The body configuration x_t is constrained to the explicitly modeled motion manifold. i.e., the mapping function γ maps from a representation of the body configuration space into the image space given mapping parameters a that are independent from the configuration. T_α represents a global geometric transformation on the appearance instance.

The manifold in the embedding space can be modeled explicitly in a function form or implicitly by points along the embedded manifold (embedded exemplars). The embedded manifold can also be modelled probabilistically using Hidden Markov Models and EM. Clearly, learning manifold representations in a low-dimensional embedding space is advantageous over learning them in the visual input space. However, our emphasis is on learning the mapping between the embedding space and the visual input space.

Since the objective is to recover body configuration from the input, it might be obvious that we need to learn mapping from the input space to the embedding space, i.e., mapping from R^d to R^e. However, learning such mapping is not feasible since the visual input is very high-dimensional, so learning such mapping will require a large number of samples in order to be able to interpolate. Instead, we learn the mapping from the embedding space to the visual input space, i.e., in a generative manner, with a mechanism to directly solve for the inverse mapping. Another fundamental reason to learn the mapping in this direction is the inherent ambiguity in 2D data. Therefore, mapping from visual data to the manifold representation is not necessarily a function, while learning a mapping from the manifold to the visual data is a function.

It is well known that learning a smooth mapping from examples is an ill-posed problem unless the mapping is constrained, since the mapping will be undefined in other parts of the space [66]. We argue that explicit modeling of the visual manifold represents a way to constrain any mapping between the visual input and any other space. Nonlinear embedding of the manifold, as was discussed in the previous section, represents a general framework to achieve this task. Constraining the mapping to the manifold is essential if we consider the existence of outliers (spatial and/or temporal) in the input space. This also facilitates learning mappings that can be used for interpolation between poses as we shall show. In what follows we explain our framework to recover the pose. In order to learn such nonlinear mapping, we use radial basis function (RBF) interpolation framework. The use of RBF for image synthesis and analysis has been pioneered by [66, 5] where RBF networks were used to learn nonlinear mappings between image space and a supervised parameter space. In our work we use RBF interpolation framework in a novel way to learn mapping from unsupervised learned parameter space to the input space. Radial basis functions interpolation provides a framework for both implicitly modeling the embedded manifold as well as learning a mapping between the embedding space and the visual input space. In this case, the manifold is represented in the embedding space implicitly by selecting a set of representative points along the manifold as the centers for the basis functions.

Let the set of representative input instances (shape or appearance) be $\mathsf{Y} = \{y_i \in R^d \quad i =$

$1, \cdots, N\}$ and let their corresponding points in the embedding space be $\mathsf{X} = \{x_i \in R^e, \; i = 1, \cdots, N\}$ where e is the dimensionality of the embedding space (e.g. $e = 3$ in the case of gait). We can solve for multiple interpolants $f^k : R^e \rightarrow R$ where k is k-th dimension (pixel) in the input space and f^k is a radial basis function interpolant, i.e., we learn nonlinear mappings from the embedding space to each individual pixel in the input space. Of particular interest are functions of the form

$$f^k(x) = p^k(x) + \sum_{i=1}^{N} w_i^k \phi(|x - x_i|), \tag{11.2}$$

where $\phi(\cdot)$ is a real-valued basic function, w_i are real coefficients, and $|\cdot|$ is the norm on R^e (the embedding space). Typical choices for the basis function includes thin-plate spline ($\phi(u) = u^2 log(u)$), the multiquadric ($\phi(u) = \sqrt{(u^2 + c^2)}$), Gaussian ($\phi(u) = e^{-cu^2}$), biharmonic ($\phi(u) = u$), and triharmonic ($\phi(u) = u^3$) splines. p^k is a linear polynomial with coefficients c^k, i.e., $p^k(x) = [1 \;\; x^\top] \cdot c^k$. This linear polynomial is essential to achieve an approximate solution for the inverse mapping as will be shown.

The whole mapping can be written in a matrix form as

$$f(x) = B \cdot \psi(x), \tag{11.3}$$

where B is a $d \times (N{+}e{+}1)$ dimensional matrix with the k-th row $[w_1^k \cdots w_N^k \;\; c^{k^T}]$ and the vector $\psi(x)$ is $[\phi(|x - x_1|) \cdots \phi(|x - x_N|) \;\; 1 \;\; x^\top]^\top$. The matrix B represents the coefficients for d different nonlinear mappings, each from a low-dimension embedding space into real numbers.

To insure orthogonality and to make the problem well posed, the following additional constraints are imposed

$$\sum_{i=1}^{N} w_i p_j(x_i) = 0, j = 1, \cdots, m \tag{11.4}$$

where p_j are the linear basis of p. Therefore the solution for B can be obtained by directly solving the linear systems

$$\begin{pmatrix} A & P \\ P^\top & 0 \end{pmatrix} B^\top = \begin{pmatrix} Y \\ 0_{(e+1)\times d} \end{pmatrix}, \tag{11.5}$$

where $A_{ij} = \phi(|x_j - x_i|), \;\; i, j = 1 \cdots N$, P is a matrix with i-th row $[1 \;\; x_i^\top]$, and Y is $(N \times d)$ matrix containing the representative input images, i.e., $Y = [y_1 \cdots y_N]^\top$. Solution for B is guaranteed under certain conditions on the basic functions used. Similarly, mapping can be learned using arbitrary centers in the embedding space (not necessarily at data points) [66, 21].

Given such mapping, any input is represented by a linear combination of nonlinear functions centered in the embedding space along the manifold. Equivalently, this can be interpreted as a form of basis images (coefficients) that are combined nonlinearly using kernel functions centered along the embedded manifold.

11.2.3 Solving for the Embedding Coordinates

Given a new input $y \in R^d$, it is required to find the corresponding embedding coordinates $x \in R^e$ by solving for the inverse mapping. There are two questions that we might need to answer:

1. What are the coordinates of point $x \in R^e$ in the embedding space corressponding to such input?

2. What is the closest point on the embedded manifold corresponding to such input?

In both cases we need to obtain a solution for

$$x^* = \underset{x}{\operatorname{argmin}} ||y - B\psi(x)|| \tag{11.6}$$

where for the second question the answer is constrained to be on the embedded manifold. In the cases where the manifold is only one dimensional (for example, in the gait case, as will be shown), only one dimensional search is sufficient to recover the manifold point closest to the input. However, we show here how to obtain a closed-form solution for x^*.

Each input yields a set of d nonlinear equations in e unknowns (or d nonlinear equations in one e-dimensional unknown). Therefore, a solution for x^* can be obtained by least square solution for the over-constrained nonlinear system in 11.6. However, because of the linear polynomial part in the interpolation function, the vector $\psi(x)$ has a special form that facilitates a closed-form least square linear approximation and therefore, avoids solving the nonlinear system. This can be achieved by obtaining the pseudo-inverse of B. Note that B has rank N since N distinctive RBF centers are used. Therefore, the pseudo-inverse can be obtained by decomposing B using SVD such that $B = USV^\top$ and, therefore, vector $\psi(x)$ can be recovered simply as

$$\psi(x) = V\tilde{S}U^T y \tag{11.7}$$

where $\tilde{S}$ is the diagonal matrix obtained by taking the inverse of the nonzero singular values in S as the diagonal matrix and setting the rest to zeros. Linear approximation for the embedding coordinate x can be obtained by taking the last e rows in the recovered vector $\psi(x)$. Reconstruction can be achieved by re-mapping the projected point.

11.2.4 Synthesis, Recovery, and Reconstruction

Given the learned model, we can synthesize new shapes along the manifold. Figure 11.4-c shows an example of shape synthesis and interpolation. Given a learned generative model in the form of Equation 11.3, we can synthesize new shapes through the walking cycle. In these examples only 10 samples were used to embed the manifold for half a cycle on a unit circle in 2D and to learn the model. Silhouettes at intermediate body configurations were synthesized (at the middle point between each two centers) using the learned model. The learned model can successfully interpolate shapes at intermediate configurations (never seen in the learning) using only two-dimensional embedding. The figure shows results for three different people.

Given a visual input (silhouette), and the learned model, we can recover the intrinsic body configuration, recover the view point, reconstruct the input, and detect any spatial or temporal outliers. In other words, we can simultaneously solve for the pose and view point, and reconstruct the input. A block diagram for recovering 3D pose and view point given learned manifold models is shown in Figure 11.4. The framework [23] is based on learning three components as shown in Figure 11.4a:

1. Learning Manifold Representation: using nonlinear dimensionality reduction we achieve an embedding of the global deformation manifold that preserves the geometric structure of the manifold as described in Section 11.2.1. Given such embedding, the following two nonlinear mappings are learned:

2. Manifold-to-input mapping: a nonlinear mapping from the embedding space into visual input space as described in Section 11.2.2.

3. Manifold-to-pose: a nonlinear mapping from the embedding space into the 3D body pose space.

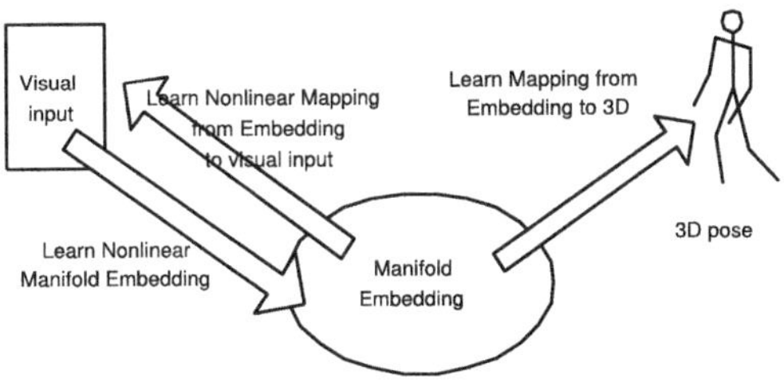

(a) Learning components

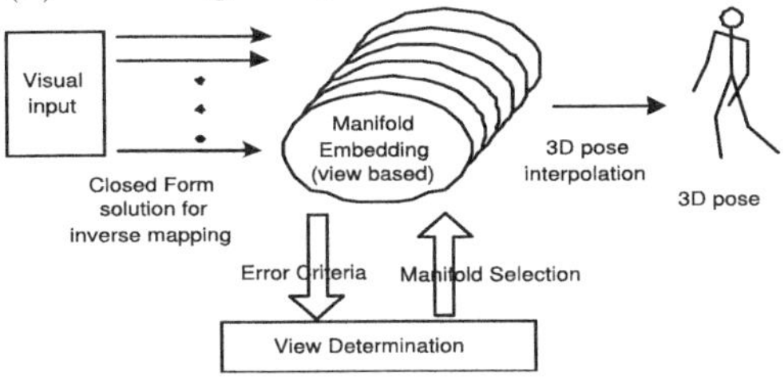

(b) Pose estimation

(c) Synthesis

Figure 11.4: (a, b) Block diagram for the learning framework and 3D pose estimation. (c) Shape synthesis for three different people. First, third, and fifth rows: samples used in learning. Second, fourth, and sixth rows: interpolated shapes at intermediate configurations (never seen in the learning).

Given an input shape, the embedding coordinate, i.e., the body configuration can be recovered in closed-form as was shown in Section 11.2.3. Therefore, the model can be used for pose recovery as well as reconstruction of noisy inputs. Figure 11.5 shows examples of the reconstruction given corrupted silhouettes as input. In this example, the manifold representation and the mapping were learned from one person's data and tested on other people's data. Given a corrupted input, after solving for the global geometric transformation, the input is projected to the embedding space using the closed-form inverse mapping approximation in Section 11.2.3. The nearest embedded manifold point represents the intrinsic body configuration. A reconstruction of the input can be achieved by projecting back to the input space using the direct mapping in Equation 11.3. As can be noticed from the figure, the reconstructed silhouettes preserve the correct body pose in each case, which shows that solving for the inverse mapping yields correct points on the manifold. Notice that no mapping is learned from the input space to the embedded space. Figure 11.6 shows examples of 3D pose recovery obtained in closed-form for different people from different view points. The training has be done using only one subject's data from five view points. All the results in Figure 11.6 are for subjects not used in the training. This shows that the model generalized very well.

Figure 11.5: Example of pose-preserving reconstruction results. Six noisy and corrupted silhouettes and their reconstructions next to them.

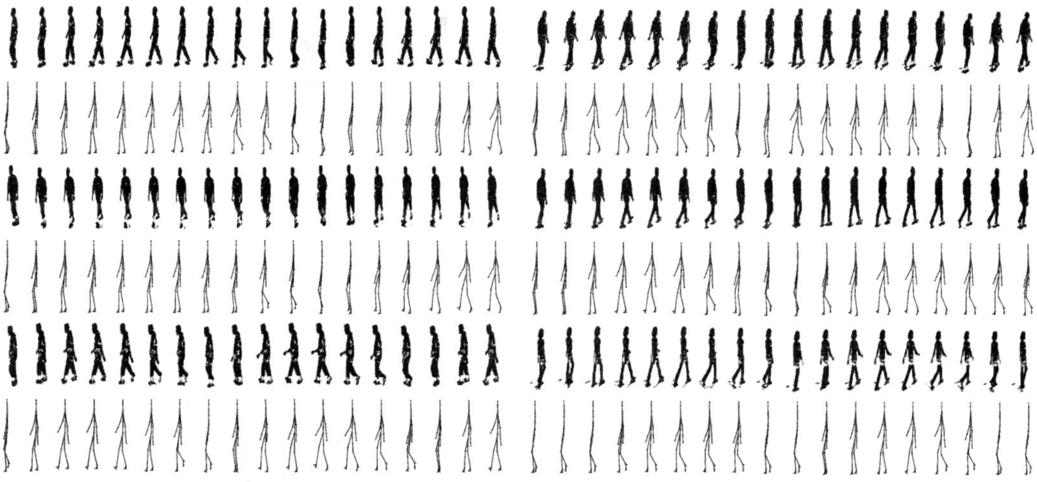

Figure 11.6: 3D reconstruction for 4 people from different views: person 70 views 1,2; person 86 views 1,2; person 76 view 4; person 79 view 4.

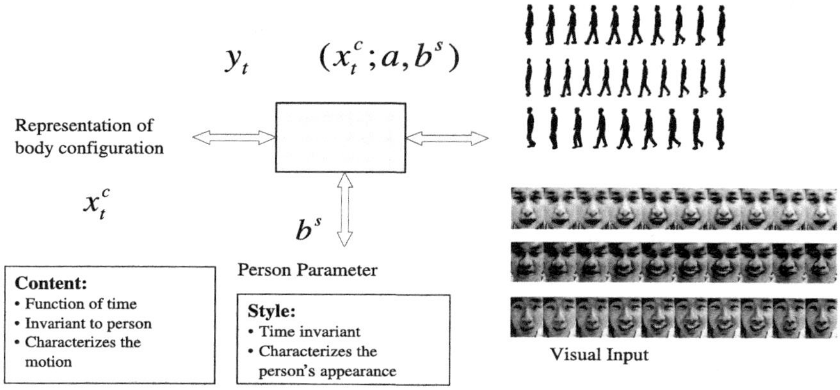

Figure 11.7: Style and content factors: Content: gait motion or facial expression. Style: different silhouette shapes or facial appearance.

11.3 Factorized Generative Models

The generative model introduced in Equation 11.8 generates the visual input as a function of a latent variable representing the body configuration constrained to a motion manifold. Obviously body configuration is not the only factor controlling the visual appearance of humans in images. Any input image is a function of many aspects such as the person's body structure, appearance, view point, illumination, etc. Therefore, it is obvious that the visual manifolds of different people performing the same activity will be different. So, how to handle all these variabilities?

To illustrate that point, we start with a single 'style' factor model and then move to the general case. Given a set of image sequences, similar to the ones in Figure 11.7, representing a motion such as gesture, facial expression, or activity, where each sequence is performed by one subject, we aim to learn a generative model that explicitly factorizes the following two factors:

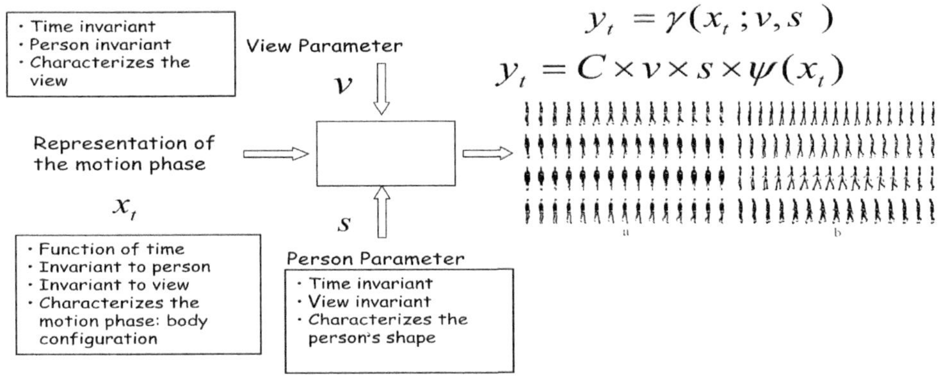

Figure 11.8: Multiple views and multiple people generative model for gait. a) Examples of training data from different views. b) Examples of training data for multiple people from the side view.

1. Content (body pose): A representation of the intrinsic body configuration through the motion as a function of time that is invariant to the person, i.e., the content characterizes the motion or the activity.

2. Style (people): A time-invariant person variable that characterizes the person's appearance or shape.

Figure 11.7 shows an example of such data where different people are performing the same activity, e.g., gait or smile motion. The content in this case is the gait motion or the smile motion, while the style is a person's shape or face appearance, respectively. On the other hand, given an observation of a certain person at a certain body pose and given the learned generative model, we aim to solve for both the body configuration representation (content) and the person's shape parameter (style).

In general, the appearance of a dynamic object is a function of the intrinsic body configuration as well as other factors such as the object appearance, the viewpoint, illumination, etc. We refer to the intrinsic body configuration as the content and all other factors as style factors. Since the combined appearance manifold is very challenging to model, given all these factors, the solution we use here utilizes the fact that the underlying motion manifold, independent of all other factors, is low in dimensionality. Therefore, the motion manifold can be explicitly modeled, while all the other factors are approximated with a subspace model. For example, for the data in Figure 11.7, we do not know the dimensionality of the shape manifold of all people, while we know that the gait is a one-dimensional manifold motion.

We describe the model for the general case of factorizing multiple style factors given a content manifold. Let $\boldsymbol{y}_t \in \mathbb{R}^d$ be the appearance of the object at time instance t, represented as a point in a d-dimensional space. This instance of the appearance is driven from a generative model in the form

$$\boldsymbol{y}_t = \gamma(\boldsymbol{x}_t, \boldsymbol{b}_1, \boldsymbol{b}_2, \cdots, \boldsymbol{b}_r; \boldsymbol{a}) \tag{11.8}$$

where the function $\gamma(\cdot)$ is a mapping function that maps from a representation of body configuration, $\boldsymbol{x}_t$ (content), at time t into the image space given variables $\boldsymbol{b}_1, \cdots, \boldsymbol{b}_r$, each representing a style factor. Such factors are conceptually orthogonal and independent of the body configuration and can be time variant or invariant. $\boldsymbol{a}$ represents the model parameters.

Suppose that we can learn a unified, style-invariant, embedded representation of the motion manifold (content) $\mathcal{M}$ in a low-dimensional Euclidean embedding space, $\mathbb{R}^e$, then we can learn a set of style-dependent nonlinear mapping functions from the embedding space into the input space, i.e., functions $\gamma_s(\boldsymbol{x}_t) : \mathbb{R}^e \rightarrow \mathbb{R}^d$ that map from embedding space with dimensionality e into the input space (observation) with dimensionality d for each style setting s. Here, a style setting is a discrete combination of style values. As described in Section 11.2.2, each such function admits a representation in the form of linear combination of basis functions [39] and can be written as

$$\boldsymbol{y}_t = \gamma_s(\boldsymbol{x}_t^c) = \boldsymbol{C}^s \cdot \psi(\boldsymbol{x}_t^c) , \tag{11.9}$$

where $\boldsymbol{C}^s$ is a $d \times N_\psi$ linear mapping and $\psi(\cdot) : \mathbb{R}^e \rightarrow \mathbb{R}^{N_\psi}$ is a nonlinear kernel map from a representation of the body configuration to a kernel induced space with dimensionality N_ψ. In the mapping in Equation 11.9 the style variability is encoded in the coefficient matrix $\boldsymbol{C}^s$. Therefore, given the style-dependent functions in the form of Equation 11.9, the style variables can be factorized in the linear mapping coefficient space using multilinear analysis of the coefficients' tensor. Therefore, the general form for the mapping function $\gamma(\cdot)$ that we use is

$$\gamma(\boldsymbol{x}_t, \boldsymbol{b}_1, \boldsymbol{b}_2, \cdots, \boldsymbol{b}_r; \boldsymbol{a}) = \boldsymbol{\mathcal{A}} \times_1 \boldsymbol{b}_1 \times \cdots \times_r \boldsymbol{b}_r \cdot \psi(\boldsymbol{x}_t) \tag{11.10}$$

where each $\boldsymbol{b}_i \in \mathbb{R}^{n_i}$ is a vector representing a parameterization of the ith style factor. $\boldsymbol{\mathcal{A}}$ is a core tensor of order $r + 2$ and of dimensionality $d \times n_1 \times \cdots \times n_r \times N_\psi$. The product operator $\times_i$ is *mode-i* tensor product as defined in [41].

The model in Equation 11.10 can be seen as a hybrid model that uses a mix of nonlinear and multilinear factors. In the model in Equation 11.10, the relation between body configuration and the input is nonlinear where other factors are approximated linearly through high-order tensor analysis. The use of nonlinear mapping is essential since the embedding of the configuration manifold is nonlinearly related to the input. The main motivation behind the hybrid model is: The motion itself lies on a low dimensional manifold, which can be explicitly modeled, while for other style factors it might not be possible to model them explicitly using nonlinear manifolds. For example, the shapes of different people might lie on a manifold; however, we do not know the dimensionality of the shape manifold and/or we might not have enough data to model such a manifold. The best choice is to represent it as a subspace. Therefore, the model in Equation 11.10 gives a tool that combines manifold-based models, where manifolds are explicitly embedded, with subspace models for style factors if no better models are available for such factors. The framework also allows modeling any style factor on a manifold in its corresponding subspace, since the data can lie naturally on a manifold in that subspace. This feature of the model was further developed in [50], where the viewpoint manifold of a given motion was explicitly modeled in the subspace defined by the factorization above.

In the following, we show some examples of the model in the context of human motion analysis with different roles of the style factors. In the following sections we describe the details for fitting such models and estimation of the parameters. Section 11.4.1 describes different ways to obtain a unified nonlinear embedding of the motion manifold for style analysis. Sections 11.4 describes learning the model. Section 11.5 describes using the model for solving for multiple factors.

11.3.1 Example 1: A Single Style Factor Model

Here we give an example of the model with a single style factor. Figure 11.7 shows an example of such data, where different people are performing the same activity as gait or smile motion. The content in this case is the gait motion or the smile motion, while the style is the person's shape or face appearance, respectively. The style is a time-invariant variable in this case. For this case, the generative model in Equation 11.10 reduces to a model in the form

$$\boldsymbol{y}_t = \gamma(\boldsymbol{x}_t^c, \boldsymbol{b}^s; \boldsymbol{a}) = \boldsymbol{\mathcal{A}} \times_2 \boldsymbol{b}^s \times_3 \psi(\boldsymbol{x}_t^c) \,, \tag{11.11}$$

where the image, $\boldsymbol{y}_t$, at time t is a function of body configuration $\boldsymbol{x}_t^c$ (content) at time t and style variable $\boldsymbol{b}^s$ that is time invariant. In this case the content is a continuous domain while style is represented by the discrete style classes that exist in the training data where we can interpolate intermediate styles and/or intermediate contents. The model parameter is the core tensor, $\boldsymbol{\mathcal{A}}$, which is a third order tensor (3-way array) with dimensionality $d \times J \times N_\psi$, where J is the dimensionality of the style vector $\boldsymbol{b}^s$, which is the subspace of the different people shapes factored out in the space of the style dependent functions in Equation 11.9.

11.3.2 Example 2: Multifactor Gait Model

As an example of a two style factor models, we consider the gait case, with multiple views and multiple people (as shown in Figure 11.8). The data set has three components: personalized shape style and view, in addition to body configuration. A generative model for walking silhouettes for different people from different view points will be in the form

$$\boldsymbol{y}_t = \gamma(\boldsymbol{x}_t, \boldsymbol{v}_t, \boldsymbol{s}; a) = \boldsymbol{\mathcal{A}} \times \boldsymbol{v}_t \times \boldsymbol{s} \times \psi(\boldsymbol{x}_t) \tag{11.12}$$

where $\boldsymbol{v}_t$ is a parameterization of the view, which is independent of the body configuration but can change over time, and also independent of the person's shape. $\boldsymbol{s}$ is a time-invariant parameterization of the shape style of the person performing the walk, independent of the body configuration and the view point. The body configuration $\boldsymbol{x}_t$ evolves along a representation of the gait manifold. In such case the tensor $\boldsymbol{\mathcal{A}}$ is a 4th-order tensor with dimensionality $d \times n_v \times n_s \times N_\psi$, where n_v is the dimensionality of the view subspace and n_s is the dimensionality of the shape style subspace.

11.3.3 Example 3: Multifactor Facial Expressions

Another example is modeling the manifolds of facial expression motions. Given dynamic facial expressions, such as sad, surprised, happy, etc., where each expression starts from a neutral pose and evolves to a peak expression, each of these motions evolves along a one-dimensional manifold. However, the manifold will be different for each person and for each expression. Therefore, we can use a generative model to generate different people's faces and different expressions using a model in the form

$$\boldsymbol{y}_t = \gamma(\boldsymbol{x}_t, \boldsymbol{e}, \boldsymbol{f}; a) = \boldsymbol{\mathcal{A}} \times \boldsymbol{e} \times \boldsymbol{f} \times \psi(\boldsymbol{x}_t) \tag{11.13}$$

where $\boldsymbol{e}$ is an expression vector (happy, sad, etc.) that is time-invariant and person-invariant, i.e., it only describes the expression type. Similarly, $\boldsymbol{f}$ is a face vector describing a person's face appearance which is time-invariant and expression-invariant. The motion content is described by $\boldsymbol{x}_t$ which denotes the motion phase of the expression, i.e., starts from neutral and evolves to a peak expression depending on the expression vector, $\boldsymbol{e}$.

11.4 Generalized Style Factorization

11.4.1 Style-Invariant Embedding

To achieve the decomposition described in the previous section, the challenge is to learn a unified and style-invariant embedded representation of the motion manifold. Several approaches can be used to achieve such a representation.

- *Unsupervised Individual Manifold Embedding and Warping:* Nonlinear dimensionality reduction can be used to obtain an embedding of each individual style-dependent manifold, as described in Section 11.2.1, and then a mean manifold can be computed as a unified representation through nonlinear warping of the manifold points. Such an approach was introduced in [24].

- *Conceptual Manifold Embedding:* In contrast to unsupervised learning of the content manifold representation described above, if the topology of the manifold is known, a conceptual topologically-equivalent representation of the manifold can be directly used. By topologically-equivalent, we mean equivalent to our notion of the underlying motion manifold. For example, since the gait is a one-dimensional closed manifold embedded in the input space, we can think of it as a unit circle twisted and stretched in the space based on the shape and the appearance of the person under consideration or based on the view. In general, all closed 1D manifolds are topologically homeomorphic to a unit circle. So we can use a unit circle as a unified representation of all gait cycles for all people for all views. The actual data is a deformed version of that conceptual manifold representation, where such deformation can be captured through the nonlinear mapping in Equation 11.9 in a generative way. The idea of conceptual manifold embedding was introduced in [22] to model image translations and rotations for tracking. It was also used in [45] to model different facial expression manifolds for different people, as will be described later. In [52] a conceptual torus manifold was used to model the joint motion and viewpoint manifold.

11.4.2 Style Factorization

To fit the model in Equation 11.10 we need image sequences at each combination of style factors, all representing the same motion (content). The input sequences do not have to be of the same length. Each style factor is represented by a set of discrete samples in the training data, i.e., a set of discrete views, discrete shape styles, discrete expressions, etc. We denote the set of discrete samples for the ith style factor by B_i and the number of these samples by $N_i = |B_i|$. A certain combination of style factors is denoted by an r-tuple $s \in B_1 \times \cdots \times B_r$. We call such a tuple "*a style setting.*" Overall, the training data needed to fit the model is of size $N_1 \times \cdots \times N_r$ sequences.

Given learned nonlinear mapping coefficients $\boldsymbol{C}^s$ for all style settings $s \in B_1 \times \cdots \times B_r$, in the form of Equation 11.9, the style parameters can be factorized by fitting a multilinear model [41, 95] to the coefficients' tensor. Higher-order tensor analysis decomposes multiple orthogonal factors as an extension of principal component analysis (PCA) (one factor), and bilinear model (two orthogonal factors). Singular value decomposition (SVD) can be used for PCA analysis and iterative SVD with *vector transpose* for bilinear analysis [88]. Higher-order tensor analysis can be achieved by higher-order singular value decomposition (HOSVD) with *matrix unfolding*, which is a generalization of SVD [41][2]

Each of the coefficient matrices $\boldsymbol{C}^s$, with dimensionality $d \times N_\psi$ can be represented as a coefficient vector $\boldsymbol{c}^s$ by column stacking (stacking its columns above each other to form a vector). Therefore, $\boldsymbol{c}^s$ is an $N_c = d \cdot N_\psi$ dimensional vector. All the coefficient vectors can then be arranged in an order $r+1$ coefficient tensor $\mathcal{C}$ with dimensionality $N_c \times N_1 \times \cdots \times N_r$. The coefficient tensor is then factorized using HOSVD as

$$\boldsymbol{\mathcal{C}} = \tilde{\boldsymbol{\mathcal{D}}} \times_1 \tilde{\boldsymbol{B}}_1 \times_2 \tilde{\boldsymbol{B}}_2 \times \cdots \times_r \tilde{\boldsymbol{B}}_r \times_{r+1} \tilde{\boldsymbol{F}} \ ,$$

[2]Matrix unfolding is an operation to reshape high order tensor array into matrix form. Given an r-order tensor $\mathcal{A}$ with dimensions $N_1 \times N_2 \times \cdots \times N_r$, the mode-$n$ matrix unfolding, denoted by $\boldsymbol{A}_{(n)} = unfolding(\boldsymbol{\mathcal{A}}, n)$, is flattening $\mathcal{A}$ into a matrix whose column vectors are the mode-n vectors [41]. Therefore, the dimension of the unfolded matrix $\boldsymbol{A}_{(n)}$ is $N_n \times (N_1 \times N_2 \times \cdots N_{n-1} \times N_{n+1} \times \cdots N_r)$.

where $\tilde{\boldsymbol{B}}_i$ is the mode-i basis of $\mathcal{C}$, which represents the orthogonal basis for the space for the i-th style factor. $\tilde{\boldsymbol{F}}$ represents the basis for the mapping coefficient space. The dimensionality of each of the $\tilde{\boldsymbol{B}}_i$ matrices is $N_i \times N_i$. The dimensionality of the matrix $\tilde{\boldsymbol{F}}$ is $N_c \times N_c$. $\mathcal{D}$ is a core tensor, with dimensionality $N_1 \times \cdots \times N_r \times N_c$, which governs the interactions (the correlation) among the different mode basis matrices.

Similar to PCA, it is desired to reduce the dimensionality for each of the orthogonal spaces to retain a subspace representation. This can be achieved by applying higher-order orthogonal iteration for dimensionality reduction [42]. The reduced subspace representation is

$$\mathcal{C} = \mathcal{D} \times_1 \boldsymbol{B}_1 \times \cdots \times_r \boldsymbol{B}_r \times_{r+1} \boldsymbol{F} \,, \tag{11.14}$$

where the reduced dimensionality for $\mathcal{D}$ is $n_1 \times \cdots \times n_r \times n_c$, for $\boldsymbol{B}_i$ is $N_i \times n_i$, and for $\boldsymbol{F}$ is $N_c \times n_c$, where $n_1, \cdots, n_r$, and n_c are the number of basis retained for each factor respectively. Since the basis for the mapping coefficients, $\boldsymbol{F}$, is not used in the analysis, we can combine it with the core tensor using tensor multiplication to obtain coefficient eigenmodes, which is a new core tensor formed by $\mathcal{Z} = \mathcal{D} \times_{r+1} \boldsymbol{F}$ with dimensionality $n_1 \times \cdots \times n_r \times N_c$. Therefore, Equation 11.14 can be rewritten as

$$\mathcal{C} = \mathcal{Z} \times_1 \boldsymbol{B}_1 \times \cdots \times_r \boldsymbol{B}_r \,. \tag{11.15}$$

The columns of the matrices $\boldsymbol{B}_1, \cdots, \boldsymbol{B}_r$ represent orthogonal basis for each style factor's subspace, respectively. Any style setting s can be represented by a set of style vectors $\boldsymbol{b}_1 \in \mathbb{R}^{n_1}, \cdots, \boldsymbol{b}_r \in \mathbb{R}^{n_r}$ for each of the style factors. The corresponding coefficient matrix $\boldsymbol{C}$ can then be generated by unstacking the vector $\boldsymbol{c}$ obtained by the tensor product

$$\boldsymbol{c} = \mathcal{Z} \times_1 \boldsymbol{b}_1 \times \cdots \times_r \boldsymbol{b}_r.$$

Therefore, we can generate any specific instant of the motion by specifying the body configuration parameter $\boldsymbol{x}_t$ through the kernel map defined in Equation 11.3. The whole model for generating image $\boldsymbol{y}_t^s$ can be expressed as

$$\boldsymbol{y}_t^s = unstacking(\mathcal{Z} \times_1 \boldsymbol{b}_1 \times \cdots \times_r \boldsymbol{b}_r) \cdots \psi(\boldsymbol{x}_t).$$

This can be expressed abstractly also by arranging the tensor $\mathcal{Z}$ into an order $r+2$ tensor $\mathcal{A}$ with dimensionality $d \times n_1 \times \cdots \times n_r \times N_\psi$. The results in the factorization in the form of Equation 11.10, i.e.,

$$\boldsymbol{y}_t^s = \mathcal{A} \times_1 \boldsymbol{b}_1 \times \cdots \times_r \boldsymbol{b}_r \cdots \psi(\boldsymbol{x}_t) \,.$$

11.5 Solving for Multiple Factors

Given a multi-factor model fitted as described in the previous section and given a new image or a sequence of images, it is desired to efficiently solve for each of the style factors, as well as the body configuration. We discriminate here between two cases: 1: *The input is a whole motion cycle.* 2: *The input is a single image.* For the first case, since we have a whole motion manifold, we can obtain a closed form analytical solution for each of factors by aligning the input sequence manifold to the model conceptual manifold representation. For the second case, we introduce an iterative solution.

Solving for Style Factors Given a Whole Sequence

Given a sequence of images representing a whole motion cycle, we can solve for the different style factors iteratively. First the sequence is embedded and aligned to the embedded content manifold. Then, the mapping coefficient matrix $\boldsymbol{C}$ is learned from the aligned embedding to the input. Given such coefficients, we need to find the optimal $\boldsymbol{b}_1, \cdots, \boldsymbol{b}_r$ factors which can generate such coefficients, i.e., minimizes the error

$$E(\boldsymbol{b}_1, \cdots, \boldsymbol{b}_r) = \|\boldsymbol{c} - \boldsymbol{\mathcal{Z}} \times_1 \boldsymbol{b}_1 \times_2 \cdots \times_r \boldsymbol{b}_r\| \tag{11.16}$$

where $\boldsymbol{c}$ is the column stacking of $\boldsymbol{C}$. If all the style vectors are known except the ith factor's vector, then we can obtain a closed-form solution for $\boldsymbol{b}_i$. This can be achieved by evaluating the product

$$\boldsymbol{\mathcal{G}} = \boldsymbol{\mathcal{Z}} \times_1 \boldsymbol{b}_1 \times \cdots \times_{i-1} \boldsymbol{b}_{i-1} \times_{i+1} \boldsymbol{b}_{i+1} \times \cdots \times_r \boldsymbol{b}_r$$

to obtain a tensor $\boldsymbol{\mathcal{G}}$. Solution for $\boldsymbol{b}_i$ can be obtained by solving the system $\boldsymbol{c} = \boldsymbol{\mathcal{G}} \times_2 \boldsymbol{b}_i$ for $\boldsymbol{b}_i$, which can be written as a typical linear system by unfolding $\boldsymbol{\mathcal{G}}$ as a matrix. Therefore, estimate of $\boldsymbol{b}_i$ can be obtained by

$$\boldsymbol{b}_i = (\boldsymbol{G}_2)^{\dagger}\boldsymbol{c} \tag{11.17}$$

where $\boldsymbol{G}_2$ is the matrix obtained by mode-2 unfolding of $\boldsymbol{\mathcal{G}}$ and $\dagger$ denotes the pseudo-inverse using SVD. Similarly, we can analytically solve for all other style factors. We start with a mean style estimate for each of the style factors since the style vectors are not known at the beginning. Iterative estimation of each of the style factors using Equation 11.17 would lead to a local minima for the error in Equation 11.16.

Solving for Body Configuration and Style Factors from a Single Image

In this case the input is a single image $\boldsymbol{y} \in \mathbb{R}^d$, it is required to find the body configuration, i.e., the corresponding embedding coordinates $\boldsymbol{x} \in \mathbb{R}^e$ on the manifold, and the style factors $\boldsymbol{b}_1, \cdots, \boldsymbol{b}_r$. These parameters should minimize the reconstruction error defined as

$$E(\boldsymbol{x}, \boldsymbol{b}_1, \cdots, \boldsymbol{b}_r) = ||\boldsymbol{y} - \boldsymbol{\mathcal{A}} \times_1 \boldsymbol{b}_1 \times \cdots \times_r \boldsymbol{b}_r \times_{r+1} \psi(\boldsymbol{x})||^2 \tag{11.18}$$

Instead of the second norm, we can also use a robust error metric and, in both cases, we end up with a nonlinear optimization problem.

One challenge is that not any point in a style subspace is a valid style vector. For example, if we consider a shape stye factor, we do not have enough data to model the class of all human shapes in this space. A training data, typically, is just a very sparse collection of the whole class. To overcome this, we assume, for all style factors, that the optimal style can be written as a convex linear combination of the style classes in the training data. This assumption is necessary to constrain the solution space. Better constraints can be achieved with sufficient training data. For example, in [50], we constrained a view factor, representing the view point by modelling the view manifold in the view factor subspace given sufficient sampled view points.

For the i-th style factor, let the mean vectors of the style classes in the training data denoted be $\bar{\boldsymbol{b}}_i^k, k = 1, \cdots, K_i$, where K_i is the number of classes and k is the class index. Such classes can be obtained by clustering the style vectors for each style factor in its subspace. Given such classes, we need to solve for linear regression weights α_{ik} such that

$$\boldsymbol{b}_i = \sum_{k=1}^{K_i} \alpha_{ik} \bar{\boldsymbol{b}}_i^k.$$

If all the style factors are known, then Equation 11.18 reduces to a nonlinear 1-dimensional search problem for the body configuration $\boldsymbol{x}$ on the embedded manifold representation that minimizes the error. On the other hand, if the body configuration and all style factors are known except the i-th factor, we can obtain the conditional class probabilities $p(k|\boldsymbol{y}, \boldsymbol{x}, \boldsymbol{s}_{/\boldsymbol{b}_i})$, which is proportional to observation likelihood $p(\boldsymbol{y} \mid \boldsymbol{x}, \boldsymbol{s}_{/\boldsymbol{b}_i}, k)$. Here, we use the notation $\boldsymbol{s}_{/\boldsymbol{b}_i}$ to denote the style factors excluding the i-th factor. This likelihood can be estimated assuming a Gaussian density centered around $\boldsymbol{\mathcal{A}} \times_1 \boldsymbol{b}_1 \times \cdots \times_i \bar{\boldsymbol{b}}_i^k \times \cdots \times_r \boldsymbol{b}_r \times \psi(\boldsymbol{x})$ with covariance Σ_{ik}, i.e.,

$$p(\boldsymbol{y} \mid \boldsymbol{x}, \boldsymbol{s}_{/\boldsymbol{b}_i}, k) \approx \mathcal{N}(\boldsymbol{\mathcal{A}} \times_1 \boldsymbol{b}_1 \times \cdots \times_i \boldsymbol{b}_i^k \times \cdots \times_r \boldsymbol{b}_r \times \psi(\boldsymbol{x}), \Sigma_{ik}).$$

Given view class probabilities, the weights are set to $\alpha_{ik} = p(k \mid \boldsymbol{y}, \boldsymbol{x}, \boldsymbol{s}_{/\boldsymbol{b}_i})$. This setting favors an iterative procedure for solving for $\boldsymbol{x}, \boldsymbol{b}_1, \cdots, \boldsymbol{b}_r$. However, wrong estimation of any of the factors would lead to wrong estimation of the others and leads to a local minima. For example, in the gait model in section 11.3.2, wrong estimation of the view factor would lead to a totally wrong estimate of body configuration and, therefore, wrong estimate for shape style. To avoid this we use a deterministic annealing-like procedure, where at the beginning the weights for all the style factors are forced to be close to uniform weights to avoid hard decisions. The weights gradually become discriminative thereafter. To achieve this, we use variable class variances which are uniform to all classes and are defined as $\Sigma_i = T\sigma_i^2 \boldsymbol{I}$ for the i-th factor. The temperature parameters T, start with large values, are gradually reduced, and in each step a new body configuration estimate is computed. We summarize the solution framework in Figure 11.9.

Input: image $\boldsymbol{y}$, style classes' means $\bar{\boldsymbol{b}}_i^k$, for all style factors $i = 1, cdots, r$, core tensor $\boldsymbol{\mathcal{A}}$

Initialization: • initialize T

- initialize α_{ik} to uniform weights, i.e., $\alpha_{ik} = 1/K_i, \forall i, k$
- Compute initial $\boldsymbol{b}_i = \sum_{k=1}^{K_i} \alpha_{ik} \bar{\boldsymbol{b}}_i^k$, $\forall i$

Iterate: • Compute coefficient $\boldsymbol{C} = \boldsymbol{\mathcal{A}} \times \boldsymbol{b}_1 \times \cdots \times \boldsymbol{b}_r$

- Estimate body configuration: 1-D search for $\boldsymbol{x}$ that minimizes $E(\boldsymbol{x}) = ||\boldsymbol{y} - \boldsymbol{C}\psi(\boldsymbol{x})||$
- For style factor $i = 1, \cdots, r$, estimate a new style factor vector $\boldsymbol{b}_i$
  - $\forall k = 1, \cdots K_i$ Compute $p(\boldsymbol{y}|\boldsymbol{x}, \boldsymbol{s}_{/\boldsymbol{b}_i}, k)$
  - $\forall k$ Update the weights $\alpha_{ik} = p(k|\boldsymbol{y}, \boldsymbol{x}, \boldsymbol{s}_{/\boldsymbol{b}_i})$
  - Estimate new factor vector as $\boldsymbol{b}_i = \sum_{k=1}^{K_i} \alpha_{ik} \boldsymbol{b}_i^k$
- Reduce T

Figure 11.9: Iterative estimation of style factors

11.6 Examples

11.6.1 Dynamic Shape Example: Decomposing View and Style on Gait Manifold

In this section we show an example of learning the nonlinear manifold of gait as an example of a dynamic shape. We used CMU Mobo gait data set [28] which contains walking people from multiple synchronized views.[3] For training we selected five people, five cycles each

[3]CMU Mobo gait data set [28] contains 25 people, about 8 to 11 walking cycles each captured from six different view points. The walkers were using a treadmill.

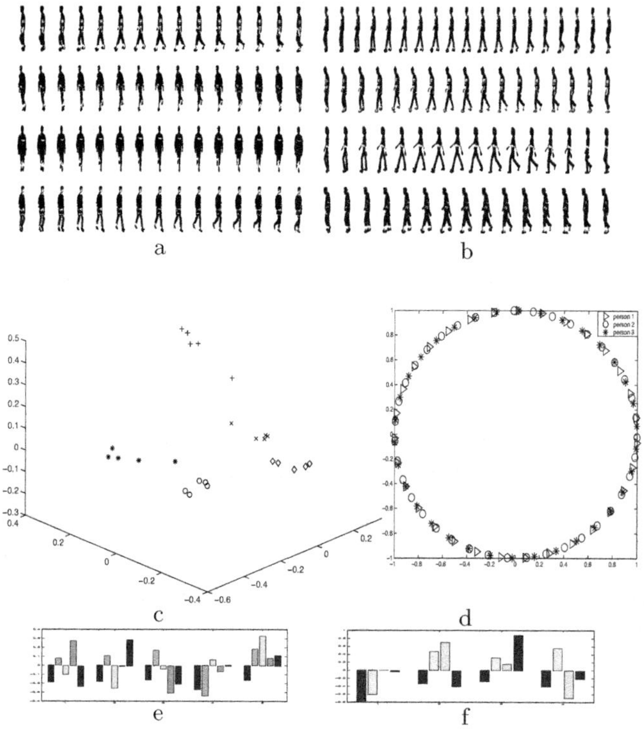

Figure 11.10: a,b) Example of training data. Each sequence shows a half cycle only. a) Four different views used for person 1 b) Side views of people 2, 3, 4, 5. c) style subspace: each person cycle has the same label. d) Unit circle embedding for three cycles. e) Mean style vectors for each person cluster. f) View vectors.

from four different views. i.e., total number of cycles for training is $100 = 5$ people $\times$ 5 cycles $\times$ 4 views. Note that cycles of different people and cycles of the same person are not of the same length. Figure 11.10a, b show examples of the sequences (only half cycles are shown because of limited space).

The data is used to fit the model as described in Equation 11.12. Images are normalized to 60×100, i.e., $d = 6000$. Each cycle is considered to be a style by itself, i.e., there are 25 styles and 4 views. Figure 11.10d shows an example of model-based aligned unit circle embedding of three cycles. Figure 11.10c shows the obtained style subspace where each of the 25 points corresponds to one of the 25 cycles used. An important thing to notice is that the style vectors are clustered in the subspace such that each persons style vectors (corresponding to different cycles of the same person) are clustered together, which indicates that the model can find the similarity in the shape style between different cycles of the same person. Figure 11.10e shows the mean style vectors for each of the five clusters. Figure 11.10f shows the four view vectors.

Figure 11.11 shows an example of using the model to recover the pose, view, and style. The figure shows samples of one full cycle and the recovered body configuration at each frame. Notice that despite the subtle differences between the first and second halves of the cycle, the model can exploit such differences to recover the correct pose. The recovery of 3D joint angles is achieved by learning a mapping from the manifold embedding and 3D joint angle from motion captured data using GRBF in a way similar to Equation 11.2. Figure 11.11c, d shows the recovered style weights (class probabilities) and view weights respectively for each frame of the cycle which shows correct person and view classification.

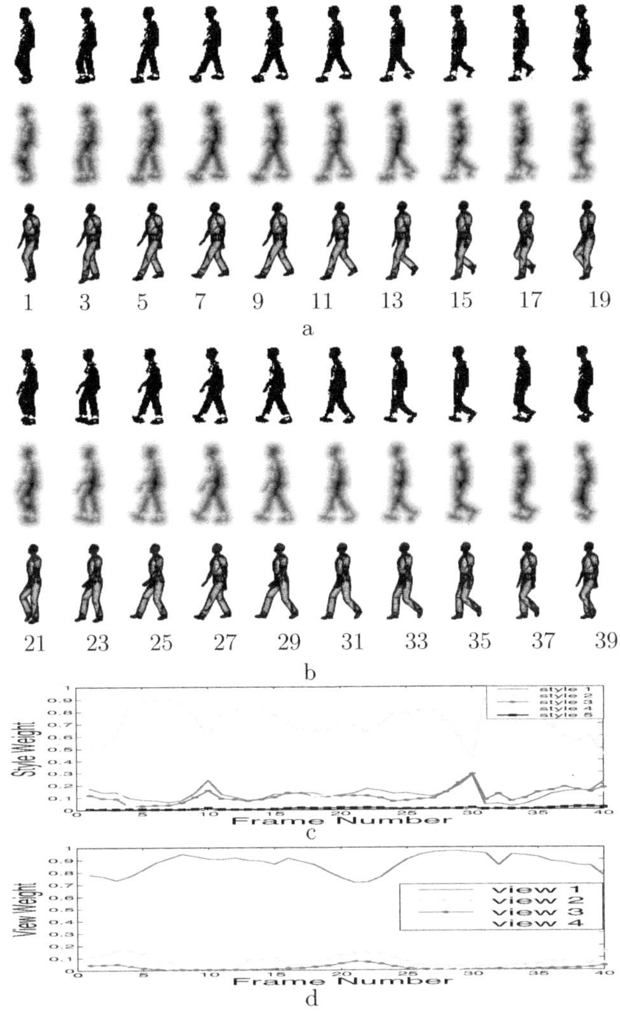

Figure 11.11: (See Color Insert.) a,b) Example pose recovery. From top to bottom: input shapes, implicit function, recovered 3D pose. c) Style weights. d) View weights.

Figure 11.12 shows examples of recovery of the 3D pose and view class for four different people, none of which was seen in training.

11.6.2 Dynamic Appearance Example: Facial Expression Analysis

We used the model to learn facial expression manifolds for different people. We used the CMU-AMP facial expression database where each subject has 75 frames of varying facial expressions. We choose four people and three expressions each (smile, anger, surprise) where corresponding frames are manually segmented from the whole sequence for training. The resulting training set contained 12 sequences of different lengths. All sequences are embedded to unit circles and aligned. A model in the form of Equation 11.13 is fitted to the data where we decompose two factors: person facial appearance style factor and expression factor besides the body configuration, which is nonlinearly embedded on a unit circle. Figure 11.13 shows the resulting person style vectors and expression vectors.

We used the learned model to recognize facial expression, and person identity at each frame of the whole sequence. Figure 11.14 shows an example of a whole sequence and the

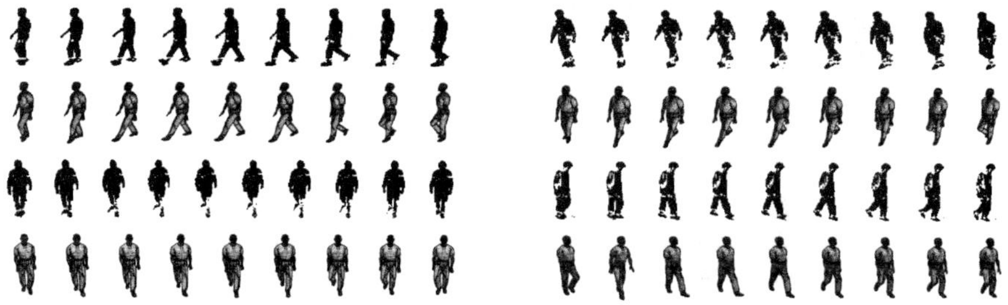

Figure 11.12: Examples of pose recovery and view classification for four different people from four views.

(a) Style plotting in 3D (b) Expression plotting in 3D

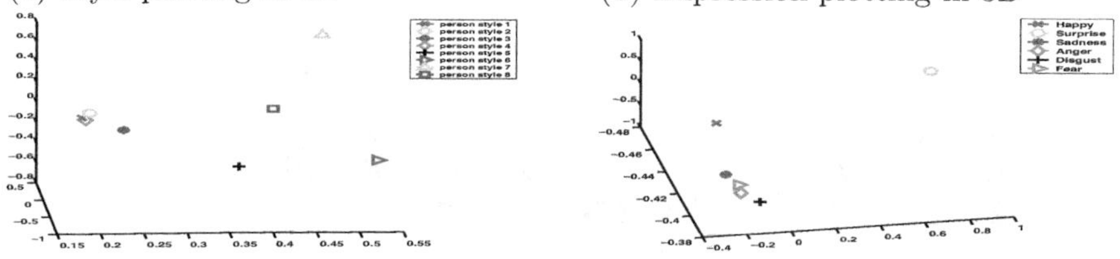

Figure 11.13: Facial expression analysis for Cohn–Kanade dataset for 8 subjects with 6 expressions and their 3D space plotting.

different expression probabilities obtained on a frame per frame basis. The figure also shows the final expression recognition after thresholding along manual expression labelling. The learned model was used to recognize facial expressions for sequences of people not used in the training. Figure 11.15 shows an example of a sequence of a person not used in the training. The model can successfully generalize and recognize the three learned expressions for this new subject.

11.7 Summary

In this chapter we focused on exploiting the underlying motion manifold for human motion analysis and synthesis. We presented a framework for learning a landmark-free, correspondence-free global representation of dynamic shape and dynamic appearance manifolds. The framework is based on using nonlinear dimensionality reduction to achieve an embedding of the global deformation manifold that preserves the geometric structure of the manifold. Given such embedding, a nonlinear mapping is learned from such embedded space into visual input space using RBF interpolation. Given this framework, any visual input is represented by a linear combination of nonlinear bases functions centered along the manifold in the embedded space. In a sense, the approach utilizes the implicit correspondences imposed by the global vector representation, which are only valid locally on the manifold through explicit modeling of the manifold and RBF interpolation where closer points on the manifold will have higher contributions than far away points. We showed how to learn a decomposable generative model that separates appearance variations from the intrinsics underlying dynamics manifold though introducing a framework for separation of style and content on a nonlinear manifold. The framework is based on decomposing multiple style factors in the space of nonlinear functions that maps between a learned unified nonlinear embedding of multiple content manifolds and the visual input space. We presented different applications of the framework for gait analysis and facial expression analysis.

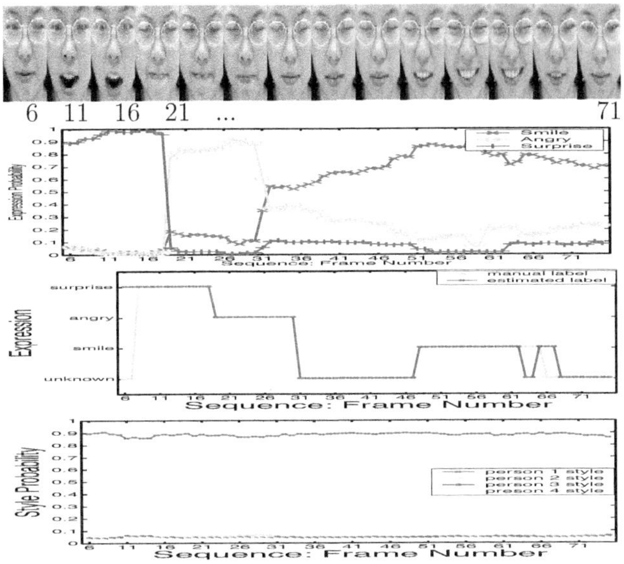

Figure 11.14: (See Color Insert.) From top to bottom: Samples of the input sequences; expression probabilities; expression classification; style probabilities.

11.8 Bibliographical and Historical Remarks

Despite the high dimensionality of both the human body configuration space and the visual input space, many human activities lie on low-dimensional manifolds. In the last few years, there has been increasing interest in exploiting this fact by using intermediate activity-based manifold representations [10, 65, 23, 85, 70, 93, 58, 57, 92]. In our earlier work [23], the visual manifolds of human silhouette deformations, due to motion, have been learned explicitly and used for recovering 3D body configuration from silhouettes in a closed-form. In that work, knowing the motion provided a strong prior to constrain the mapping from the shape space to the 3D body configuration space. However, the approach proposed in [23] is a view-based approach; a manifold was learned for each of the discrete views. In contrast, in [50, 52] the manifold of both the configuration and view is learned in a continuous way. In [85], manifold representations learned from the body configuration space were used to provide constraints for tracking. In both [23] and [85] learning an embedded manifold representation was decoupled from learning the dynamics and from learning a regression function between the embedding space and the input space. In [92], coupled learning of the representation and dynamics was achieved through introducing Gaussian Process Dynamic Model [99] (GPDM), in which a nonlinear embedded representation and a nonlinear observation model were fitted through an optimization process. GPDM is a very flexible model since both the state dynamics and the observation model are nonlinear. The problem of simultaneously estimating a latent state representation coupled with a nonlinear dynamic model was earlier addressed in [78]. Similarly, in [57], models that coupled learning dynamics with embedding were introduced.

Manifold-based representations of the motion can be learned from kinematic data, or from visual data, e.g., silhouettes. The former is suitable for generative model-based approaches and provides better dynamic-modeling for tracking, e.g., [85, 93]. Learning motion manifolds from visual data, as in [23, 16, 58], provides useful representations for recovery and tracking of body configurations from visual input without the need for explicit body models. The approach introduced in [50] learns a coupled representation for both the vi-

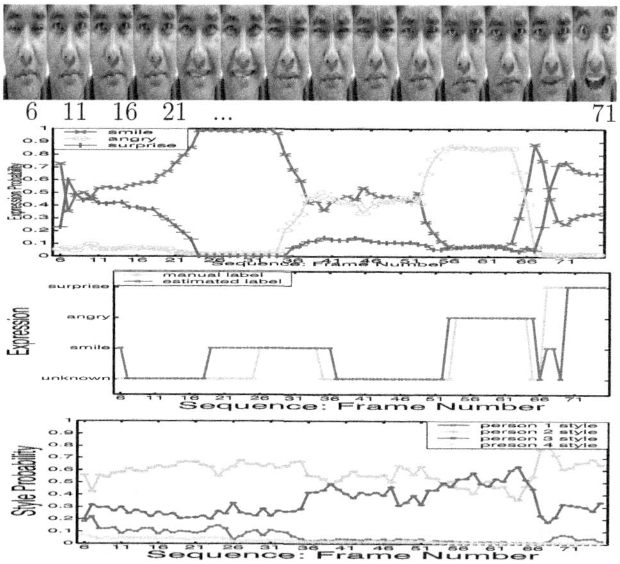

Figure 11.15: (See Color Insert.) Generalization to new people: expression recognition for a new person. From top to bottom: Samples of the input sequences; Expression probabilities; Expression classification; Style probabilities

sual manifold and the kinematic manifold. Learning a representation of the visual motion manifold can be used in a generative manner as in [23] or as a way to constrain the solution space for discriminative approaches as in [89].

The use of a generative model in the framework presented in this chapter is necessary since the mapping from the manifold representation to the input space will be well defined in contrast to a discriminative model where the mapping from the visual input to manifold representation is not necessarily a function. We introduced a framework to solve for various factors such as body configuration, view, and shape style. Since the framework is generative, it fits well in a Bayesian tracking framework and it provides separate low dimensional representations for each of the modelled factors. Moreover, a dynamic model for configuration is well defined since it is constrained to the 1D manifold representation. The framework also provides a way to initialize a tracker by inferring about body configuration, view point, and body shape style from a single or a sequence of images.

The framework presented in this chapter was basically applied to one-dimensional motion manifolds such as gait and facial expressions. One-dimensional manifolds can be explicitly modeled in a straightforward way. However, there is no theoretical restriction that prevents the framework from dealing with more complicated manifolds. In this chapter we mainly modeled the motion manifold while all appearance variability is modeled using subspace analysis. Extension to modeling multiple manifolds simultaneously is very challenging. We investigated modeling both the motion and the view manifolds in [49, 50, 52, 51]. The proposed framework has been applied to gait analysis and recognition in [44, 46, 53, 47]. It was also used in analysis and recognition of facial expressions in [45, 48].

Acknowledgment

This research is partially funded by NSF award IIS-0328991 and NSF CAREER award IIS-0546372

Bibliography

[1] P. N. Belhumeur, J. Hespanha, and D. J. Kriegman. Eigenfaces vs. fisherfaces: Recognition using class specific linear projection. In *ECCV (1)*, pages 45–58, 1996.

[2] M. Belkin and P. Niyogi. Laplacian eigenmaps for dimensionality reduction and data representation. *Neural Comput.*, 15(6):1373–1396, 2003.

[3] Y. Bengio, O. Delalleau, N. Le Roux, J.-F. Paiement, P. Vincent, and M. Ouimet. Learning eigenfunctions links spectral embedding and kernel pca. *Neural Comp.*, 16(10):2197–2219, 2004.

[4] Y. Bengio, J.-F. Paiement, P. Vincent, O. Delalleau, N. L. Roux, and M. Ouimet. Out-of-sample extensions for lle, isomap, mds, eigenmaps, and spectral clustering. In *Proc. of NIPS*, 2004.

[5] D. Beymer and T. Poggio. Image representations for visual learning. *Science*, 272(5250), 1996.

[6] M. J. Black and A. D. Jepson. Eigentracking: Robust matching and tracking of articulated objects using a view-based representation. In *ECCV (1)*, pages 329–342, 1996.

[7] A. Bobick and J. Davis. The recognition of human movement using temporal templates. *IEEE Transactions on Pattern Analysis and Machine Intelligence*, 23(3):257–267, 2001.

[8] R. Bowden. Learning statistical models of human motion. In *IEEE Workshop on Human Modelling, Analysis and Synthesis*, 2000.

[9] M. Brand. Shadow puppetry. In *International Conference on Computer Vision*, volume 2, page 1237, 1999.

[10] M. Brand. Shadow puppetry. In *Proc. of ICCV*, volume 2, pages 1237–1244, 1999.

[11] M. Brand and K. Huang. A unifying theorem for spectral embedding and clustering. In *Proc. of the Ninth International Workshop on AI and Statistics*, 2003.

[12] C. Bregler and S. M. Omohundro. Nonlinear manifold learning for visual speech recognition. In *Proc. of ICCV*, pages 494– 499, 1995.

[13] L. W. Campbell and A. F. Bobick. Recognition of human body motion using phase space constraints. In *ICCV*, pages 624–630, 1995.

[14] Z. Chen and H. Lee. Knowledge-guided visual perception of 3-d human gait from single image sequence. *IEEE SMC*, 22(2):336–342, 1992.

[15] C. M. Christoudias and T. Darrell. On modelling nonlinear shape-and-texture appearance manifolds. In *Proc. of IEEE CVPR*, volume 2, pages 1067–1074, 2005.

[16] C. M. Christoudias and T. Darrell. On modelling nonlinear shape-and-texture appearance manifolds. In *Proc. of CVPR*, pages 1067–1074, 2005.

[17] T. F. Cootes, C. J. Taylor, D. H. Cooper, and J. Graham. Active shape models: Their training and application. *CVIU*, 61(1):38–59, 1995.

[18] T. Cox and M. Cox. *Multidimensional scaling.* London: Chapman & Hall, 1994.

[19] R. Cutler and L. Davis. Robust periodic motion and motion symmetry detection. In *Proc. IEEE CVPR*, 2000.

[20] T. Darrell and A. Pentland. Space-time gesture. In *Proc IEEE CVPR*, 1993.

[21] A. Elgammal. Nonlinear manifold learning for dynamic shape and dynamic appearance. In *Workshop Proc. of GMBV*, 2004.

[22] A. Elgammal. Learning to track: Conceptual manifold map for closed-form tracking. In *Proc. of CVPR*, June 2005.

[23] A. Elgammal and C.-S. Lee. Inferring 3d body pose from silhouettes using activity manifold learning. In *Proc. of CVPR*, volume 2, pages 681–688, 2004.

[24] A. Elgammal and C.-S. Lee. Separating style and content on a nonlinear manifold. In *Proc. of CVPR*, volume 1, pages 478–485, 2004.

[25] R. Fablet and M. J. Black. Automatic detection and tracking of human motion with a view-based representation. In *Proc. ECCV 2002, LNCS 2350*, pages 476–491, 2002.

[26] D. Gavrila and L. Davis. 3-d model-based tracking of humans in action: a multi-view approach. In *IEEE Conference on Computer Vision and Pattern Recognition*, 1996.

[27] R. Goldenberg, R. Kimmel, E. Rivlin, and M. Rudzsky. 'Dynamism of a dog on a leash' or behavior classification by eigen-decomposition of periodic motions. In *Proceedings of the ECCV'02*, pages 461–475, Copenhagen, May 2002. Springer-Verlag, LNCS 2350.

[28] R. Gross and J. Shi. The cmu motion of body (mobo) database. Technical Report TR-01-18, Pittsburgh: Carnegie Mellon University, 2001.

[29] J. Ham, D. D. Lee, S. Mika, and B. Schölkopf A kernel view of the dimensionality reduction of manifolds. In *Proceedings of ICML*, page 47, 2004.

[30] I. Haritaoglu, D. Harwood, and L. S. Davis. W4: Who ? when ? where? what? a real time system for detecting and tracking people. In *3rd International Conference on Face and Gesture Recognition*, 1998.

[31] D. Hogg. Model-based vision: a program to see a walking person. *Image and Vision Computing*, 1(1):5–20, 1983.

[32] N. R. Howe, M. E. Leventon, and W. T. Freeman. Bayesian reconstruction of 3d human motion from single-camera video. In *Proc. NIPS*, 1999.

[33] H. S.Seung and D. D. Lee. The manifold ways of perception. *Science*, 290(5500):2268–2269, December 2000.

[34] I. T. Jolliffe. *Principal Component Analysis.* New York: Springer-Verlag, 1986.

[35] J. O'Rourke and N. Badler. Model-based image analysis of human motion using constraint propagation. *IEEE PAMI*, 2(6), 1980.

[36] S. X. Ju, M. J. Black, and Y. Yacoob. Cardboard people: A parameterized model of articulated motion. In *International Conference on Automatic Face and Gesture Recognition*, pages 38–44, Killington, Vermont, 1996.

[37] I. A. Kakadiaris and D. Metaxas. Model-based estimation of 3D human motion with occlusion based on active multi-viewpoint selection. In *Proc. IEEE Conf. Computer Vision and Pattern Recognition, CVPR*, pages 81–87, Los Alamitos, California, U.S.A., 18–20 1996. IEEE Computer Society.

[38] A. Kapteyn, H. Neudecker, and T. Wansbeek. An approach to n-model component analysis. *Psychometrika*, 51(2):269–275, 1986.

[39] G. S. Kimeldorf and G. Wahba. A correspondence between Bayesian estimation on stochastic processes and smoothing by splines. *The Annals of Mathematical Statistics*, 41:495–502, 1970.

[40] K. Grauman, G. Shakhnarovich, and T. Darrell. Inferring 3d structure with a statistical image-based shape model. In *ICCV*, 2003.

[41] L. D. Lathauwer, B. de Moor, and J. Vandewalle. A multilinear singular value decomposiiton. *SIAM Journal on Matrix Analysis and Applications*, 21(4):1253–1278, 2000.

[42] L. D. Lathauwer, B. de Moor, and J. Vandewalle. On the best rank-1 and rank-(r1, r2, ..., rn) approximation of higher-order tensors. *SIAM Journal on Matrix Analysis and Applications*, 21(4):1324–1342, 2000.

[43] N. Lawrence. Gaussian process latent variable models for visualization of high dimensional data. In *Proc. of NIPS*, 2003.

[44] C.-S. Lee and A. Elgammal. Gait style and gait content: Bilinear model for gait recogntion using gait re-sampling. In *Proc. of FGR*, pages 147–152, 2004.

[45] C.-S. Lee and A. Elgammal. Facial expression analysis using nonlinear decomposable generative models. In *IEEE Workshop on AMFG*, pages 17–31, 2005.

[46] C.-S. Lee and A. Elgammal. Style adaptive Bayesian tracking using explicit manifold learning. In *Proc. of British Machine Vision Conference*, pages 739–748, 2005.

[47] C.-S. Lee and A. Elgammal. Gait tracking and recognition using person-dependent dynamic shape model. In *Proc. of FGR*, pages 553–559. IEEE Computer Society, 2006.

[48] C.-S. Lee and A. Elgammal. Nonlinear shape and appearance models for facial expression analysis and synthesis. In *Proc. of ICPR*, pages 497–502, 2006.

[49] C.-S. Lee and A. Elgammal. Simultaneous inference of view and body pose using torus manifolds. In *Proc. of ICPR*, pages 489–494, 2006.

[50] C.-S. Lee and A. Elgammal. Modeling view and posture manifolds for tracking. In *Proc. of ICCV*, 2007.

[51] C.-S. Lee and A. Elgammal. Coupled visual and kinematics manifold models for human motion analysis. *IJCV*, July 2009.

[52] C.-S. Lee and A. Elgammal. Tracking people on a torus. *IEEE Trans. PAMI*, March 2009.

[53] C.-S. Lee and A. M. Elgammal. Towards scalable view-invariant gait recognition: Multilinear analysis for gait. In *Proc. of AVBPA*, pages 395–405, 2005.

[54] A. Levin and A. Shashua. Principal component analysis over continuous subspaces and intersection of half-spaces. In *ECCV, Copenhagen, Denmark*, pages 635–650, May 2002.

[55] J. Magnus and H. Neudecker. *Matrix Differential Calculus with Applications in Statistics and Econometrics.* John Wiley & Sons, New York, New York, 1988.

[56] D. Marimont and B. Wandell. Linear models of surface and illumination spectra. *J. Optical Society of America*, 9:1905–1913, 1992.

[57] K. Moon and V. Pavlovic. Impact of dynamics on subspace embedding and tracking of sequences. In *Proc. of CVPR*, volume 1, pages 198–205, 2006.

[58] V. I. Morariu and O. I. Camps. Modeling correspondences for multi-camera tracking using nonlinear manifold learning and target dynamics. In *Proc. of CVPR*, pages 545–552, 2006.

[59] P. Mordohai and G. Medioni. Unsupervised dimensionality estimation and manifold learning in high-dimensional spaces by tensor voting. In *Proceedings of International Joint Conference on Artificial Intelligence*, 2005.

[60] G. Mori and J. Malik. Estimating human body configurations using shape context matching. In *European Conference on Computer Vision*, 2002.

[61] M. Turk and A. Pentland. Eigenfaces for recognition. *Journal of Cognitive Neuroscience*, 3(1):71–86, 1991.

[62] H. Murase and S. Nayar. Visual learning and recognition of 3D objects from appearance. *International Journal of Computer Vision*, 14:5–24, 1995.

[63] R. C. Nelson and R. Polana. Qualitative recognition of motion using temporal texture. *CVGIP Image Understanding*, 56(1):78–89, 1992.

[64] S. Niyogi and E. Adelson. Analyzing and recognition walking figures in xyt. In *Proc. IEEE CVPR*, pages 469–474, 1994.

[65] D. Ormoneit, H. Sidenbladh, M. J. Black, T. Hastie, and D. J. Fleet. Learning and tracking human motion using functional analysis. In *Proc. IEEE Workshop on Human Modeling, Analysis and Synthesis*, pages 2–9, 2000.

[66] T. Poggio and F. Girosi. Network for approximation and learning. *Proceedings of the IEEE*, 78(9):1481–1497, 1990.

[67] R. Polana and R. Nelson. Low level recognition of human motion (or how to get your man without finding his body parts). In *IEEE Workshop on Non-Rigid and Articulated Motion*, pages 77–82, 1994.

[68] R. Polana and R. C. Nelson. Qualitative detection of motion by a moving observer. *International Journal of Computer Vision*, 7(1):33–46, 1991.

[69] R. Polana and R. C. Nelson. Detecting activities. *Journal of Visual Communication and Image Representation*, June 1994.

[70] A. Rahimi, B. Recht, and T. Darrell. Learning appearance manifolds from video. In *Proc. of IEEE CVPR*, volume 1, pages 868–875, 2005.

[71] J. M. Rehg and T. Kanade. Visual tracking of high DOF articulated structures: an application to human hand tracking. In *ECCV (2)*, pages 35–46, 1994.

[72] J. M. Rehg and T. Kanade. Model-based tracking of self-occluding articulated objects. In *ICCV*, pages 612–617, 1995.

[73] J. Rittscher and A. Blake. Classification of human body motion. In *IEEE International Conferance on Computer Vision*, 1999.

[74] K. Rohr. Towards model-based recognition of human movements in image sequence. *CVGIP*, 59(1):94–115, 1994.

[75] R. Rosales, V. Athitsos, and S. Sclaroff. 3D hand pose reconstruction using specialized mappings. In *Proc. ICCV*, 2001.

[76] R. Rosales and S. Sclaroff. Inferring body pose without tracking body parts. Technical Report 1999-017, 1, 1999.

[77] R. Rosales and S. Sclaroff. Specialized mappings and the estimation of human body pose from a single image. In *Workshop on Human Motion*, pages 19–24, 2000.

[78] S. Roweis and Z. Ghahramani. An EM algorithm for identification of nonlinear dynamical systems. In S. Haykin, editor, *Kalman Filtering and Neural Networks*.

[79] S. Roweis and L. Saul. Nonlinear dimensionality reduction by locally linear embedding. *Sciene*, 290(5500):2323–2326, 2000.

[80] B. Schölkopf and A. Smola. *Learning with Kernels: Support Vector Machines, Regularization, Optimization and Beyond.* The MIT Press, Cambridge, Massachusetts, 2002.

[81] G. Shakhnarovich, P. Viola, and T. Darrell. Fast pose estimation with parameter-sensitive hashing. In *ICCV*, 2003.

[82] A. Shashua and A. Levin. Linear image coding of regression and classification using the tensor rank principle. In *Proc. of IEEE CVPR, Hawaii*, 2001.

[83] H. Sidenbladh, M. J. Black, and D. J. Fleet. Stochastic tracking of 3d human figures using 2d image motion. In *ECCV (2)*, pages 702–718, 2000.

[84] H. Sidenbladh, M. J. Black, and L. Sigal. Implicit probabilistic models of human motion for synthesis and tracking. In *Proc. ECCV 2002, LNCS 2350*, pages 784–800, 2002.

[85] C. Sminchisescu and A. Jepson. Generative modeling for continuous non-linearly embedded visual inference. In *Proceedings of ICML*, pages 96–103. ACM Press, 2004.

[86] Y. Song, X. Feng, and P. Perona. Towards detection of human motion. In *IEEE Computer Society Conference on Computer Vision and Pattern Recognition (CVPR 2000)*, pages 810–817, 2000.

[87] J. Tenenbaum. Mapping a manifold of perceptual observations. In *Advances in Neural Information Processing*, volume 10, pages 682–688, 1998.

[88] J. Tenenbaum and W. T. Freeman. Separating style and content with bilinear models. *Neural Computation*, 12:1247–1283, 2000.

[89] T.-P. Tian, R. Li, and S. Sclaroff. Articulated pose estimation in a learned smooth space of feasible solutions. In *Proc. of CVPR*, page 50, 2005.

[90] K. Toyama and A. Blake. Probabilistic tracking in a metric space. In *ICCV*, pages 50–59, 2001.

[91] L. Tucker. Some mathematical notes on three-mode factor analysis. *Psychometrika*, 31:279–311, 1966.

[92] R. Urtasun, D. J. Fleet, and P. Fua. 3D people tracking with Gaussian process dynamical models. In *Proc. of CVPR*, pages 238–245, 2006.

[93] R. Urtasun, D. J. Fleet, A. Hertzmann, and P. Fua. Priors for people tracking from small training sets. In *Proc. of ICCV*, pages 403–410, 2005.

[94] M. A. O. Vasilescu. An algorithm for extracting human motion signatures. In *Proc. of IEEE CVPR, Hawaii*, 2001.

[95] M. A. O. Vasilescu and D. Terzopoulos. Multilinear analysis of image ensebles: Tensorfaces. In *Proc. of ECCV, Copenhagen, Denmark*, pages 447–460, 2002.

[96] R. Vidal and R. Hartley. Motion segmentation with missing data using powerfactorization and gpca. In *Proceedings of IEEE CVPR*, volume 2, pages 310–316, 2004.

[97] R. Vidal, Y. Ma, and S. Sastry. Generalized principal component analysis (gpca). In *Proceedings of IEEE CVPR*, volume 1, pages 621–628, 2003.

[98] H. Wang and N. Ahuja. Rank-r approximation of tensors: Using image-as-matrix representation. In *Proceedings of IEEE CVPR*, volume 2, pages 346–353, 2005.

[99] J. Wang, D. J. Fleet, and A. Hertzmann. Gaussian process dynamical models. In *Proc. of NIPS*, 2005.

[100] K. W. Weinberger and L. K. Saul. Unsupervised learning of image manifolds by semidefinite programming. In *Proc. of CVPR*, volume 2, pages 988–995, 2004.

[101] C. R. Wern, A. Azarbayejani, T. Darrell, and A. P. Pentland. Pfinder: Real-time tracking of human body. *IEEE Transaction on Pattern Analysis and Machine Intelligence*, 1997.

[102] Y. Yacoob and M. J. Black. Parameterized modeling and recognition of activities. *Computer Vision and Image Understanding: CVIU*, 73(2):232–247, 1999.

Index

I

9781439871096